S0-AAI-127

Radio over Fiber Technologies for Mobile Communications Networks

The Artech House Universal Personal Communications Series

Ramjee Prasad, Series Editor

CDMA for Wireless Personal Communications, Ramjee Prasad

IP/ATM Mobile Satellite Networks, John Farserotu and Ramjee Prasad

OFDM for Wireless Multimedia Communications, Richard van Nee and Ramjee Prasad

Radio over Fiber Technologies for Mobile Communications Networks, Hamed Al-Raweshidy and Shozo Komaki, editors

Third Generation Mobile Communication Systems, Ramjee Prasad, Werner Mohr, and Walter Konhäuser, editors

Towards a Global 3G System: Advanced Mobile Communications in Europe, Volume 1, Ramjee Prasad, editor

Towards a Global 3G System: Advanced Mobile Communications in Europe, Volume 2, Ramjee Prasad, editor

Universal Wireless Personal Communications, Ramjee Prasad

WCDMA: Towards IP Mobility and Mobile Internet, Tero Ojanperä and Ramjee Prasad, editors

Wideband CDMA for Third Generation Mobile Communications, Tero Ojanperä and Ramjee Prasad, editors

WLAN Systems and Wireless IP for Next Generation Communications, Neeli Prasad and Anand Prasad, editors

For further information on these and other Artech House titles, including previously considered out-of-print books now available through our In-Print-Forever® (IPF®) program, contact:

Artech House
685 Canton Street
Norwood, MA 02062
Phone: 781-769-9750
Fax: 781-769-6334
e-mail: artech@artechhouse.com

Artech House
46 Gillingham Street
London SW1V 1AH UK
Phone: +44 (0)20 7596-8750
Fax: +44 (0)20 7630-0166
e-mail: artech-uk@artechhouse.com

Find us on the World Wide Web at: www.artechhouse.com

Radio over Fiber Technologies for Mobile Communications Networks

Hamed Al-Raweshidy
Shozo Komaki
Editors

Artech House
Boston • London
www.artechhouse.com

Library of Congress Cataloging-in-Publication Data
Radio over fiber technologies for mobile communications networks/Hamed
Al-Raweshidy, Shozo Komaki, editors.
    p. cm. — (Artech House universal personal communications series)
Includes bibliographical references and index.
ISBN 1-58053-148-2 (alk. paper)
    1. Optical communications.   2. Fiber optics.   3. Radio frequency modulation.
    I. Al-Raweshidy, Hamed.   II. Komaki, Shozo.   III. Series.
TK5103.59 .R34 2002
0621.3845'6—dc21   2001056567

British Library Cataloguing in Publication Data
Radio over fiber technologies for mobile communications networks. — (Artech House
universal personal communications series)
1. Mobile communication systems   2. Optical fibers
I. Al-Raweshidy, Hamed   II. Komaki, Shozo
621.3'8456

ISBN 1-58053-148-2

Cover design by Igor Valdman

© **2002 ARTECH HOUSE, INC.**
685 Canton Street
Norwood, MA 02062

All rights reserved. Printed and bound in the United States of America. No part of this book
may be reproduced or utilized in any form or by any means, electronic or mechanical,
including photocopying, recording, or by any information storage and retrieval system, without
permission in writing from the publisher.
    All terms mentioned in this book that are known to be trademarks or service marks have
been appropriately capitalized. Artech House cannot attest to the accuracy of this information.
Use of a term in this book should not be regarded as affecting the validity of any trademark
or service mark.

International Standard Book Number: 1-58053-148-2
Library of Congress Catalog Card Number: 2001056567

10 9 8 7 6 5 4 3 2 1

To the memory of my late brother Rashid, and to Father and Mother, and all of my family who have helped me complete this book
—H. Al-Raweshidy

Contents

Preface

He said that there was one only good, namely, knowledge; and one only evil, namely, ignorance.

—Diogenes Laërtius, *Socrates*

With the development of the third generation of mobile communications, IMT 2000, radio over fiber technology will play a significant role in solving problems facing this technology. Envisioning a global village, people could transmit and receive "any time, anywhere, and anything." In addition, the explosive growth in Internet applications, such as the World Wide Web, demonstrates the tremendous increase in bandwidth and low power that the coming world of multimedia interactive applications will require from future networks. Radio over fiber systems have many advantages such as enhanced microcellular coverage, higher capacity, lower cost, lower power, and easier installation. Therefore, radio over fiber technology is the most suitable candidate for indoor applications such as airport terminals, shopping centers, and large offices, in addition to outdoor applications such as those underground, tunnels, narrow streets (i.e., dead zone areas), and highways.

The next generation of cellular mobile phone systems will make extensive use of microcells. This will permit a large increase in the numbers of users and will also allow a significant increase in the available channel bandwidth, so that broadband services can be offered, in addition to the voiceband services offered with current systems. The introduction of large numbers of microcells will result in the need to interconnect huge numbers of cells and microcells, and this can be carried out effectively using optical fiber, which offers a high transmission capacity at low cost. The transmission of radio signals over fiber, with simple optical-to-electrical conversion, followed by radiation at remote antennas, which are connected to a central station, has been proposed as a method of minimizing costs. The reduction in cost is brought about in two ways. First, the remote antenna or radio distribution point needs to perform only simple functions, and it is small in size and low in cost. Second, the resources provided by the central stations can be shared among many antenna sites. This technique of modulating the RF subcarrier onto an optical carrier for distribution over a fiber network is known as radio over fiber (RoF) technology. In addition to the advantages of potential low cost, RoF technology has the further benefit that transferring the RF frequency allocation to a central station can allow flexible network channel allocation and rapid response to variations in traffic demand.

It is very clear that the synergy of wireless and optical fiber communications is an important issue, taking into consideration future requirements for the fourth generation, such as software radio and intelligent networks, in which RoF technology will play a significant role through its macrodiversity.

This book provides a comprehensive introduction to RoF technology. It provides the technical background necessary to understand how RoF systems are designed, starting from components and system requirements and ending with analysis and applications. It also provides a proposal for a simulation model for a RoF system in third generation (3G) and fourth generation (4G).

Chapter 1 of this book introduces various aspects of components, such as laser diode behavior, losses, and noise with comprehensive analysis.

Chapter 2 describes and presents results on single-mode fiber optic transmission of microwave QAM signals, and how to improve system performance. In addition, theoretical and some experimental bit error rate performance results are described for two types of 256-QAM signals.

Chapter 3 introduces the optical fiber link in mobile communication systems. The chapter starts with a system-level analysis and continues with considerations about requirements for cellular base station remote antenna links. We then show that these requirements can be fulfilled with direct, intensity-modulated analog optical fiber links. A comparison of conventional coaxial links and optical fiber links, considering technical as well as economical aspects, and hybrid fiber/coaxial solutions, are also presented in this chapter.

Chapter 4 covers the UMTS RoF and the 4G RoF such as HiperLAN2. This chapter suggests an integrated RoF simulation model and discusses some benefits of RoF systems, such as macrodiversity. A full analysis for UMTS/WCDMA performance and voice activities is also presented.

In Chapter 5, RoF for in-building coverage is described, and the use of RoF for cellular in-building distributed antenna systems. It examines the problems associated with these issues and assesses the competing technologies and architectures that can be used. Both current and emerging RoF technologies are included.

Fiber optic radio networking and radio highways are discussed in Chapter 6. This includes software radio principles, with three main classifications: software radio terminals, software radio base stations, and software radio networks. In addition, various types of link configuration and photonic multiplexing schemes, such as SCMA, photonic FDMA or WDMA, photonic TDMA, photonic CDMA, and chirp multiple transform access and routing methods, are also discussed.

Chapter 7 explains the historical aspects of this technology's development and the evolution of key factors in more recent applications, such as that used by Tekmar Sisterni in the 2000 Olympics in Sydney, and also Bluewater by Tekmar. The different scenarios for RoF market issues are also discussed in this chapter.

Finally, the applications of RoF in intelligent transport systems and multiple service wireless systems based on RoF, such as CATV and fixative, are discussed in Chapter 8.

This book is intended for everyone involved in the field of mobile communications, optical fiber technologies, and internal and outdoor coverage. It provides treatment of the subject matter at different levels for system designers, researchers, managers, and graduate students. The majority of chapters provide both a general review for the beginner and a more thorough approach for the reader interested in technical details. The views expressed in this book are from both academia and industries in Europe, Japan, and the United States.

The contributors have tried their best to make each chapter quite self-contained. We hope that this book will help to stimulate many new research avenues and solutions for future mobile data communications problems.

We cannot claim that this book is errorless. Any remarks to improve the readability and to correct any errors would be highly appreciated.

Acknowledgments

The material in this book originates from several research institutes in Europe, Japan, and the United States, and I would like to thank all the contributors of this book.

I would like to thank everyone in my group, especially Habib Ali, Ammar Daher, Nazim Khashjori, Stephen Ampem, Kamiran Haider, Adrin Sivarajah, Mudar Barry, Maswood Jahromi, and Haider Ali, for their kind help on this book.

Hamed Al-Raweshidy
Kent, England
February 2002

1

Basic Microwave Properties of Optical Links: Insertion Loss, Noise Figure, and Modulation Transfer

Istvan Frigyes

1.1 Introduction

Virtually all of the optical links transmitting microwave signals apply intensity modulation of light. Essentially, three different methods exist for the transmission of microwave signals over optical links with intensity modulation. In direct intensity modulation an electrical parameter of the light source is modulated by the information-bearing microwave or, more generally, the radio-frequency (RF) signal. In practical links, this is the current of the laser diode, serving as the optical transmitter. A second method applies an unmodulated light source and an external light intensity modulator. In a third method, microwaves or millimeter waves are optically generated in most cases via *remote heterodyning*, that is, a method in which more than one optical signal is generated by the light source, one of which is modulated by the information-bearing signal, and these are mixed or heterodyned by the photodetector or by an external mixer to form the output microwave/millimeter-wave signal. All of these have different loss and noise properties. In this chapter we investigate the first and second of these. Methods to generate microwave or millimeter-wave carriers are rather widespread; to deal separately with all of them would not be of much interest.

Of course, direct intensity modulation is, in principle, by far the simplest of the three solutions. So it is used everywhere that it can be used. One limiting phenomenon to its use is the modulation bandwidth of the laser. Relatively simple lasers can be modulated to frequencies of several gigahertz, say, 5–10 GHz. Although there are reports of direct intensity modulation lasers operating at up to 40 GHz or even higher [1], these diodes are rather expensive or nonexistent in commercial form.

That is why at higher microwave frequencies, say, above 10 GHz, external modulation rather than direct modulation is applied. In entering into the millimeter band a new adverse effect, the nonconvenient transfer function of the transmission medium (i.e., the fiber) is observed. It turns out that fiber dispersion *and* coherent mixing of the sidebands of modulated light may cause transmission zeros, even in the case of rather moderate lengths of fiber. For example, a standard fiber having a 1-km length has a transmission zero at 60 GHz if 1,550-nm wavelength light is intensity modulated. Due to this phenomenon, optical generation rather than transmission of the RF signal is preferable.

Because the number of base stations in a wireless network is high, simple and inexpensive components must be used. Therefore, in the *uplink* of an RoF cellular system (i.e., base station to central station), it is convenient to use direct intensity modulation with cheap lasers; this may require downconversion of the uplink RF signal received at the base station. In the downlink there is *one* light source; economical factors are less important there, so either more expensive lasers or external modulators can be applied.

Noise figure and insertion loss can be regarded as basic microwave characteristics of the optical links. However, some of the optical components, in particular modulators and optical fibers, have characteristics that cause performance degradation that cannot be deduced from noise figure or insertion loss. Phenomena like linear and nonlinear distortion and phase noise fall into this category. These are also discussed in Sections 1.5.3 to 1.5.5.

1.2   Insertion Loss and Noise Figure Concepts

From the point of view of the microwave signal to be transmitted, the fiber optic link is a simple transducer (like a waveguide or transmission-line section, a filter, or an amplifier). Therefore, for its primary characterization, the same parameters may be used as those used for the latter. The basic parameters of a transducer are its insertion loss and its noise figure.

Note that for characterizing a transducer inserted into the transmitter part of a communication system, it is the insertion loss that is of most interest; for a transducer in the receiver part, the noise figure is usually the most important parameter.

The purpose of this section is to provide definitions of these parameters. They are given for sake of completeness only because these are ubiquitous parameters in everyday use by engineers dealing with microwave or other RF transmission systems. However, in optical communication systems, which are more conventional than those discussed in this book, insertion loss is a not very important characteristic, whereas noise figure is a nonexistent, or at least not very useful, characteristic. Therefore, these parameters are not widely used in optics.

1.2.1 Insertion Loss

Figure 1.1 shows a microwave transmission system. The generator is represented as the series connection of a voltage source and a source impedance; e_G is the source voltage and Z_G is the source impedance. Impedance Z_G might be a positive resistance or any *passive* complex impedance. The load is represented by its impedance Z_L, which is also a real or complex passive impedance.

The loss (or gain) caused by inserting the *microwave transducer* between the generator and the load is defined as

$$L \triangleq \frac{P_a}{P_L} = \frac{|e_G|^2/4R_G}{|I_2^2|R_L} \tag{1.1}$$

where P_a is the generator available power, P_L is the power delivered to the load, and R_G and R_L are the real parts of the generator and load impedance,

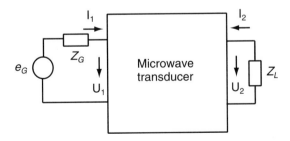

Figure 1.1 A microwave transmission system.

respectively. Note that the transducer produces gain rather than loss, if $L < 1$; $L = 1$ for lossless matched transducers. Note also that the generator voltage can be expressed in the form

$$e_G = 2\sqrt{P_a R_L} \qquad (1.2)$$

a formula to be used several times in the following sections.

1.2.2 Noise Figure

The noise figure concept is a notion in principle related to thermal noise. Take into account that a resistor—due to thermal effects, such as the Brownian motion of charge carriers, thermal radiation, or others—produces an ac voltage. This is a randomly varying voltage, and thus can be modeled as a stochastic process, being stationary, at least in the short term. The single-sided spectral density of the available power $N_0(f)$ can be given as [2]

$$N_{0T}(f; T) = \frac{hf}{\exp\left[\dfrac{hf}{k_B T}\right] - 1} \qquad (1.3a)$$

where:

$h = 6.62 \times 10^{-34}$ W sec^2 is the Planck constant;
f is frequency (in hertz);
T is temperature (in Kelvin);
$k_B = 1.38 \times 10^{-23}$ W sec/K is the Boltzmann constant; subscript T stands for *thermal*.

To proceed, take limiting cases into account. If $\dfrac{hf}{k_B T} \ll 1$, the first two terms of the Taylor expansion of the exponential expression yield a good approximation:

$$N_{0T}(f; T) \approx \frac{hf}{1 + \dfrac{hf}{k_B T} - 1} = k_B T \qquad (1.3b)$$

while for $\dfrac{hf}{k_B T} \gg 1$

$$N_{0T}(f; T) \approx hf \exp\left[-\frac{hf}{k_B T}\right] \approx 0 \tag{1.3c}$$

(Consider these numerical examples: For f = 300 GHz, T = 30K, being really limiting values, $N_0 = k_B \cdot T - 1.1$ dB; for f = 200 THz, T = 290K, $N_0 = k \cdot T - 129$ dB.)

Thus, (1.3b) gives an excellent approximation up to the highest "electrical" frequencies, say, up to 300 GHz. That is, in the electrical domain thermal noise can be regarded as *white*; this approximation will be applied in what follows for electrical frequency bands. However, (1.3b) is a good approximation in the whole optical band, and so in this domain thermal noise is virtually nonexistent. (This latter statement shows why the noise figure concept is not a generally used notion in optical techniques.)

Taking (1.3b) and (1.2) into account, a resistor can be represented as a voltage generator (see Figure 1.2), with generator voltage of

$$e_N = 2\sqrt{k_B TBR} \tag{1.4}$$

where B is the noise bandwidth of the system.

Of course, because this voltage is randomly varying, e_N should be regarded as an rms magnitude.

In Figure 1.2 the noise source is given with its Thévenin equivalent; of course, it also can be represented by its Norton equivalent, in which case there is a noise current source, having a source current k_N of

$$k_N = 2\sqrt{k_B TBG} \tag{1.5}$$

where $G = 1/R$ is the source conductance.

According to Figure 1.2, a *real* generator with resistive impedance can be regarded, for example, as one with two voltage sources: a signal source and a noise source, as shown in Figure 1.3.

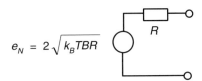

$$e_N = 2\sqrt{k_B TBR}$$

Figure 1.2 A resistor as noise generator.

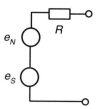

Figure 1.3 A generator with noisy resistance.

After this introduction, the noise figure of any two-port network, such as the transducer of Figure 1.1, is defined as

$$F \triangleq \frac{(S/N)_{in}}{(S/N)_{out}} \qquad (1.6a)$$

that is, as the ratio of the signal-to-noise power ratio (S/N) at the input to the S/N at the output. More precisely, to be unique, the generator temperature T_g must be specified. This conventional temperature designated by T_0 and called *standard temperature* is usually taken as 290K. Thus, the final definition of the noise figure is

$$F = \frac{(S/N)_{in}}{(S/N)_{out}} \Bigg|_{T_g = T_0} \qquad (1.6b)$$

In practical cases, F is always greater than 1; this is due to the fact that in the general case additional noise power is generated within the two-port network, but loss of the two-port network also causes $F > 1$. This additional noise can result from very different physical phenomena, such as increased thermal noise, shot noise in optical transmission laser intensity, and phase noise, and so forth.

Taking the noise figure into account, a noisy two-port network is equivalent to a noiseless one, and the noise produced by this two-port network is reduced to the input. In the simplest, impedance-matched case, one noise generator can be applied as in Figure 1.2, and the spectral density of the input white noise is

$$N_0 = N_{0T} \cdot F = F k_B T_0 \qquad (1.7)$$

In a more general case, two noise generators are needed [3]; between the two possibilities, for our purposes, the design shown in Figure 1.4 is more appropriate.

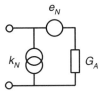

Figure 1.4 Noise generators reduced to the input of a noisy two-port network for a general case; G_A is the input conductance of the network.

The noise sources are characterized by two parameters: the minimal noise figure of the two port, F_0, and the corresponding resistance, $R_{A0} = 1/G_{A0}$. The following relationships apply:

$$k_N^2 = 2(F_0 - 1)k_B T_0 B G_{A0} \tag{1.8a}$$

and

$$e_N^2 = 2(F_0 - 1)k_B T_0 B R_{A0} \tag{1.8b}$$

To conclude this introductory section, let us determine the noise figure of a matched dissipative attenuator of loss L and temperature T_0. To apply the definition of (1.6a), take into account that signal power at the output is S_{in}/L, whereas the output noise power is the same as the input noise power (because the temperature of the whole system is T_0). Thus, the noise figure is

$$F = \frac{S/N}{\dfrac{S/L}{N}} = L \tag{1.9}$$

Note that in deriving this relationship it was tacitly assumed that the attenuator (regarded as a black box) operates at *electrical* frequencies. We will see later that if the signal is converted to *optical* frequencies, (1.9) may not hold.

1.3 Direct-Modulated Optical Links

The direct intensity modulation of laser diodes can be explained in a very simple manner. It is based on the fact that electrons flowing through the semiconductor diode generate photons. Thus, by modulating the current, for example, via a modulated microwave/millimeter-wave signal, the intensity of the emitted light will be modulated in the same way. In Figure 1.5 laser diode characteristics are shown together with characteristic quantities. For a more detailed discussion of direct modulation of a laser diode, see [4, 5].

Output optical power versus current can be given as

$$P_{\text{opt}}(I) = \frac{hf}{e}\,\eta_L(I - I_{th}); \quad I_S \geq I \geq I_{th} + \delta I \tag{1.10}$$

with η_L the laser quantum efficiency, that is, the average number of generated photons per electron. The system to be discussed next is shown in Figure 1.6. Structures such as those shown in this figure are discussed in [6–10]. Our discussion mirrors that of [6].

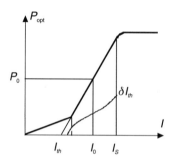

Figure 1.5 Schematic optical power versus current characteristic of a laser diode: I_{th}, threshold current; I_0, bias current; I_S, saturation current; and P_0, average optical power.

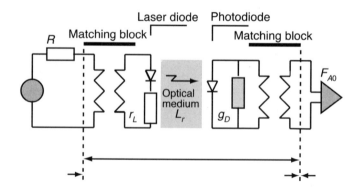

Figure 1.6 Direct-modulated optical microwave link.

According to the figure, the optical link is regarded as a passive microwave transducer. It is assumed that both the impedance of the laser diode and the admittance of the photodiode are real. This can be eventually achieved by adding an appropriate reactance and susceptance in series and in parallel, respectively. Note that problems of wideband matching of these links are not addressed here, and so this approach can be justified. Wideband matching will be discussed briefly in Section 1.3.5.

The real parts of the laser impedance r_L and the photodiode admittance g_D can be adjusted virtually to any magnitude, via the transformers; these transformed magnitudes are designated as R_L and G_D, respectively. The transformers shown in Figure 1.6 can represent any lossless matching circuit. Of course, *practical* transforming networks are not lossless; in our discussion it is tacitly assumed that r_L and g_D include losses of these matching networks.

The optical power, as usual, can be detected either by a p-i-n photodiode or by an avalanche photodiode. As in the present study, the optical link is regarded as a passive microwave component and no optical amplifier is assumed within this link. However, there is a post-photodetector electrical amplifier, characterized by its optimal noise figure F_{A0} and the corresponding generator resistance R_{A0} according to Figure 1.4 and (1.8a) and (1.8b). Note that the amplifier noise figure depends on the generator impedance—in the present case, on the impedance of the photodiode. This may well be different from the noise matching impedance, which will be taken into account later.

1.3.1 Loss Characteristics

According to (1.1), the insertion loss is defined as the available power of the generator divided by the power delivered to the postdetection amplifier input conductance G_A. In subsequent calculations the following convention will be applied: dc currents will be designated by an uppercase I, possibly together with some subscript; high-frequency currents will be designated with a lowercase i; and similarly, constant or average optical power will be designated by P_{opt}, whereas variable or modulated-intensity optical power will be designated by p_{opt}. Concerning electrical power, this distinction is not needed, so an uppercase P will be used.

The high-frequency current flowing through the laser diode, according to (1.2), is

$$i = \frac{2}{R + R_L} \sqrt{\frac{P_a R R_L}{r_L}} \qquad (1.11)$$

where R is the generator resistance, which is normally equal to the characteristic impedance of the connecting transmission line.

The high-frequency component of the optical power generated by the laser is given as

$$p_{opt} = \frac{h \cdot f}{e} \eta_L i \tag{1.12}$$

where h is Planck's constant; f is the optical carrier frequency; e is the charge of the electron; and η_L is the quantum efficiency of the laser diode.

The microwave component of the received optical power that is input to the photodiode is

$$p_{opt,rec} = \frac{p_{opt}}{L_r} \tag{1.13}$$

where L_r is the combined (resulting) loss of the optical components, including coupling loss into and out of the optical fiber.

The source current of the detector is

$$k = \eta_Q M \frac{e}{hf} p_{opt,rec} \tag{1.14}$$

where η_Q is the quantum efficiency of the photodiode and M is the current multiplication factor of the avalanche photodiode at the microwave carrier frequency. In the case of a p-i-n photodetector, $M = 1$.

The output (electrical) power is

$$P_{out} = k^2 \frac{1}{\left(1 + \dfrac{G_D}{G_A}\right)^2} \frac{G_D}{G_A g_D} \tag{1.15}$$

where G_A, as seen, is the amplifier input conductance.

To understand the derivation of (1.15), you should take into account the fact that the output power is $P_{out} = \dfrac{k^2 n^2}{(G_D + G_A)^2} G_A$ where n is the turns ratio of the transformer of Figure 1.6; furthermore, $n^2 = G_D / g_D$. Also note that in (1.15) it is assumed that $g_D > 0$; if $g_D = 0$, which in practical cases might well be approximated, any transformer at the output can be applied, and independently $P_{\text{out}} = k^2 n^2 G_A$. So, finally, the total loss can be expressed as

$$L = bL_r^2 \qquad (1.16)$$

with

$$b = \frac{RG_A}{4(M\eta_L\eta_Q)^2}\left(1 + \frac{R_L}{R}\right)^2\left(1 + \frac{G_D}{G_A}\right)^2 \frac{r_L g_D}{R_L G_D} \qquad (1.17)$$

Equations (1.16) and (1.17) lead to some interesting conclusions. First, note that the electrical loss is proportional to the *square* of the optical loss, that is, one "optical" decibel is equivalent to two "electrical" decibels. Physically, this can be understood by taking into account the fact that *photons* are related to *electrons* (i.e., optical power to detected electric current), and power is proportional to current-squared. Second, the fact that laser impedance is in practical cases low, of the order of a few ohms, while photodiode resistance is high, say, several kilo-ohms, can lead to a significant decrease of microwave loss, or, in principle, to a microwave gain even in the case of a significant optical loss. We will return to this point later.

Before that, however, let us see some matching conditions. Loss is minimal if the matching circuits transform generator and load impedance to the appropriate (matched) impedances, that is, if $R_L = R$, $G_D = G_A$. Then we obtain

$$L_0 = \left(\frac{2L_r}{M\eta_L\eta_Q}\right)^2 r_L g_D \qquad (1.18)$$

Note that *wideband matching* cannot be achieved with reactant circuits in the case of high-Q impedances, restricting the applicability of (1.18) to narrow frequency bands only.

If laser impedance r_L is equal to that of the generator and photodiode admittance g_D to the input admittance of the amplifier, and both input and output are matched (i.e., if $r_L = R = R_L$ and $g_D = G_A = G_D$), the microwave power loss, designated L_1, is

$$L_1 = \left(\frac{2L_r}{M\eta_L\eta_Q}\right)^2 RG_A \qquad (1.19)$$

This can be regarded as the case of *resistive matching*. Transformers are omitted and passive resistors are connected in series and in parallel, respectively, to the diodes to fulfill the matching conditions. Resistive matching is often applied, for the sake of simplicity and also for easing wideband matching requirements. Note that under ideal circumstances ($\eta_L = \eta_Q = 1$, $L_r = 1$, $M = 1$) a loss of 6 dB occurs, due to the power dissipated in the matching resistors.

In the idealized case, in which both $r_L/R \ll 1$ and $g_D/G_A \ll 1$ and transformers are not applied (i.e., laser and photodiode are directly connected to the generator and postdetection amplifier, respectively), resulting in $r_L = R_L$ and $g_D = G_D$,

$$L_2 = \left(\frac{L_r}{2M\eta_L\eta_Q} \right)^2 RG_A \tag{1.20a}$$

Thus, L_2 is, by 12 dB less than L_1, valid in the resistive matched case.

As a last example, let $R_L = r_L$ and $G_D = G_A$ with $r_L \ll R$ (i.e., the laser is connected directly, without a transformer, to the generator while the photodiode is matched to the amplifier). Then

$$L_3 = \left(\frac{L_r}{M\eta_L\eta_Q} \right)^2 Rg_D \tag{1.20b}$$

Let us briefly investigate very low loss links or links with gain. Assume that optimal matching is applied with lossless transformers to result in the insertion loss L_0 of (1.18). Further assume a p-i-n photodiode, that is, $M = 1$. Take $r_L = 5\Omega$ and $1/g_D = 3,000\Omega$; then the link is amplifying as long as the term representing optical loss, $L_r/\eta_L\eta_Q$, is less than 11 dB. The question can be asked: As the power gain needs the power supply, what is the source in this case? To answer this trivial question, we must take into account the fact that a photodetector for correct operation must be reverse biased. If the photocurrent and a high load resistance $1/g_D$ are present, high-bias dc voltage is needed for maintaining the appropriate bias conditions, yielding also supply power for the gain. Similarly, the bias source of the laser diode is the power supply for light generation.

As a numerical example of a low-loss link, assume for a direct modulated link an operational wavelength $\lambda = 1.3 \ \mu$m; let $\eta_L = \eta_Q = 0.8$ and the resulting optical loss = 10 dB (being composed of fiber exciting loss = 3 dB; fiber loss = 3 dB, which corresponds to about a 12-km fiber length; loss of connectors, splices = 2 dB; and illumination loss of the photodetector = 2 dB). Then, according to (1.18), $L_0 = 1.63 = 2.1$ dB, a really very low loss. The corresponding magnitudes, assuming that $R_A = 1/G_D = 50\Omega$, would be $L_1 = 30$ dB (resistive matching) and $L_2 = 18$ dB (nonmatched), respectively.

1.3.2   Noise Figure of Passive Optical Microwave Links

Coming now to the evaluation of the noise figure, there are several noise sources; these will be characterized by noise current generators, as shown in Figure 1.7.

The mean square values of these, respectively, are as follows: For generator noise, k_g:

$$k_g^2 = \frac{4k_B T_0 B}{(R + R_L)} \left(\frac{\eta_L \eta_Q M}{L_r}\right)^2 \frac{R_L}{r_L} \frac{G_D}{g_D} \tag{1.21}$$

Equation (1.21) follows directly from (1.11) through (1.14) if one applies the noise-source-voltage of a resistance R, (1.4).

For shot noise, k_s (denoting by I_d the diode dark current and detected current caused by possible optical background noise; background noise may be present in free-space optical links):

$$k_s^2 = 2eB\left[(I_0 - I_{th})F_M \frac{\eta_L \eta_Q M^2}{L_r} + I_d\right] \frac{G_D}{g_D} \tag{1.22}$$

For laser intensity noise, k_i:

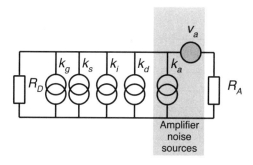

Figure 1.7 Noise sources of an optical link.

$$k_i^2 = (I_0 - I_{th})^2 \text{RIN}(\omega_c)\left(\frac{\eta_L \eta_Q M}{L_r}\right)^2 B \frac{G_D}{g_D} \qquad (1.23)$$

For thermal noise of the photodiode:

$$k_d^2 = 4k_B T_0 B G_D \qquad (1.24)$$

Finally, for amplifier noise, reduced to the input of the amplifier as shown in Figure 1.4 and (1.8a) and (1.8b):

$$k_a^2 = 2(F_{A0} - 1)\frac{k_B T_0 B}{R_{A0}} \qquad (1.25)$$

and

$$e_a^2 = 2(F_{A0} - 1)\, k_B T_0 B R_{A0} \qquad (1.26)$$

where it is assumed that the voltage and current sources are uncorrelated. Relating to k_a and e_a of (1.25) and (1.26), note that we are investigating *non-impedance-matched* situations. Therefore, the simplest, though usually applied, formula of the noise figure of cascaded noisy networks is not applicable—this assumes *matched* networks. This justifies our dealing with a general description of a noisy amplifier (i.e., reducing the amplifier noise into two noise generators placed at the input). However, because our aim is not to investigate noisy electrical amplifiers in detail, the situation is simplified by not taking into account the possible correlation of these two noise sources.

Here, apart from the quantities defined before, F_M is the avalanche photodiode noise figure, usually $2 < F_M < M$; B is the receiver bandwidth; and $\text{RIN}(\omega_c)$ is the laser relative intensity noise measured at the microwave carrier angular frequency ω_c. It is assumed that B is low enough to assume RIN constant in this band. This is a not very important assumption and it is only made for the sake of simplicity.

A few comments should be made with regard to (1.21) through (1.26). The noise temperature of the photodiode can be very different from T_0; for the sake of simplicity T_0 was applied in (1.24); a deeper investigation of this problem is out of the scope of this study.

With regard to (1.22), note that shot noise has in this case two components: *quantum* noise is one of them and photodiode dark current the other one. Further shot-noise-current-squared spectral density is in general $2qI$ [2], with q the elementary charge and I the average current in the quantum noise component. Elementary charge q equals e if a p-i-n photodiode is applied; however, in the case of an avalanche photodiode, the average q is $q = Me$ as one photon generates M electrons on average. Also, in (1.22), (1.10) is taken into account and the average optical power is expressed via the laser dc current. This results in the first term of (1.22), comprising $M^2 \eta_L \eta_Q / L_r$.

Similarly, (1.10) is applied in (1.23) and fluctuation of the optical power intensity is expressed as fluctuation of the laser dc current. This is similar to reducing amplifier noise to the input as was done in (1.7).

As usual, it is reasonable to assume statistical independence of the noise sources. Thus, individual noise contributions are power additive, and the output noise power can be written as

$$N_{\text{out}} = k_N^2 \frac{1}{(1 + G_D/G_A)^2 G_A} \tag{1.27}$$

with

$$k_N^2 = k_g^2 + k_s^2 + k_i^2 + k_d^2 + k_a^2 + v_a^2 G_D^2 \tag{1.28}$$

Taking all of these into account, the noise figure can be written as a second-order polynomial of optical loss L_r:

$$F = 1 + \frac{R_L}{R} + \frac{(R + R_L)^2 r_L}{RR_L}(a_0 + a_1 L_r + a_2 L_r^2) \tag{1.29}$$

with the coefficients of

$$a_0 = \frac{(I_0 - I_{th})^2}{4k_B T_0} \text{RIN}(\omega_c) \tag{1.30}$$

$$a_1 = \frac{e(I_0 - I_{th})}{k_B T_0} \frac{F_M}{\eta_L \eta_Q} \tag{1.31}$$

$$a_2 = \frac{1}{(\eta_L \eta_Q M)^2} \left[\frac{e}{k_B T_0} I_d + g_D + g_D \varphi(F_{A0} - 1) \right] \tag{1.32}$$

$$\varphi = \frac{1}{2} \left(R_{A0} G_D + \frac{1}{R_{A0} G_D} \right) \tag{1.33}$$

The physical reason for the dependence as described in (1.29) is as follows. Noise produced in the generator side is attenuated together with the signal; therefore, in the noise figure expression this does not depend on optical loss. Noise from the generator resistance and laser intensity noise belong to this category. The mean square value of the shot noise current is proportional to the photocurrent. Because this latter is proportional to the optical power, it is attenuated like the optical signal power. However, because the electrical signal power is inversely proportional to the optical loss *squared*, the shot noise contribution to the noise figure [i.e., signal/shot noise at the output = (constant/L_r^2)/(constant/L_r)] is inversely proportional to optical loss, that is, noise contribution is proportional to the optical loss. Finally, the contribution of the unattenuated noise sources (those situated at the receiver end) increases in the same manner in which signal power is attenuated: proportional to the optical loss squared. Noise of the photodiode resistance and amplifier noise, reduced to the input of the amplifier, fall into this category.

The rather complicated dependence of the noise figure on the optical power is schematically shown for various situations in Figure 1.8.

Also, by applying (1.16), one can express the noise figure by the electric loss

$$F = 1 + \frac{R_L}{R} + \frac{(R + R_L)^2 r_L}{RR_L}\left(a_0 + \frac{a_1}{\sqrt{b}}\sqrt{L} + \frac{a_2}{b}L\right) \qquad (1.34)$$

where b is given in (1.17).

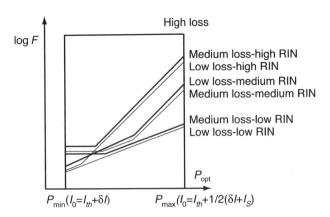

Figure 1.8 Noise figure versus optical power in various situations.

1.3.3 Noise Figure Behavior Under Ideal or Near-Ideal Circumstances

To envision the ultimate possibilities, let us make the following—of course, idealized—assumptions:

- No dark current is assumed: $I_d = 0$.
- No RIN is assumed: $\text{RIN}(\omega_c) = 0$.
- No optical loss is assumed: $L_r = \eta_L = \eta_Q = 1$.

We also assume that a p-i-n photodiode is applied: $M = F_M = 1$. In this case the noise figure can be written as

$$F = 1 + \frac{R_L}{R} + \frac{(R + R_L)^2 r_L}{R R_L} \left\{ \frac{e(I - I_{th})}{k_B T_0} + g_D [1 + (F_A - 1)\varphi] \right\}$$
(1.35a)

To idealize further, let the shot noise also be very low—this is the case if the optical power is very low:

$$P_{\text{opt}} \approx 0, \text{ that is, } (I - I_{th}) \approx 0$$

and the amplifier ideal: $F_{A0} = 1$. The noise figure in this case is

$$F = 1 + \frac{R_L}{R} + \frac{(R + R_L)^2 r_L g_D}{R R_L}$$
(1.35b)

In this overidealized case, F can approach 1 if $R_L = r_L$ is chosen and both r_L and g_D are low. But, as already seen, this is the case in real situations: Laser diodes have very low resistance, say, 3–5Ω, while that of photodiodes is very high (of the order of kilo-ohms or even more). That is,

$$F = 1 + \frac{r_L}{R} + g_D R \ (\approx 1)$$
(1.35c)

Equation (1.35c) conflicts with some statements in the literature [7] according to which $F \geq 2$. As seen, in reality it can be less than 2 and can approach 1. However, if the laser diode must be matched to the source resistance (i.e., $R_L = R$ is required), then

$$F = 2 + 4r_L g_D \ (\approx 2) \tag{1.35d}$$

As seen in the matched situation, the $F > 2$ limit holds. A rather interesting result: Although matching is optimal for minimizing loss, it can be suboptimal for minimizing the noise figure; of course, if link loss is of a determining magnitude, that is, if the bracketed term in (1.34) is high, loss has to be kept low to result in a low noise figure. By comparing L_0 and L_3 of (1.17) and (1.20b), we can see that the latter is higher by a factor of $R/2r_L$.

1.3.4 Comparison of the Noise Figure Behavior of Electrical and More Realistic Optical Links

It is well known that in a system containing only electrical components at reference temperature, the noise figure is always higher than or equal to the link loss. As shown at the end of Section 1.2, equality holds in the passive matched case. The question can be asked if this is true for systems that also contain optical elements.

Furthermore, a theoretical minimum noise figure in such a link is also of interest. To get a first impression on this problem, assume an ideal situation in which only the shot noise is nonnegligible. This would be possible in a RIN-free and high-loss case if the postdetection amplifier has a very low noise figure and g_D is very low. In this case, that is, if $a_2 = a_0 = 0$ in a high-loss link, the noise figure is proportional to the *square root* of the (electrical) loss; therefore, above a certain link loss, the noise figure *must* be less than the loss itself. Consequently, in contradiction to electric circuits, the theoretical noise figure in an optical link *can* be less than the electric link loss. As already stated, physically this is possible because parts of the *electrical* link operate at *optical* frequencies and in this band noise temperature is 0; see (1.38).

To discuss in detail, under the assumptions made and assuming further that $\eta_L = 1$ (i.e., the number of photons generated in the laser is equal to the number of exciting electrons), then

$$F = 1 + \frac{r_L}{R} + \frac{(R + r_L)^2}{R} \frac{a_1}{\sqrt{b}} \sqrt{L} \approx R \frac{a_1}{\sqrt{b}} \sqrt{L}; \tag{1.36}$$

$$a_1 = \frac{e(I_0 - I_{th})}{k_B T_0} = \frac{e^2 P_{opt}}{hf \cdot k_B T_0}; \ b = \frac{1}{4}$$

Figure 1.9 shows the result of a calculation in an idealized case: $\lambda = 1.55$ μm, $R = 1/G_A = 50\Omega$, $r_L = R_L \ll R$, $g_D = G_D \ll G_A$, and a laser optical power of 1 mW was assumed. In the region of high attenuation the application of the above formulas leads to:

$$F^{[dB]} \cong \frac{1}{2} L^{[dB]} + 6.8 \qquad (1.37)$$

Thus, in this ideal case, $F < L$ if $L > 13.6$ dB.

If none of the noise sources is neglected, there are three regions where different causes are predominant as depicted in Figure 1.10. As long as the loss is low, RIN and/or generator noise is predominant and thus noise figure is constant. Increasing the loss decreases these contributions, and loss-dependent ones become significant. If the noise is low, shot noise is the main factor at low loss-magnitudes and thus noise figure is proportional to square-root loss. By increasing the loss, the noise figure increases linearly. However, if noise is high, the central part of the curve is missing because shot noise is suppressed by RIN and by thermal noise.

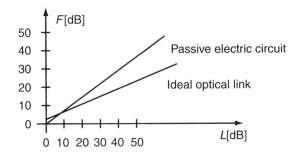

Figure 1.9 Noise figure versus link loss in an ideal situation.

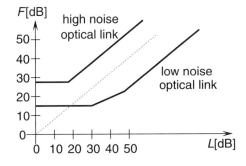

Figure 1.10 Qualitative form of the noise figure versus link loss.

In real cases, of course, a_2 is nonzero and therefore the last term of (1.29) becomes predominant if the optical loss is high. In such situations the noise figure is less than the link loss if the following condition

$$F_{A0} < 1 + \frac{\frac{1}{4}\left(\sqrt{\frac{G_A}{G_D}} + \sqrt{\frac{G_D}{G_A}}\right)^2 - 1}{\frac{1}{2}\left(R_{A0}G_D + \frac{1}{R_{A0}G_D}\right)} \tag{1.38}$$

is fulfilled, where the dark current is taken to be negligible for simplicity; in practical cases a few decibels of noise figure can be gained by means of an appropriate choice of amplifiers and appropriate matching conditions.

This latter result can be significant, for example, in optical distribution networks of fixed or mobile microwave or millimeter-wave communication systems. Link loss in these systems can be rather high in, among others, the downlink of picocellular systems. Reducing the noise figure can be a very important task.

To understand the numerical magnitudes, let us look at the high-impedance case again. Take $G_D = g_D << G_A$ and take also $g_D << 1/R_{A0}$. The approximate form of (1.38) is then

$$F_{A0} < 1 + \left(\frac{1}{4}\frac{G_A}{g_D} - 1\right)2R_{A0}g_D \approx 1 + \frac{1}{2}R_{A0}G_A$$

If $R_{A0}G_A = 1$ is taken, the above formula yields $F_{A0} < 1.5$ dB, a rather favorable noise figure. Applying, say, $F_{A0} = 1$ dB $= 1.25$ into (1.34), we get $F = L/2$. The noise figure is 3 dB less than the insertion loss. Also, applying the appropriate equation of (1.20a), noise figure in this example is

$$F = \frac{1}{8}\left(\frac{L_r}{M\eta_L\eta_Q}\right)^2 RG_A; \quad L >> 1$$

1.3.5   Some Points of Practical Interest: Bandwidth, Dynamic Range, and Measurement Results

Two parameters of a transmission system not discussed yet are its bandwidth and its dynamic range. In this section, although *modulation bandwidth* of the laser diode will be mentioned, we do not deal with bandwidth of the optical part of the system. In doing so we assume that these are of much wider bandwidth than required by the wireless system in which fiber radio is applied. As will be seen in Section 1.5, this is not always true, due to the particular adverse effects of the dispersion of the optical fiber.

These are particularly pronounced if the RF modulating light is rather high, say, 60 GHz or so, being a special candidate for future wireless systems. However, these effects are not related to basic microwave characteristics, which are the subject of this chapter. Our concern now is with the achievable *bandwidth* of the matching circuits. Up to now these were assumed to be simple transformers, as shown in Figure 1.6. Further, both laser and photodiode impedances were assumed to be *real*, although this was not a realistic assumption. Our discussion of the achievable bandwidth is very much simplified: The simplest models of the laser diode and of the photodiode are used, that is, circuit equivalents containing one reactant parasitic element only. In photodiodes we assume that parasitics are the only reason for bandwidth limitations. More detailed discussions can be found in [11–14].

Note that in this application, in contrast to digital optical transmission, bandpass rather than lowpass matching is needed. Because the high-resistance photodiode always contains a parasitic shunt capacitor and the low-resistance laser diode always contains a parasitic series inductor, the simplest method of matching is to apply a parallel or a series resonant circuit, respectively, together with the transformer shown in Figure 1.6. It is well known that the 3-dB bandwidth of a resonant circuit is

$$B = \frac{g}{2\pi C}$$

$$B = \frac{r}{2\pi l}$$
(1.39)

for parallel and series circuits, respectively. In (1.39) g is the reciprocal load resistance of the parallel circuit and r is the resistance of the series resonant circuit; C and l are capacitance and inductance, respectively. We also know that more complicated reactant matching of circuits does not allow for much gain in bandwidth [15]. Thus, parasitics of the light source and the photodetector put limits on the minima of r_L and g_D, and therefore also on the minimal loss of the fiber optic link. Usually $r_L/L \gg g_D/C$. Therefore, to achieve the appropriate bandwidth, g_D must be increased by adding a parallel resistor to the diode. Taking this into account and applying the first equation of (1.39) in (1.18) we get for the minimal link loss, valid in the case of lossless matching,

$$L_0 = \left(\frac{2L_r}{M\eta_L\eta_Q}\right)^2 2\pi CBr_L \qquad (1.40a)$$

As a numerical example, let us deal again with the situation examined at the end of Section 1.3.1. If capacitance of the photodiode is 2 pF and a bandwidth of 100 MHz is desired, photodiode resistance has to be decreased from 3 kΩ to about 800Ω (i.e., it has to be shunted by 800Ω). This leads to an increase of the link loss by 5.8 dB; the computed loss is $L_0 = 7.9$ dB.

As already mentioned, the situation we are dealing with differs significantly from the more usual one in which digital signals are transmitted by the optical link. In this latter case, a lowpass match is needed while we can deal now with bandpass matching. For example, in a digital link carrying 10 Gbps, the useful bandwidth extends from nearly 0 up to about 5 or 10 GHz, depending on the signaling format applied. In an optical microwave link operating at 10 GHz microwave frequency and carrying a 156 Mbps QPSK signal, the useful bandwidth is about 10 GHz ± 40 MHz.

Consequently, as also seen from the above numerical example, even at reasonable bandwidths relatively high load impedance of the photodiode can be used. What can cause a problem is the *relative* bandwidth, or, in other words, the loaded Q factor versus the unloaded Q of the applied transmission line. If the loaded Q is too high, a nonnegligible additional loss will be caused by the matching circuit. It is well known that the loss of a resonant circuit can be expressed as

$$L_m = \frac{1}{(1 - Q_L/Q_U)^2} \qquad (1.40b)$$

where Q_L and Q_U represent the loaded and unloaded Q, respectively, and the subscript m designates the loss of the matching circuit. For example, for $L_m < 0.5$ dB, the loaded Q must be at least 20 times lower than the unloaded one.

Coming now to the problem of dynamic range, the maximum power of the transmitted signal is limited by laser saturation (and, of course, minimum power by noise). With a sinusoidal current superimposed on a laser bias current, the rms amplitude of this sinusoidal current is limited by

$$i \le (I_0 - I_{th} - \delta I_{th})/\sqrt{2} \qquad (1.40c)$$

$$I_0 \le (I_S - I_{th} - \delta I_{th})/2$$

with currents as defined in Figure 1.5.

Bias current should be chosen very carefully because link performance depends on it and, consequently, on the optical power in a rather complicated way. This is particularly the case if RIN is not negligible (in most cases, it is not). The dependence is through several parameters: RIN, noise figure, and modulation bandwidth depend on I_0. As discussed in the literature, RIN depends on frequency and on optical power [16]. A typical plot is shown in Figure 1.11. The peak of the RIN curve coincides approximately with the relaxation oscillation frequency of the laser.

Dependence of the noise figure on bias current can be seen in (1.30) and (1.31): The RIN contribution is proportional to $(I_0 - I_{th})^2$, while the shot noise contribution is proportional to $(I_0 - I_{th})$. Thus, noise figure is decreasing in some cases if bias is set not much higher than threshold current.

However, when bias current is decreasing, we must take into account that besides decreasing the dynamic range, modulation bandwidth is also decreased.

In Tables 1.1 and 1.2, computed parameters of links with four different RIN lasers are given. In Table 1.1 noise figures are given in various matching situations; F_0 stands for the matched case with a lossless matching network; F_1 for resistive match; and F_2 for the nonmatched version. The following parameters are applied in the Table 1.1 and later in Table 1.2: $r_L = 5\,\Omega$; $F_{A0} = 1$ dB, $1/g_D = 3{,}000\,\Omega$; $G_A = 1/R_0$; $L_r/\eta_L\,\eta_Q = 13$ dB; and $B = 100$ MHz.

Parameters used in Table 1.1 are such that optical loss is of a realistic, but not too low, magnitude. In laser 1 RIN is very high; in fact, it is of determining magnitude. Therefore, the noise figure increases with the applied optical power: Although RIN is decreasing [16], there is the coefficient of I^2. RIN and shot noise have nearly equal weight in lasers 2 and 3. RIN is negligible in laser 4; shot noise is of determining magnitude in that case.

While for relatively high-loss links, the noise figure will be minimal if lossless matching is applied, an optimal mismatch of the laser occurs in the

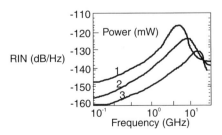

Figure 1.11 RIN versus frequency at a few optical power levels.

Table 1.1
Noise Figure for Different RIN Lasers

Laser Number	$I_0 - I_{th}^{mA}$	$RIN(\omega_c)^{dB/Hz}$	F_0^{dB}	F_1^{dB}	F_2^{dB}
1	1	−105	46	56	50
	10	−111	60	70	64
2	1	−135	18	32	22
	10	−141	31	41	35
3	1	−144	14.5	31	19
	10	−150	24.5	36	28
4	1	−150	13	23	17
	10	−180	22	33	28
	40	−190	28	38	33

Table 1.2
Maximal Attainable S/N Values in Fiber Optic Links of Table 1.1

Laser Number	$I_0 - I_{th}^{mA}$	$S/N_{max}^{dB} (F_0)$	$S/N_{max}^{dB} (F_1)$
1	1	16	6
	10	21	11
2	1	43	29
	10	50	40
3	1	46.5	30
	10	56.5	44.5

case of low-loss links. For example, if shot noise is the determining factor of noise figure (as in the case of laser 4 in Table 1.1), it is easy to compute that noise figure is minimized with

$$\frac{R_L}{R} = \sqrt{\frac{A}{A+1}}; \quad A = r_L \frac{e(I_0 - I_{th})}{k_B T_0} \frac{F_M L_r}{\eta_L \eta_D}$$

In principle, $F < 3$ dB can be achieved. Take again an example: Assume laser 4 of Table 1.1, with $I_0 - I_{th} = 1$ mA, $r_L = 3\Omega$, and $L_r = 1$ (of course, this is an unrealistic assumption); then the optimal $R_L = 16\Omega$ and $F = 2$ dB.

The double-limited form of laser characteristic (i.e., it is peak and minimum-limited) results in the maximal attainable S/N for an optical link:

Noise is determined by the noise figure and signal is maximized by laser characteristic. In Table 1.2 computed maximum S/N values are given, related to F_0 and F_1 of Table 1.1. Note that dynamic range (DR) is given as

$$DR = \frac{(S/N)_{max}}{(S/N)_{min}}$$

where S/N$_{min}$ is the minimal S/N needed and the S/N$_{max}$ values are given in Table 1.2. We can see that while noise figure is higher with higher optical power (or I_0), S/N$_{max}$ and, consequently, dynamic range are also increasing. To conclude this section, some measurements are given, after [17]. Lasers corresponding to lasers 1 and 2 were used, with link parameters listed above, except that bandwidth was lower (50 MHz). Resistive matching was used. The link loss agreed within 0.5 dB with L_1 of (1.19) and the basic noise figure was in good agreement with the values of Table 1.1. Noise figure versus additional optical attenuation is shown in Figures 1.12 and 1.13. As seen, the behavior of the link with a low-RIN laser follows well the lower curve of Figure 1.10, whereas the link of the high-RIN laser follows the higher link.

Other measured results are given in [7, 8].

1.3.6 Improving Performance with a Low-Noise Preamplifier

As seen earlier, when applying optical links for the transmission of microwave signals (among others) three problems arise: Dynamic range can be too low;

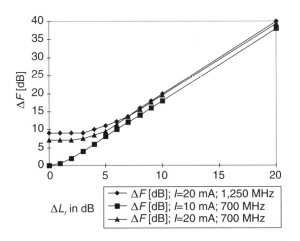

Figure 1.12 Noise figure versus additional optical loss for laser 1.

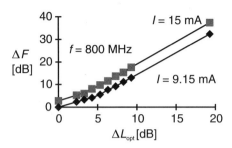

Figure 1.13 Noise figure versus additional optical loss for laser 2.

S/N can be too low; and even if these are acceptable, sensitivity can be insufficient. The latter means that electrical power needed to attain appropriate S/Ns can be too high. By applying a low-noise preamplifier (see Figure 1.14), all of these can be facilitated. Of course, the *maximal* attainable S/N *is* defined by F_2 and the maximal modulating power S_{max} and cannot be increased. As a matter of fact, it is somewhat *decreased* if a preamplifier is used; however, this decrease is not significant if $F_1 \ll F_2$.

In the following, we assume that the laser is matched to the preamplifier, either by a lossless transforming network ($R_L = R \gg r_L$) or by a series resistor (when $R_L = R = r_L$); in a nonmatched case, load and generator resistances and their noise and power have to be taken into account in a similar way as was done in Sections 1.3.1 and 1.3.2. It is well known that resulting noise figure F of two cascaded matched two-ports can be computed as

$$F = F_1 + \frac{F_2 - 1}{G} \tag{1.41}$$

where F_1 and G are noise figure and gain of the first stage, respectively, and F_2 is the noise figure of the second stage. System sensitivity can significantly be improved with a constant-gain low-noise amplifier, at the expense of a (usually not too large) decrease in dynamic range. In designing this link, two of the three parameters—the minimal signal to noise, S/N_{min}, dynamic

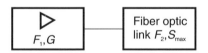

Figure 1.14 Fiber optic link with preamplifier; parameters are assumed to be given.

range, DR, and minimum acceptable input signal, S_{1min}—are specified. S_{1min} is given as

$$S_{1min} = \frac{S_{max}}{G \cdot DR} = F k_B T_0 \left(\frac{S}{N}\right)_{min} \tag{1.42}$$

Putting (1.41) into (1.42) and making elementary transformations, we obtain for the upper limit of the gain of the preamplifier

$$G \le \frac{1}{F_1}\left[\frac{S_{max}}{k_B T_0 B}\frac{1}{DR(S/N)_{min}} - (F_2 - 1)\right]; \quad DR < \frac{S_{max}}{F_2 k_B T_0 (S/N)_{min}} \tag{1.43}$$

If besides improving sensitivity, an increase in the dynamic range is also required, a preamplifier with automatic gain control and a constant output power of S_{max} should be applied. Then the gain is

$$G_{max} \ge G \ge G_{min} \ge 1$$

There exists an absolute maximum of the gain and thus an absolute minimum of S_{1min} determined by the noise figures and S_{max}. Noise figure is minimal if gain is maximal [see (1.41)], and so the lowest input level that fulfills specifications is given as

$$S_{1min} = F_{min} k_B T_0 (S/N)_{min} \tag{1.44}$$

gain being limited, however, by

$$G_{max} S_{1min} = S_{max} \tag{1.45}$$

Equations (1.41), (1.44), and (1.45) result in an upper limit for the maximal gain of

$$G_{max} \le \frac{1}{F_1}\left[\frac{S_{max}}{k_B T_0 B(S/N)_{min}} - (F_2 - 1)\right] \tag{1.46}$$

a lower limit for S_{1min}

$$S_{1\min} \geq \frac{F_1 k_B T_0 B (\text{S/N})_{\min}}{1 - (F_2 - 1) \dfrac{k_B T_0 B (\text{S/N})_{\min}}{S_{\max}}} \tag{1.47}$$

and an upper limit for the dynamic range (DR) of

$$\text{DR} = \frac{S_{\max}}{S_{1\min}} = G_{\max} \leq (G_{\max})_{\max} \tag{1.48}$$

Finally, note that if the input signal is maximal (i.e., equal to $S_{1\min}$) and so $G = 1$, noise figure F is higher than F_2; thus the *maximal* S/N is somewhat decreased due to the presence of the amplifier, namely,

$$F = F_1' + F_2 - 1 \tag{1.49}$$

where $F_1' > F_1$ is the preamplifier noise figure if G is low. For example, if the amplifier is composed of n identical stages, each of which has a noise figure f, then

$$F_1' = nf - n + 1 \tag{1.50}$$

1.3.7 Application of Optical Amplifiers

Until now, the microwave fiber optic link itself was regarded as a passive electrical transducer followed by and possibly also preceded by an electrical amplifier. As in point-to-point optical transmission systems, the fact that this applied configuration does not take optical amplifiers into account is justified. However, in optical *distribution* systems the application of an optical amplifier is more or less a must. The reason for this is the particular dependence of the electrical loss on optical loss, that is, that electrical insertion loss is proportional to the optical loss squared; see (1.16). If optical power is distributed among several users, the result is a particular *distribution loss.*

Assume first a single input/single output link of insertion loss L; if the optical power is divided into n equal parts (e.g., by the application of a lossless star coupler), transmitted to n detectors via n identical links, the detected electrical current at one output will be decreased by a factor n and the electrical power by a factor n^2. So the sum-electrical-power (n times the power of an individual receiver) is $n/n^2 = 1/n$ times lower than in the single-output link; hence, a distribution loss of $1/n$ is produced. To compensate for that, an optical amplifier (of gain at least n) should be applied.

In recent years a tremendous amount of research has been done in the field of optical amplifiers. From the system point of view, an erbium-doped fiber amplifier operating in the 1.55-μm frequency band proved itself to be the most appropriate. Our aim is not to deal too much with optical amplifiers, so we refer to the literature [16, 18]. From our point of view, as with in electrical amplifiers, the two main parameters are their gain (G_{oa}) and noise figure (F_{oa}), where the subscript means *optical amplifier.*

The role of the gain is self-evident: If an amplifier is applied in all of the formulas of the present and the next section, optical loss L_r has to be replaced by L_r/G_{oa}. However, although the basic definition of noise figure is the same as in electrical networks [see (1.6)], it must be kept in mind that in this case the input S/N is not determined by thermal noise but rather by shot noise. Therefore, the noise figure of an optical amplifier is

$$F \triangleq \frac{(S/N)_{in}}{(S/N)_{out}} = \frac{P_{opt,rec}}{2hfB \cdot (S/N)_{out}}$$

The nonideal noise behavior of an optical amplifier ($F_{oa} > 1$) has the only formal consequence in our discussion of replacing F_M by $F_M F_{oa}$ in every relevant formula.

1.4 Optical Links Operating with External Intensity Modulators

Optical intensity modulation can be achieved in the easiest way by applying *direct modulation,* which formed the basis of our discussions in the previous section. There are, however, a few reasons that could rule out its application. The first one is bandwidth: Although lasers with modulation cutoff frequencies up to about 40 GHz were reported [1], commercially available laser diodes usually have cutoff frequencies of about 10 GHz or below.

Further, it can be shown that a change in the laser current also results in a change in the optical frequency [19]. This phenomenon, called *chirping,* results in frequency modulation superimposed on the (wanted) intensity modulation. This may or may not be harmful. However, fields modulated by an external modulator are virtually free from chirping.

A third advantage of the external modulator over direct modulation could be the higher achievable gain. In Section 1.3.7, we saw that a microwave optical link might have gain; however, in direct modulation links this is restricted to low values and to rather special situations. As will be shown later, gain of an externally modulated link can be increased by increasing the optical power. The disadvantages of external modulators are twofold: (1) Being a separate component, its application makes the design more complex, and (2) it causes optical loss.

External optical modulator functioning is based on at least three different principles: electro-optical, electroabsorption, and interferometric modulators. To operate the latter, however, one of the former effects is needed. Basic designs or their refined variants can be operated well up to the millimeter frequency band or above, to 100 GHz, 300 GHz, or even 1 THz. More or less detailed descriptions of their operation, in particular the physical basis of operation, can be found in [4, 12, 16]. The loss and noise characteristics of these types of microwave optical links are investigated in [6, 8, 20].

In spite of some of its disadvantages, the modulator based on a Mach-Zehnder interferometer is the most widely applied. In this section first the operating principle and the most important characteristics of this type of modulator are described and then a loss/gain and noise figure analysis similar to that of Section 1.3 is presented. For the most part, our analysis will follow that of [6].

1.4.1 The Mach-Zehnder Interferometric Optical Modulator

A simplified diagram of a Mach-Zehnder modulator is shown in Figure 1.15. Light is propagating in a plane optical waveguide realized on a substrate made of an electro-optic material (usually a lithiumniobate), the refraction index of which is controllable by an applied electric field.

At cross section A the input optical field is divided into two halves. If the modulating signal is applied on the electrodes, the phase shift of the optical field in the upper branch of the structure will vary according to the temporal variation of the signal, while that of the lower branch will remain

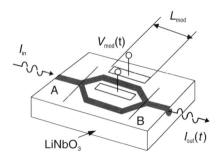

Figure 1.15 Schematic representation of a Mach-Zehnder modulator.

constant. At cross section B the two optical fields interfere with each other, resulting in a modulated amplitude and consequently in a modulated intensity.

From the electrical point of view, the electrodes can be regarded either as the two plates of a capacitor or as a coplanar transmission line, depending on the length of the electrodes relative to the wavelength. Both approaches will be dealt with briefly in the sequel.

Note that the design shown in Figure 1.15 is the simplest one and only this will be dealt with here. Applying different and more complex structures for electrodes having various special characteristics can be achieved: push-pull operation, single-sideband modulation, and others. These will not be discussed here, but reference is made to [11, 21, 22].

To see the functioning of this type of modulator, assume that the structure is lossless and symmetric. The phase constant of the *unbiased* optical waveguide is β_0, while it is $\beta_m(t)$ if a time-varying voltage is applied on the electrodes. Then the rms complex envelope of the optical field strength in point B is

$$E(t) = \frac{E_{\text{in}} e^{j\phi}}{\sqrt{2}} (e^{-j\beta_0 L_{\text{mod}}} + e^{-j\beta_m(t) L_{\text{mod}}}) \qquad (1.51)$$

$$= \frac{E_{\text{in}} e^{j(\phi - j\beta_0 L_{\text{mod}})}}{\sqrt{2}} (1 + e^{j[\beta_0 - \beta_m(t)] L_{\text{mod}}})$$

where E_{in} is the rms value of the optical field strength at A, the input Y junction; and ϕ is the phase shift caused by the connecting lines, which is insignificant in the present discussion.

Let us first assume that the length of the electrodes is much less than the wavelength of the modulating microwave signal. The phase constant along the modulator plates is then constant as driving voltage is constant. The optical power can be written as

$$P_{\text{opt}}(t) = |E|^2(t) = \frac{E_{\text{in}}^2}{2} \{1 + \cos[(\beta_0 - \beta_m(t)) L_{\text{mod}}]\} \qquad (1.52)$$

As seen, the argument of the cosine function is the phase difference of field strengths arriving along the upper and the lower branches of the interferometer.

Assume that the change of the phase constant is proportional to the modulating voltage. Then β_m can be written in the following form:

$$\beta_m(t) = \beta_0 - \frac{\pi}{L_{\text{mod}}} \frac{u(t)}{V_\pi} \tag{1.53a}$$

and so the output optical power of this intensity modulator is as follows:

$$P_{\text{opt}}(t) = \frac{P_{\text{opt,in}}}{2} \left(1 + \cos \pi \frac{u(t)}{V_\pi} \right) \tag{1.53b}$$

where:

$P_{\text{opt,in}} = E_{\text{in}}^2$ is the input power to the modulator;
$u(t)$ is the modulating voltage;
V_π is the voltage needed to cause a differential phase shift of π in the upper arm.

Of course, if $u(t) = V_\pi$, there will be 0 output power. Note that V_π is inversely proportional to the modulator length.

Thus, the optical (intensity) *transfer function* of the modulator is

$$\frac{P_{\text{opt}}(t)}{P_{\text{opt,in}}} = \frac{1}{2} \left(1 + \cos \pi \frac{u(t)}{V_\pi} \right) \tag{1.54}$$

Note that the transfer function is a periodic function of the voltage, with period $2V_\pi$.

Let the modulating voltage be a dc bias plus a sinusoid, which, for the sake of simplicity, is unmodulated. Optical intensity will then be

$$P_{\text{opt}}(t) = P_{\text{opt,in}}/2 \left[1 + \cos \frac{\pi}{V_\pi} (U + \sqrt{2} u_0 \cos \omega_c t) \right] \tag{1.55}$$

where:

U is the dc bias voltage;
u_0 is the rms amplitude of the modulating sinusoid;
ω_c is the RF angular frequency.

It is not our intention to discuss in detail the properties of a Mach-Zehnder modulator. We should mention, however, that there is a nonlinear relationship between the output intensity and the modulating signal, and the behavior of the modulator depends very much on the bias voltage relative to V_π. If $U = V_\pi/2$, the response may be close to linear, in particular if the modulation index $u_0\pi/V_\pi$, is small; if $U = V_\pi$, the optical carrier is suppressed, and in the output intensity the modulating signal exhibits only even harmonics; with $U = 2V_\pi$, the carrier is maximum and, again, there are no odd harmonics. These characteristics are utilized in various applications; in our present investigations, we assume a quasilinear regime, with bias voltage $U = V_\pi/2$.

In this case the RF modulated optical power p_{opt} can be written as

$$p_{opt}(t) = \frac{P_{opt,in}}{2}[1 - \sin(m\cos\omega_c t)] \approx \frac{P_{opt,in}}{2}(1 - m\cos\omega_c t)$$

$$(1.56)$$

with $m = \sqrt{2}\pi u_0/V_\pi$. The approximate equation is valid for $m \ll 1$.

For the sake of completeness, we give the Fourier expansion of the output optical intensity in the general case. For any m:

$$p_{opt}(t) = \frac{P_{opt,in}}{2}\left[1 - \sum_{k=0}^{\infty}(-1)^k J_{2k+1}(m\pi)\cos(2k+1)\omega_c t\right]$$

$$(1.57a)$$

with J_s being the sth-order Bessel function of the first kind.

From the electrical point of view, the modulator electrodes are regarded basically as a capacitor. This can be, however, lossy, and a parallel matching resistance can be applied. Further, a series inductor can be applied in order to tune the capacitor to resonance. Thus, the general electrical block schematic of an optical link with a Mach-Zehnder modulator is as shown in Figure 1.16. As can be seen in this figure, the "optical medium" and the receiver part are the same as those shown for the direct modulated link in Figure 1.6.

Because we are dealing with links operating up to the millimeter-wave band, the RF wavelength can be rather low and, hence, electrode length L_{mod} can be of the order of this wavelength. In this case the electrodes form the two conductors of a transmission line (actually an asymmetric coplanar line) and in order to be efficient a *traveling*-wave modulator should be applied. The electrical equivalent of the link in this case is shown in Figure 1.17. No tuning coil is needed and r_M is equal to the characteristic impedance of the coplanar transmission line.

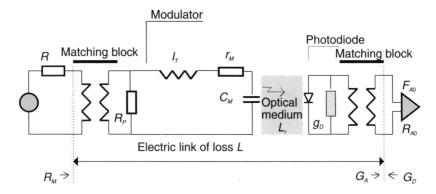

Figure 1.16 Externally modulated optical link: l_t, tuning inductor; r_M, C_M, resistance and capacitance of the modulator electrodes, respectively; and R_P, possible matching resistor.

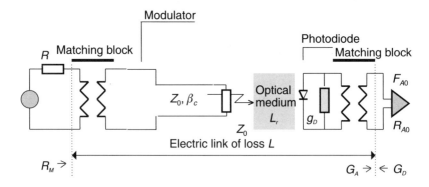

Figure 1.17 Optical microwave link with traveling-wave external modulator.

The modulating voltage along the coplanar line in this case can be written as

$$u(x, t) = \sqrt{2}u_0 \cos\left(\omega_c t - \frac{\omega_c n_c x}{c}\right)$$

where:

n_c is the refractive index at the RF frequency;

x is the spatial coordinate along the modulator;

c is the velocity of light.

To determine the (variable) phase shift, let us deal with an optical wave, the instantaneous magnitude of which is $E_{opt}(x = 0, t_0)$; then the modulating voltage $u(x_1, t)$ is effective at instant $t_0 + x_1 \cdot n/c$, where n is the refractive index at optical frequencies. (The optical wave needs $x_1 \cdot n/c$ seconds to reach point $x = x_1$. Note that in the general case $n \neq n_c$.) Taking this into account, the phase shift in the upper arm of the modulator can be written as

$$\varphi = \beta_0 L_{mod} + \frac{\pi}{V_\pi} \left[U + \int_0^{L_{mod}} \sqrt{2} u_0 \cos \omega_c \left(t_0 - \frac{n_c - n}{c} x \right) dx \right]$$

leading, after a few trigonometrical manipulations, to [21], the intensity transfer function

$$\frac{P_{opt}(t)}{P_{opt,in}} = \frac{1}{2}$$

(1.57b)

$$\left\{ 1 + \cos \frac{\pi}{V_\pi} \left[U + \sqrt{2} u_0 \frac{\sin \frac{\omega_c L_{mod}(n_c - n)}{2c}}{\frac{\omega_c L_{mod}(n_c - n)}{2c}} \cos \left(\omega_c t - \frac{\omega_c L_{mod}(n_c - n)}{2c} \right) \right] \right\}$$

Note that the sensitivity of the modulator is at a maximum if the velocities of the optical and RF waves are equal (i.e., $n = n_c$). As before, in this case V_π is inversely proportional to L_{mod}, whatever L_{mod} is. Otherwise (and usually), $n \neq n_c$. Due to the factor $\sin x/x$, the required bandwidth determines a maximal L_{mod}. In turn, this determines V_π, the sensitivity of the modulator.

As in the case of a lumped-capacitor modulator, taking $U = V_\pi/2$, the modulator is quasilinear and we can write:

$$P_{opt}(t) \approx \frac{P_{opt,in}}{2} [1 - m' \cos(\omega_c t - \psi)]$$

(1.57c)

with

$$\psi = \beta_c L_{mod} \frac{n_c - n}{2}; \quad m' = m \frac{\sin \psi}{\psi}$$

(1.57d)

where β_c is the free-space phase constant at the RF carrier frequency.

Note that methods exist, based on the appropriate design of the electrode's shape, to decrease the *effective* difference $n - n_c$, below its true magnitude [9]; these are, however, outside of the scope of this chapter.

1.4.2 Loss in Externally Modulated Optical Microwave Links

A link with a modulator of lumped capacitance will be considered first. In this situation, coupling from the electrical voltage source to the optical medium is the most intensive if the RF voltage on the capacitor is maximal. This occurs if the current flowing through the capacitor is maximal. Thus, in the case of so-called series match R_P is omitted and the series resonant circuit tuned to resonance at the RF carrier, ω_c. The current flowing through the resonant circuit is then

$$i = \frac{2}{R + R_M} \sqrt{\frac{P_a R R_M}{r_M}} \tag{1.58}$$

and the voltage of the capacitor

$$u_0 = -j \cdot i r_M Q; \quad Q = 1/\omega_c r_M C_M \tag{1.59}$$

with Q as the Q factor of the series resonant circuit. Thus, the mean optical power [see (1.56)] is

$$P_{\text{opt}} = \frac{P_{\text{opt,in}} u_0}{V_\pi} = \frac{2 P_{\text{opt,in}} Q}{V_\pi} \frac{\sqrt{P_a R R_M r_M}}{R + R_M} \tag{1.60}$$

The received optical power (i.e., that at the input of the photodetector) is

$$P_{\text{opt,r}} = \frac{2 P_{\text{opt,in}} Q}{V_\pi L_r} \frac{\sqrt{P_a R R_M r_M}}{R + R_M} \tag{1.61}$$

where, as before, L_r is the resulting optical loss. In the present case, this contains the loss of the modulator (usually this is a significant loss, of the order of 5–7 dB), the loss of the fiber itself, together with its connectors and splices, and the illumination loss of the photodetector. The RF photocurrent generator is then

$$k = \eta_Q M \frac{e}{hf} \frac{2P_{\text{opt,in}} Q}{V_\pi L_r} \frac{\sqrt{P_a R R_M r_M}}{R + R_M} \tag{1.62}$$

and the output electrical power is

$$p_{\text{out}} = \left(2\eta_Q M \frac{e}{hf}\right)^2 \left(\frac{P_{\text{opt,in}} Q}{V_\pi L_r}\right)^2 \frac{R_M r_M}{R\left(1 + \frac{R_M}{R}\right)^2} \frac{G_D}{G_A g_D \left(1 + \frac{G_D}{G_A}\right)^2} P_a \tag{1.63}$$

In these formulas resistances are as shown in Figure 1.16 and

f is the optical frequency;
f_c and ω_c are RF carrier frequency and angular frequency, respectively;
h and e are Planck's constant and the electronic charge, respectively;
η_Q is the quantum efficiency of the photodetector;
M is the multiplication factor of the avalanche diode, if this is applied.

With (1.63) for the insertion loss of an externally modulated optical link with a series match applied for the modulator, we have again, as in (1.16),

$$L = b \cdot L_r^2 \tag{1.64}$$

and now

$$b = \left(\frac{hf}{2\eta_Q Me} \frac{V_\pi}{P_{\text{opt,in}} Q}\right)^2 \left(1 + \frac{R_M}{R}\right)^2 \left(1 + \frac{G_D}{G_A}\right)^2 \frac{R G_A g_D}{R_M r_M G_D} \tag{1.65}$$

To get the final form of the insertion loss formula, note that V_π (the voltage needed for a phase change of π) is inversely proportional to the length L_{mod} of the modulator electrodes, whereas the capacitance C_M is proportional to that length. Thus, V_π is inversely proportional to the capacitance, or, in other words, $V_\pi \cdot C_M$ is constant, being a figure of merit of the modulator. With (1.59), we can write:

$$V_\pi/Q = S\omega_C r_M; \quad S \triangleq V_\pi C_M$$

and so finally

$$b = \left(\frac{hf}{2\eta_Q Me} \frac{S}{P_{opt,in}} \right)^2 \left(1 + \frac{R_M}{R} \right)^2 \left(1 + \frac{G_D}{G_A} \right)^2 \frac{RG_A g_D r_M}{R_M G_D} \omega_c^2$$

(1.66)

Note that (1.66) is formally equivalent to (1.17) if laser diode efficiency η_L is replaced by $(2e \cdot P_{opt,in}/hfS\omega_c)$ in (1.17).

Very much like the direct modulated case, various types of matching can be applied. Again loss is minimal if (virtually) lossless matching is applied (i.e., $R_M = R$, $G_D = G_A$):

$$L_0 = \left(\frac{2hf}{\eta_Q Me} \frac{SL_r}{P_{opt,in}} \omega_c \right)^2 g_D r_M$$

(1.67)

In the case of resistive matching (i.e., $R_M = R = r_M$, $G_D = G_A = g_M$), then

$$L_1 = \left(\frac{2hf}{\eta_Q Me} \frac{SL_r}{P_{opt,in}} \omega_c \right)^2 G_A R$$

(1.68)

Finally, if no matching is applied while $R_M = r_M \ll R$, $G_D = g_M \ll G_A$, then

$$L_2 = \left(\frac{hf}{2\eta_Q Me} \frac{SL_r}{P_{opt,in}} \omega_c \right)^2 G_A R$$

(1.69)

In comparing direct modulation to lumped-electrode external modulation [see (1.17) through (1.20a) and (1.66) through (1.69)], besides the similarities, there are two significant differences. First, to decrease loss (or to increase gain) there is a very important additional degree of freedom, the magnitude of the optical power. In principle, any magnitude of gain can be achieved by increasing the optical power and this is limited only by such secondary factors as nonlinear effects in the fiber or the price of the light source. (Nonlinear effects are discussed in Section 1.5.) Secondly, there is a pronounced frequency dependence in that the loss is proportional to the RF carrier frequency squared. In the case of direct modulation, there is no frequency dependence, at least as long as it is not limited by optical factors; that is, as long as ω_c is well below the relaxation oscillation angular frequency of the laser diode.

Although this dependence on carrier frequency exists, the loss (or gain) and *bandwidth* are virtually independent of each other. However, bandwidth and modulator sensitivity are related to each other. We can write, for example, for the 3-dB bandwidth (in hertz):

$$B_{3\text{dB}} = 2f_c(1 - \sqrt{1 - 1/Q}); \quad Q > 1 \tag{1.70}$$

and, as we have already seen, Q and V_π are related to each other.

The situation changes somewhat if a traveling-wave modulator is used. There is no resonant circuit in this case and modulator resistance equals the characteristic impedance of the coplanar transmission line (see Figure 1.17). To repeat the analysis of (1.58) and (1.59), now

$$i = \frac{2}{R + R_M}\sqrt{\frac{P_a R R_M}{Z_0}} \tag{1.71}$$

the voltage along the transmission line is

$$u_0 = \frac{2}{R + R_M}\sqrt{P_a R R_M Z_0} \tag{1.72}$$

and the received optical power, based on (1.61), is

$$p_{\text{opt,r}} = \frac{2P_{\text{opt,in}}}{L_r V_\pi} \frac{\sin\psi}{\psi} \frac{\sqrt{P_a R R_m Z_0}}{R + R_M} \tag{1.73}$$

Comparing (1.61) and (1.73), note that Q in the former is replaced by $\sin\psi/\psi$ in the latter. Thus, the insertion loss can still be given by (1.65), with a change in coefficient b:

$$b = \left(\frac{hf}{2\eta_Q Me}\frac{V_\pi}{P_{\text{opt,in}}}\frac{\psi}{\sin\psi}\right)^2\left(1 + \frac{R_M}{R}\right)^2\left(1 + \frac{G_D}{G_A}\right)^2\frac{R G_A g_D}{R_M Z_0 G_D} \tag{1.74}$$

In the present case b is also increasing with increasing frequency; as in the lumped-capacitor design, however, this increase is much slower.

Furthermore, note that in contrast to the previous cases, b is *inversely* proportional to Z_0; thus, a high-impedance transmission line is more advantageous than a low-impedance one. Of course, in the choice of the characteristic impedance, the possibilities are not very widespread. For example, if an asymmetrical coplanar line is applied, taking reasonable dimensions, characteristic impedances of 20–50Ω can be achieved [21].

The special cases discussed earlier were reactant matching:

$$L_0 = \left(\frac{2hf}{\eta_Q Me} \frac{V_\pi}{P_{\text{opt,in}}} \frac{\psi}{\sin\psi} \right)^2 \frac{g_D}{Z_0}$$

and resistive matching (in the present case on the input side this means that the characteristic impedance of the modulator line is equal to R, the generator impedance):

$$L_1 = \left(\frac{hf}{2\eta_Q Me} \frac{V_\pi}{P_{\text{opt,in}}} \frac{\psi}{\sin\psi} \right)^2 \frac{G_A}{R}$$

1.4.3   Noise Figure of Externally Modulated Links

Because the link and detector are the same in direct and externally modulated optical links, Figure 1.7 may also serve in the present case as the basis for calculating the noise figure; the same designations will be used for the various noise current generators as used there. First, the lumped-capacitor design is considered. Taking noise sources one by one, for generator noise, k_g, we have:

$$k_g^2 = \frac{4k_B T_0 B}{R + R_M} \left(\eta_Q M \frac{e}{hf} \frac{1}{\omega_c} \right)^2 \left(\frac{P_{\text{opt,in}}}{SL_r} \right)^2 \frac{R_M G_D}{r_M g_D} \qquad (1.75)$$

For shot noise, k_S:

$$k_s^2 = 2eB \left(P_{\text{opt,in}} F_M \eta_Q \frac{M^2 e}{hf \cdot L_r} + I_d \right) \frac{G_D}{g_D} \qquad (1.76)$$

For laser intensity noise, k_i:

$$k_i^2 = P_{\text{opt,in}}^2 \, \text{RIN}\,(\omega_c) \left(\frac{\eta_Q M e}{L_r h f} \right)^2 B \frac{G_D}{g_D} \tag{1.77}$$

Finally, sources of noise generated at the output are the same as in the case of direct modulation; we repeat them for the sake of completeness. For thermal noise of the photodiode (1.24):

$$k_d^2 = 4 k_B T_0 B G_D$$

and for amplifier noise (1.25),

$$k_a^2 = 2(F_{A0} - 1) \frac{k_B T_0 B}{R_{A0}}$$

and (1.26)

$$e_a^2 = 2(F_{A0} - 1) k_B T_0 B R_{A0}$$

Applying (1.27) and (1.28), we obtain for the noise figure the same expression as (1.29):

$$F = 1 + \frac{R_M}{R} + \frac{(R + R_M)^2 r_M}{R R_M} (a_0 + a_1 L_r + a_2 L_r^2) \tag{1.78}$$

however, with somewhat different coefficients:

$$a_0 = \frac{(\omega_c S)^2}{4 k_B T_0} \text{RIN}(\omega_c) \tag{1.79}$$

$$a_1 = \frac{h f (\omega_c S)^2}{2 k_B T_0 P_{\text{opt,in}}} \tag{1.80}$$

$$a_2 = \left(\frac{h f \cdot \omega_c S}{e \cdot \eta_Q M P_{\text{opt,in}}} \right)^2 \left[\frac{e}{2 k_B T_0} I_d + g_D [1 + \varphi(F_{A0} - 1)] \right] \tag{1.81}$$

Looking for the traveling-wave modulator, the detailed computations are neglected; only final parameters are given. The noise figure can be expressed as

$$F = 1 + \frac{R_M}{R} + \frac{(R + R_M)^2}{RR_MZ_0}(a_0 + a_1L_r + a_2L_r^2) \qquad (1.82)$$

with

$$a_0 = \frac{(V_\pi\psi/\sin\psi)^2}{4k_BT_0}\,\text{RIN}(\omega_c) \qquad (1.83)$$

$$a_1 = \frac{hf(V_\pi\psi/\sin\psi)^2}{2k_BT_0P_{\text{opt,in}}} \qquad (1.84)$$

$$a_2 = \left(\frac{hf \cdot V_\pi\psi/\sin\psi}{e \cdot \eta_QMP_{\text{opt,in}}}\right)^2\left\{\frac{e}{2k_BT_0}I_d + g_D[1 + \varphi(F_{A0} - 1)]\right\} \qquad (1.85)$$

As seen here, the main difference between (1.79), (1.80), and (1.81) compared to (1.83), (1.84), and (1.85) and (1.30) through (1.32) is the different dependence from the optical power: In the case of direct modulation, the noise figure is independent of or increasing with increasing power, whereas in the case of both types of external modulators, it is independent of or decreasing with increasing optical power. The quantitative form of loss and RIN dependence of power is shown in Figure 1.18.

To conclude this section, we should mention that calculations similar to those in Sections 1.3.5 and 1.3.6 on dynamic range, the effect of low-noise amplifiers preceding the fiber optic link, and so on, could be made for the external modulators also. We leave these as exercises for the reader. However, the question of dynamic range will be briefly discussed in relation to QAM microwave/millimeter-wave transmission.

1.5 Modulation Transfer in Microwave Fiber Optic Links

Basic microwave transducer characteristics are, as discussed in Sections 1.3 and 1.4, insertion loss and noise figure. In determining their magnitude, we

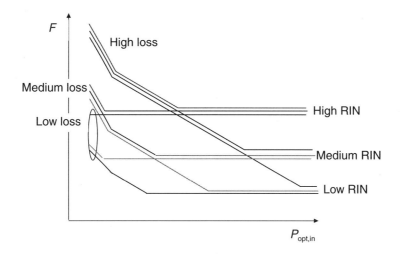

Figure 1.18 Dependence of the noise figure of an externally modulated link on optical power in various situations.

have assumed that the only optical adverse effects are power loss and intensity noise. This is a suitable assumption if microwave frequency is rather low, the optical power is rather low, and the fiber is not very long. However, in the general case, fiber dispersion and nonlinearity of the fiber and of other optical components must be taken into account. These can cause additional loss and also distortion of the modulating signal. The present section gives an overview of the most significant phenomena resulting from these characteristics. Intensity modulation will be assumed as previously; however, to describe the effects considered, electrical field strength rather than intensity will be needed in most cases. We assume throughout that light is modulated by an external single-electrode-pair Mach-Zehnder modulator.

1.5.1 Effect of Fiber Dispersion on the Transmission of a Microwave Carrier

Applying the modulating voltage of (1.55),

$$u = U + \sqrt{2}u_0 \cos\omega_c t \tag{1.86}$$

in (1.51) and taking (1.53) into account, we get for the complex envelope of the optical field strength at the fiber input (at $z = 0$)

$$E(0, t) = \sqrt{P_0}\left(1 + e^{j(U+\sqrt{2}u_0 \cos\omega_c t)}\right) \tag{1.87}$$

with P_0 the optical power that has been input to the fiber.

We restrict our investigation to the quasilinear regime of operation of the modulator ($U = V_\pi/2$). In several applications, a different bias is used for various purposes: Suppressed carrier modulation can be produced, harmonics of the microwave signal can be generated, and so forth. We do not deal with these, however; instead, see [21, 23–25].

In the quasilinear regime

$$E(0, t) = \sqrt{P_0}(1 + je^{jm\cos\omega_c t}); \quad m \triangleq \frac{\sqrt{2}}{V_\pi}u_0 \tag{1.88}$$

Taking the well-known Fourier series of $\cos(m \cos x)$ and $\sin(m \cos x)$ into account [26], we can write

$$E(0, t) = \sqrt{P_0}[1 + jJ_0(m) - 2J_1(m)\cos\omega_c t + \text{HH}] \tag{1.89}$$

$$\approx \sqrt{P_0}[1 + jm_0 - m_1\cos\omega_c t]; \quad m_0 \triangleq J_0(m); \quad m_1 \triangleq 2J_1(m)$$

Here HH refers to higher harmonics and is insignificant in our present discussion.

The optical field of (1.89) is transmitted by an optical fiber. Assuming that the wavelength of the optical carrier does not correspond to the zero-dispersion point of the fiber (the light of a 1,550-μm wavelength fulfills this condition on a standard fiber), the propagation constant can be written as

$$\gamma = \alpha + j\beta = \alpha + j\left[\beta_0 + \beta_1(\omega - \omega_0) + \frac{\beta_2}{2}(\omega - \omega_0)^2\right]; \quad \beta_i = \frac{d^i\beta}{d\omega^i} \tag{1.90}$$

The field strength at the cross section z of the fiber can be written in the following form:

$$E(z, t) = E(0, t)e^{-\alpha z}e^{-j\beta z} = \sqrt{P_0}e^{-\alpha z}r(z, \tau); \quad \tau \triangleq t - \frac{z}{v_g} \tag{1.91}$$

with $v_g = 1/\beta_1$ being the group velocity.

Then it is $r(z, \tau)$ that characterizes the signal distortion while propagating along the fiber:

$$r(z, \tau) = \mathbf{F}^{-1}\left\{ R(0, \omega) \exp\left[j\frac{\beta_2 z}{2}\omega^2 \right] \right\} \qquad (1.92)$$

($\mathbf{F}^{-1}$ designates the inverse Fourier transform and $R = \mathbf{F}\{r\}$.) Magnitude $r(0, \tau)$ is a constant + sinusoid; thus,

$$R(0, \omega) = (1 + jm_0)\delta(\omega) - \frac{m_1}{2}[\delta(\omega - \omega_c) + \delta(\omega + \omega_c)] \quad (1.93)$$

With this,

$$R(z, \omega) = \left\{ (1 + jm_0)\delta(\omega) - \frac{m_1}{2}[\delta(\omega - \omega_c) + \delta(\omega + \omega_c)] \right\} (1.94)$$

$$\exp\left[j\frac{\beta_2 z}{2}\omega^2 \right]$$

Forming the inverse Fourier transform of R for r, we obtain

$$r(z, \tau) = 1 + jm_0 - m_1 \exp\left[j\frac{\beta_2 z}{2}\omega_c^2 \right] \cos\omega_c\tau \qquad (1.95)$$

Take the fiber length $z = L_f$. The detected photocurrent in the photodetector (placed at the end of the fiber) is proportional to the absolute-squared value of r:

$$|r(L_f, \tau)|^2 = \left(1 - m_1 \cos\omega_c\tau \cdot \cos\frac{\beta_2 L_f \omega_c^2}{2} \right)^2$$

$$+ \left(m_0 - m_1 \cos\omega_c\tau \cdot \sin\frac{\beta_2 L_f \omega_c^2}{2} \right)^2 \qquad (1.96)$$

$$= -2m_1 \cos\omega_c\tau \left(\cos\frac{\beta_2 L_f \omega_c^2}{2} + m_0 \sin\frac{\beta_2 L_f \omega_c^2}{2} \right)$$

$$+ \text{ dc} + \text{2nd harmonic}$$

If m is low, then $m_0 \approx 1$ and $m_1 \approx m$. Then the ratio of the ω_c component of the photocurrent to the dc current is

$$\frac{i_{RF}}{I_0} = -\sqrt{2}m \cos\omega_c \tau \cdot \cos\left(\frac{\beta_2 L_f \omega_c^2}{2} - \frac{\pi}{4}\right) \qquad (1.97)$$

The two last equations show the extremely adverse effect of fiber dispersion (more precisely of chromatic dispersion of single-mode fibers), in particular if higher microwaves or millimeter waves are applied. It also shows that investigating optical intensity may not be sufficient and optical field strength has to be discussed instead; the investigation of modulation transfer needs this deepened form of discussion.

For discussing modulation transfer, note that modulation of the light will be completely suppressed if the argument of the second cosinusoid in (1.97) equals $\pi/2$ or odd multiples of it. Otherwise,

$$\frac{\beta_2 L_f \omega_c^2}{2} = \frac{3\pi}{4} + k\pi; \quad k = 0, 1, 2, \ldots \qquad (1.98)$$

This limits drastically the span length of an optical microwave link operating at a high carrier frequency. To understand the numerical values: In the case of standard fiber β_2 is about 20 ps^2/km in the 1,550-μm optical wavelength band. Thus, the first transmission zero is at about the 1.5-km fiber length if the carrier frequency is 60 GHz, so the fiber length must be significantly shorter than 1.5 km. Also note that from this point of view a fiber of length 0 km is not the optimum: A 3-dB increase in the fundamental harmonic of the detected current is produced if the argument of the cosinusoid in (1.97) is $\pi/4$.

Further, the situation is similar with numerical differences only if the approximations on Bessel functions are not applied: m_0 in (1.96) is less than 1; critical values of the fiber length are then shorter. Note also that by applying a modified electrode design in the modulator, its behavior is somewhat different. For example, in the case of push-pull electrodes, the term $-\pi/4$ is missing from the cosine argument; thus, the length of transmission zero is by a factor of 1.5 shorter in the case of a 60-GHz RF carrier. Essentially, however, the behavior in this case is very similar to that already discussed.

1.5.2 Countermeasures to the Dispersion-Induced Suppression of Modulation

Several countermeasures are available for the really adverse effect of modulation suppression discussed in Section 1.5.1. In one of them, single-sideband rather than double-sideband modulation is applied. In this case one of the sidebands of (1.93) is suppressed. This can be done via special electrode design in the modulator [22], via appropriate (optical) filters [27], via balanced photodetectors, or by other means [28]. Suppression of one of the sidebands results in

$$R(0,\ \omega) = (1 + jm_0)\delta(\omega) - \frac{m_1}{2}\delta(\omega - \omega_c) \tag{1.99}$$

Again with this

$$R(z,\ \omega) = \left\{(1 + jm_0)\delta(\omega) - \frac{m_1}{2}\delta(\omega - \omega_c)\right\} \exp\left[j\frac{\beta_2 z}{2}\omega^2\right] \tag{1.100}$$

and its inverse Fourier transform, as before, is

$$r(z,\ \tau) = 1 + jm_0 + m_1 \exp\left[j\left(\omega_c\tau + \frac{\beta_2 z}{2}\omega_c^2\right)\right] \tag{1.101}$$

while the absolute-squared at the end of the fiber is

$$|r(L_f,\ \tau)|^2 = \left[1 - m_1\cos\left(\omega_c\tau + \frac{\beta_2 L_f \omega_c^2}{2}\right)\right]^2$$

$$+ \left[m_0 - m_1\sin\left(\omega_c\tau + \frac{\beta_2 L_f \omega_c^2}{2}\right)\right]^2 \tag{1.102}$$

$$= -2m_1\cos\left(\omega_c\tau + \frac{\beta_2 L_f \omega_c^2}{2}\right) - 2m_0 m_1\sin\left(\omega_c\tau + \frac{\beta_2 L_f \omega_c^2}{2}\right)$$

$$+ \text{dc} + \text{2nd harmonic}$$

If, as in (1.97), we again take a low m, the relative magnitude of the RF current can be written as

$$\frac{i_{\text{RF}}}{I_0} = -\sqrt{2}m \cos\left(\omega_c \tau + \frac{\beta_2 L_f \omega_c^2}{2} - \frac{\pi}{4} \right) \qquad (1.103)$$

Comparing (1.103) and (1.97), we realize the significant difference: The frequency-dependent loss in the RF signal, characteristic in the latter case, is replaced by a frequency-dependent phase shift. In the present case, that is, if we are transmitting an unmodulated sinusoid, this phase shift is not harmful at all. We briefly return to the investigation of this phenomenon in Section 1.5.3 when linear distortion is discussed.

In the most widely proposed solution microwave/millimeter-wave optical *generation* rather than simple optical *transmission* is applied. Optical millimeter-wave (or microwave) generation is usually based on the principle of *remote heterodyning*: Two or more optical signals are simultaneously transmitted and are heterodyned in the receiver. One or more of the heterodyning products is the required RF signal. Heterodyning can be realized by the photodetector itself or the optical signals can be detected separately and then converted in an electrical (RF) mixer. In a complete (duplex) system, the photodetector can be replaced by an electroabsorptive transceiver [29].

Because phase noise is a key problem in digital microwave transmission, care must be taken to produce a small phase noise only by the heterodyned signals. This can be achieved if the two (or more) optical signals are phase coherent; in turn, this can be realized if the different frequency optical signals are somehow deduced from a common source or they are phase-locked to one *master* source. (Due to the characteristics of light sources, it would be extremely difficult to ensure an acceptable phase noise after heterodyning two *individual* sources.) A rather complete listing of the possibilities is given in [30] and some of the individual solutions can be found in [28, 31–34].

Of course, *digital modulation* must also be introduced. This can be done by modulating an appropriate intermediate frequency with the information-bearing (digital) signal and modulating one of the optical signals with this.

The advantage of optical RF generation via remote heterodyning over modulating an optical carrier with the RF signal is apparent: In mixing *two* (rather than three) optical signals, no effect like the suppression of modulation occurs. Taking into account that according to (1.98) frequency *squared* determines the critical length, we can assume that intermediate frequency is low enough not to cause modulation suppression in the case of reasonable fiber lengths. [As an example, fiber length where complete suppression of the modulation occurs, as already seen, is about 1.5 km if the RF is 60

GHz; with an intermediate frequency (IF) of 2 GHz, this length is 900 times longer, that is, about 1,350 km.]

Although there are several solutions (some of them referred to above), one typical design is shown in a somewhat simplified way in Figure 1.19; it was introduced in [35].

The functioning of this system is as follows: The master laser's intensity is modulated by the unmodulated RF reference signal; several harmonics of the reference signal and consequently several sidebands are generated. The *reference laser* is injection locked by one of these and the *signal laser* by another one in such a way that the difference of their frequencies corresponds to the (millimeter-wave) "local oscillator" frequency. (Local oscillator is shown in quotation marks because it is not actually *local* relative to the receiver, that is, the photodetector.) And, as seen, the optical field generated by the signal laser is also modulated by the information-bearing intermediate frequency signal.

So the complex envelope of the signal laser optical field strength is

$$E_1(t) = \sqrt{P_1}\left(1 + j \cdot e^{jm\cos\omega_{\mathrm{IF}}\tau}\right) = \sqrt{P_1}\left(1 + jm_0 - m_1\cos\omega_{\mathrm{IF}}\tau + \mathrm{HH}\right) \tag{1.104}$$

and that of the reference laser

$$E_2(t) = \sqrt{P_1}\,e^{j\omega_{\mathrm{LO}}t} \tag{1.105}$$

where P_1 and P_2 refer to the powers of the signal laser and the reference laser, respectively; ω_{IF} and ω_{LO} are the intermediate frequency and local

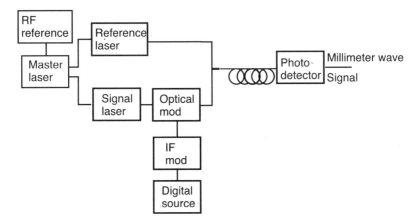

Figure 1.19 A remote heterodyning optical/millimeter-wave transmission system.

oscillator frequency, respectively, and m_0 and m_1 have the same meaning as in (1.90).

Taking $E_1 + E_2$ and producing the photocurrent—being proportional to the absolute-squared of the summed field strength—there will be dc terms, terms of ω_{LO}, and terms of $\omega_{LO} \pm \omega_{IF}$. There are also higher-order terms, but these were not taken into account in (1.104) and so they are neglected here as well.

One of $\omega_{LO} \pm \omega_{IF}$ is taken as RF and selected by an optical or an RF filter. Thus, the RF signal (actually, the RF current) can be written as

$$i_{RF}(t) \approx \sqrt{P_1 P_2}\, m \cos(\omega_{LO} + \omega_{IF})t \qquad (1.106)$$

where again the approximation $m_0 \approx 1$, $m_1 \approx m$ was made. As seen from the last formulas, fiber chromatic dispersion has in this case a similar, but from the practical point of view insignificant, effect as in single-sideband modulation.

To conclude this section, note that the term *fiber dispersion* was used to identify one characteristic of single-mode fibers: the frequency dependence of the phase constant, β. To be specific, the term *chromatic dispersion* is sometimes used instead. However, it is well known that β depends on the field polarization as well, a phenomenon called *birefringence* of a dielectric material. The polarization mode dispersion effect of birefringence can result in similar phenomena, such as modulation suppression, as discussed in the previous section. We do not discuss these phenomena in a quantitative detail, mainly because they are rather slight effects, at least at reasonable (at most a few tens of kilometers) fiber lengths. To understand the effects, note that in RF transmission two or three optical signals are heterodyned; and there is a power loss in the heterodyne product if polarization of the two signals is different. Now due to birefringence, in the photodetector (at $z = L_f$) polarization *can* be different, resulting in power loss. In a particularly adverse case, polarization of the fields to be heterodyned can become orthogonal, resulting in an infinite loss. For details, see [36–38].

1.5.3   Enhancement of the Effect of Fiber Dispersion: Linear Distortion

In Sections 1.5.1 and 1.5.2, suppression of the modulation in fiber optic links due to fiber chromatic dispersion as well as countermeasures to this effect were discussed. The electrical signal modulating light was an unmodulated sinusoid. In the present subsection, we assume that a digital signal is modulating the RF carrier. Fiber dispersion causes in-band linear distortion of the signal. Depending on the RF modulation applied this will more or less degrade signal quality.

Present-day mobile and other wireless systems operate with binary modulation: minimum shift keying (MSK), Gaussian MSK (GMSK), or similar continuous-phase modulation, perhaps binary phase shift keying (BPSK). To increase capacity, future systems are likely to take advantage of higher order modulation. Some of the producers of mobile equipment have developed mobile systems applying 64-QAM. Quadrature amplitude modulation (QAM) systems are much more sensitive to distortions, both linear and nonlinear, than MSK or BPSK. Therefore in what follows we deal mainly with QAM systems, with reference to trellis-coded modulation (TCM), which is considered the coded version of QAM.

The complex envelope of an M-ary QAM signal is

$$s(t) = \sum_{k=-\infty}^{\infty} A(a_k + jq_k)b(t - kT); \quad a_k, q_k = \frac{\pm 2i - 1}{\sqrt{M} - 1}; \, i = 1, \ldots, \frac{\sqrt{M}}{2} \tag{1.107}$$

and the symbol error probability (if additive white Gaussian noise is the only disturbing effect) is given as [39]

$$P_E = 2(1 - 1/\sqrt{M}) \, \text{erfc} \, \frac{d}{\sqrt{N_0}}; \quad d = \frac{\sqrt{E_{\text{peak}}}}{\sqrt{2}(\sqrt{M} - 1)} \tag{1.108}$$

where:

$b(t)$ is the signaling pulse;

d is the minimal distance of signal vectors from the closest decision boundary in the signal space;

N_0 is the additive noise spectral density;

$E_{\text{peak}} = A^2 \cdot T$ is the peak energy of the signal constellation;

T is the symbol for time.

Now light is modulated (via a Mach-Zehnder modulator) by the QAM-modulated RF signal. Repeating for this case the analysis of Section 1.5.1, the optical field strength is

$$E(0, t) = \sqrt{P_0}(1 + je^{jm(t)\cos\omega_c t}); \quad m(t) \triangleq \frac{s(t)\pi}{V_\pi} \tag{1.109}$$

With transformations similar to those made in (1.89) through (1.97), we can write for the detected photocurrent

$$\frac{i_{RF}}{I_0} = -\sqrt{2}m(L_f, \tau)\cos\omega_c\tau \cdot \cos\left(\frac{\beta_2 L_f \omega_c^2}{2} - \frac{\pi}{4}\right) \qquad (1.110)$$

Here $m(L_f, \tau)$ is the distorted version of $m(t)$; this distortion is the linear distortion due to the fiber dispersion. Time function $m(L_f, \tau)$ can be determined by the application of the somewhat generalized form of (1.94) through (1.96). Based on these, a *modulation transfer function* $K(z, \omega)$ can be defined:

$$K(z, \omega) = \cos\left(\frac{\beta_2 z}{2}\omega^2 - \frac{\pi}{4}\right) \qquad (1.111)$$

and so

$$m(L_f, \tau) = \mathbf{F}^{-1}[M(0, \omega) \cdot K(L_f, \omega)] \qquad (1.112)$$

with $M(0, \omega)$ the Fourier transform of $M(0, \tau)$.

Looking at (1.109) through (1.112), we recognize that besides causing significant loss of the signal-bearing component of the detected light, an in-band power difference is also caused by the fiber dispersion. QAM transmission is extremely sensitive to any form of linear distortion, including in-band power difference. Figure 1.20 shows simulated error probability versus E_b/N_0 (signal energy over noise spectral density) curves for the case of 64-QAM if the RF signal is transmitted through fibers of various lengths. In the simulation, carrier-frequency of 5 GHz was assumed, as was a standard fiber and an optical wavelength of 1,550 nm. Note that the enhancement of loss due to in-band power difference in one of the examples is about 12 dB.

Investigating (1.111), an approximate formula for in-band power difference (IBPD) can be given:

$$IBPD \approx -20\log\left(2\pi B\frac{\partial}{\partial\omega}|K(z, \omega)_{\omega=\omega_c}|\right) \qquad (1.113)$$

where B is the signal bandwidth ($\approx 1/T$, where T represents time). Performing the calculations of (1.113), we get:

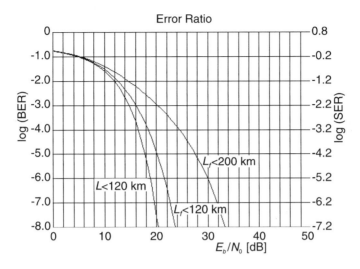

Figure 1.20 Simulated error probabilities: 64-QAM, 5-GHz RF frequency, 1,550-nm optical wavelength, and a standard fiber ($\beta_2 = 20$ ps^2/km).

$$\text{IBPD} \approx -20 \log [1 + \beta_2 L_f(a, \omega_r) \omega_r^2 \sqrt{a(\omega = \omega_c)} - 1 B/f_c]$$

$$(1.114)$$

Here $a(\omega = \omega_c) \triangleq \dfrac{1}{K^2(L_f, \omega_c)}$, ω_r is an optionally chosen reference frequency, and $L_f(a, \omega_r)$ is a fictitious fiber length: that length at which $a = a(\omega = \omega_c)$ at the reference frequency.

We can see that the in-band power difference is high if the relative bandwidth B/f_c is rather high and a at ω_c is also rather high. Note, however, that modulation systems such as 32-TCM, 64-QAM, or 128-TCM are extremely sensitive to all sorts of linear distortion, including in-band power difference; a few decibels can cause significant degradation. An equalizer, adaptive or not, can counteract this distortion.

Single-sideband modulation is more favorable from this point of view also. Based on (1.101) and (1.103), for this case the modulation transfer function can be written as

$$K(z, \omega) = \exp \left[j \frac{\beta_2 z}{2} \omega^2 \right]$$

$$(1.115)$$

As seen, fiber now causes a phase shift proportional to the frequency squared, that is, a frequency-linear group delay distortion:

$$D(z, \omega) = \frac{\partial}{\partial \omega}\left(\frac{\beta_2 z \omega^2}{2}\right) = \beta_2 z \omega \qquad (1.116)$$

Thus, the total group-delay variation within the signal bandwidth B at the photodetector output is

$$\Delta D = 2\pi \beta_2 L_f B \qquad (1.117)$$

The ΔD is negligible even in the worst case (i.e., $L_f = 200$ km). In this case, $\Delta D = 6.3$ ps, being really negligible as symbol time is about 4 ns.

1.5.4   Some Nonlinear Effects: Modulator

In Sections 1.5.1 and 1.4.1, we have seen that there is a nonlinear relationship between light intensity at the modulator output and modulating electrical signal [see (1.56)]. This nonlinearity has different effects in different cases. If the electrical signal is composed of multiplexed subcarriers, digital performance is degraded by intermodulation noise [40]. In the case of single-channel QAM transmission, modulator nonlinearity causes AM compression, also increasing error probability. This effect is discussed next.

To deal with this problem, (1.109) is written in a somewhat modified form:

$$s(t) = \sum_{(k)} A_k b(t - kT) e^{j\varphi_k} \qquad (1.118)$$

introducing

$$a_k + j q_k = A_k e^{j\varphi_k}$$

With this, the detected current (if the photodetector is placed right at the output of the modulator) is

$$\frac{i(t)}{I_0} = 1 - \sin\left[m \operatorname{Re} \sum_{(k)} A_k e^{j\varphi_k} e^{j\varphi_c t}\right]; \quad m \triangleq \frac{\sqrt{2}U_0 \pi}{V_\pi} \qquad (1.119)$$

where U_0 is now the rms magnitude of the maximal RF amplitude ($a_k = q_k = 1$) and ω_c is the RF carrier angular frequency.

QAM error probability—not taking into account modulator distortion—is given in (1.107). To investigate the effect of this distortion, see the signal space constellation as depicted in Figure 1.21 (for $M = 16$). As can be seen, due to the distortion signal vector absolute values are decreased and, hence, they come closer to the decision boundaries. The decrease is characterized by the *describing function* of the sinusoidal nonlinearity (defined as the ratio of the output signal fundamental harmonic amplitude to the amplitude of the input signal). Taking (1.89) into account, this is

$$T_{\sin}(U_0) = \frac{J_1(\sqrt{2}U_0\pi/V_\pi)}{\sqrt{2}U_0\pi/V_\pi} \tag{1.120}$$

With the designations of Figure 1.21, a rather tight bound for the penalty due to the sinusoidal nonlinearity can be written as

$$L_{\sin,\text{QAM}} < 20\log(d/d') = -20\log[(\sqrt{M}-1)\cdot T(u_0) - (\sqrt{M}-2)] \tag{1.121}$$

The computed loss for 4-, 16-, 64-, and 256-QAM systems is shown in Figure 1.22.

Specifying the tolerable penalty, say, 3 dB, the dynamic range of the system can be determined: Maximal input RF voltage and consequently RF power are determined by the curves of Figure 1.22; minimal RF power is determined by the noise figure as computed in Section 1.4. However, this type of distortion can easily be compensated for, and so the dynamic range is increased by the application of a *predistorter*; this should have a describing function that is the inverse to that of (1.120):

$$T_{\text{pred}}(x) = \frac{x}{J_1(x)}; \quad x < \frac{\pi}{2} \tag{1.122}$$

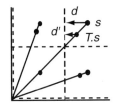

Figure 1.21 16-QAM vector constellation in the signal space showing distortion penalty.

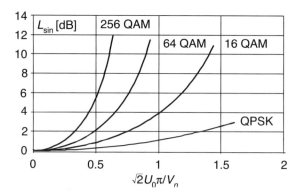

Figure 1.22 Loss due to the sinusoidal nonlinearity of the modulator for various QAM systems.

1.5.5   More Nonlinear Effects: Fiber Nonlinear Refraction

Because the cross section of an optical fiber is very small, field intensity is very high, even if overall power is relatively low, say, 1–10 mW. Therefore, an optical fiber can be modeled as a nonlinear transmission line. Nonlinear transmission lines, in particular those operating in the optical frequency band, are rather complicated components. They produce several unusual effects such as nonlinear scattering and nonlinear refraction. Dispersive nonlinear transmission lines are even more complicated due to the interaction of dispersion and nonlinearity; examples of these effects are modulation instability (while transmitting a dc signal) and soliton propagation. Fiber nonlinear characteristics are discussed in much detail in [41]. Among the several nonlinear effects, nonlinear refraction is investigated here along with self-phase modulation and the cross-phase modulation resulting from it. A QAM scheme is assumed.

As will be shown, for remote heterodyning systems cross-phase modulation can be one of the effects that sets a limit on the optical power; this is not the case in links intensity modulated by the microwave or millimeter-wave signal. This is a rather important difference and very likely serves as an argument in choosing one or the other system. With regard to that, remember that the noise figure decreases as optical power is increased. However, note also that cross-phase modulation is not the only nonlinear effect possibly limiting the optical power. Brillouin scattering is another phenomenon that is harmful above 1–10 mW [41].

Nonlinear refraction is the term used for the phenomenon in which the phase constant of the fiber depends on the optical power. In this section we investigate the effects of nonlinear refraction in a *dispersion-free* fiber. The results can be regarded as valid in the 1.3-μm band in standard fibers or at 1.55 μm in dispersion-shifted fibers (i.e., close to the zero-dispersion point of the fiber).

In a dispersion-free nonlinear transmission line—fiber or waveguide—the propagation constant can be written as

$$\gamma(z) = \alpha + j[\beta_0 + \beta_1(\omega - \omega_o) + \gamma^1 P(z)] \qquad (1.123)$$

where:

α is the attenuation constant;
ω_0 is the optical angular frequency;
γ^1 is the nonlinear part of the phase constant;
$P(z)$ is the power along the line.

Note that due to the nonzero α, the power (and, consequently, also γ^1) is a function of the distance z.

Applying the same designations as in (1.91) and in the following equations we get the differential equation for $r(z, \tau)$:

$$\frac{\partial r(z, \tau)}{\partial z} = \frac{-j}{\gamma^1 P_0} e^{-\alpha z} |r(z, \tau)|^2 r(z, \tau) \qquad (1.124)$$

Because there is no dispersion—as mentioned, at present we are dealing with nonlinearity only—we can assume that $|r(z, \tau)| = |r(0, \tau)|$, and so we can try to obtain the solution to (1.124) in this form:

$$r(z, \tau) = r(0, \tau) \exp[j\Phi_{NL}(z, \tau)] \qquad (1.125)$$

Using (1.125) in (1.124), we obtain for the nonlinear phase shift Φ_{NL}

$$\Phi_{NL} = |r(0, \tau)|^2 \cdot \frac{z_{eff}}{L_{NL}}; \quad L_{NL} = \frac{1}{\gamma^1 P_0}; \quad z_{eff} = \frac{1 - e^{-\alpha z}}{\alpha}$$
$$(1.126)$$

Here:

L_{NL} is the so-called nonlinear length, characteristic to the fiber length above which nonlinear phenomena can be of significance;

P_0 is the peak optical power at $z = 0$;
α is the attenuation constant of the fiber; 0.25 dB/km = 0.0575 Np/km can be taken for 1,300 nm;
z_{eff} is called the effective length; note that $z_{eff} < z$ and also $z_{eff} \leq 1/\alpha$.

As seen in (1.124), note first that the optical carrier will be chirped due to self-phase modulation: Phase is varying, following QAM power variation due to the pattern of the transmitted symbol sequence, and phase variation results in a nonconstant, chirped optical frequency. But note also that this will have no effect on link performance if light is modulated by the (modulated) RF carrier, as $|r(z, \tau)|^2 = |r(0, \tau)|^2$.

The situation is completely different in remote heterodyned links. Note that in this case two signals copropagate along the fiber; according to (1.104) and (1.105), one of them contains the digitally modulated intermediate frequency and the other the unmodulated local oscillator signal. Then cross-phase modulation has to be taken into account instead of self-phase modulation (i.e., the nonlinear phase shift caused in one channel due to the intensity modulation of the other). Phase shift in channel 1 and 2 can then be given [41] (self-phase modulation and cross-phase modulation, respectively):

$$\Phi_{NL1} = \Phi_{NL}; \quad \Phi_{NL2} = 2\Phi_{NL} \qquad (1.127)$$

with Φ_{NL} as given in (1.125).

According to this, (1.104) and (1.105) must be corrected to include the nonlinear phase shifts: E_1 must be multiplied by $\exp[j\Phi_{NL1}]$ and E_2 by $\exp[j\Phi_{NL2}]$. This results in a detected current at $z = L_f$ of

$$\frac{i_{RF}(L_f, \tau)}{I_0} = A_2 m(t) \cos(\omega_{LO} \tau + \omega_{IF} \tau + \Phi_{NLk}) \qquad (1.128)$$

With subscript k it is explicitly stated that Φ_{NL} depends on symbol k. That is, cross-phase modulation and self-phase modulation cause amplitude modulation (AM)-to-phase modulation (PM) conversion, which increases the error probability.

Note that the phase recovered by the carrier synchronizing system coincides with the average phase. Thus, signal vectors with higher-than-average power are shifted on a constant amplitude circle in one sense and those of lower-than-average power in the other one. The situation for 16-QAM is visualized in Figure 1.23, which shows the original d and the distortion-caused d'. Taking this into account and using an elementary calculation,

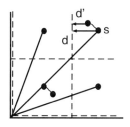

Figure 1.23 Signal-space constellation showing AM-to-PM conversion penalty.

similar to that resulting in (1.121), we get an upper bound of the loss due to the AM/PM conversion:

$$L_{AM/PM} < 20 \log(d/d') \tag{1.129}$$
$$= -20 \log[(\sqrt{M} - 1)\sqrt{2} \cos(\pi/4 + b\Phi_{NL}) - (\sqrt{M} - 2)]$$

with b a function of M: $b = 0.57$ in 64-QAM, $b = 0.44$ in 16-QAM, and, of course, $b = 0$ in QPSK.

In Figure 1.24, the computed penalty due to self-phase modulation/cross-phase modulation–induced AM-to-PM conversion is shown for 16-QAM and 64-QAM.

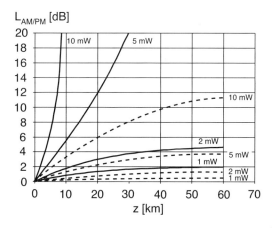

Figure 1.24 AM-to-PM loss in remote heterodyning transmission; solid lines: 64-QAM; dotted lines: 16-QAM; $P_1 = P_2$, $\gamma = 3/W$ km.

1.6 Conclusions

Microwave characteristics, which determine the properties of the *physical layer* of an optical link transmitting microwave or millimeter-wave signals, were presented; intensity modulation of light was assumed throughout. These characteristics are essentially different both from those of wireless links and from those of optical links transmitting baseband (digital) signals. The differences, which are partly responsible for determining the overall system design, are seen in loss/gain, noise figure, and modulation transfer.

Here we list again the most spectacular of these differences:

- *Concerning insertion loss:* electrical loss is proportional to the optical loss *squared,* and also a *gain* is possible in spite of the fact that these links are essentially *passive.* The power needed for this gain is supplied by the light source (and, of course, by its electrical power supply) and by the bias source of the photodetector.

- *Concerning noise figure:* In principle, the noise figure *can* be lower than the insertion loss and matching conditions for minimal loss (maximal gain) and minimal noise figure may or may not be identical.

- *Concerning signal transmission:* Loss characteristics, although uniquely determined by matching characteristics and by optical loss in links of lower RF frequencies and short lengths, are strongly influenced by fiber chromatic dispersion in the case of higher frequencies and longer lengths. As a result of dispersion and the *coherent character* of the two sidebands in intensity modulation, it turns out that modulation can significantly be suppressed or even become completely extinct. Modulation extinction length is inversely proportional to the dispersion parameter of the fiber and to RF frequency squared. In addition to the RF loss, fiber dispersion also causes in-band power difference in the transmission signal, decreasing further signal performance. Taking these effects into account, optical *generation* rather than optical transmission of the RF signals is in many cases more favorable. Modulator nonlinearity can significantly decrease the dynamic range of a link; the application of a predistorter can improve this situation. Fiber nonlinear refraction can be an adverse characteristic, particularly in the case of QAM (or TCM) transmission; it has different effects in links *transmitting* RF signals and in links *generating* the RF carrier. In the former of these two it has virtually no effect, whereas in the latter it sets a limit to the applicable optical power. This should be a significant point when choosing between possible solutions.

Phenomena were not discussed with equal detail, mainly due to restrictions in the available size. Some important properties, such as phase noise and nonlinearity-induced intermodulation noise, were merely mentioned with a note on which references to check to obtain more information on the topic. Some important topics have not yet been sufficiently clarified to be included in a book; an example of this is the simultaneous effect of fiber dispersion and nonlinearity. Some of the topics are discussed here for the first time (as far as this author knows) or at least discussed in such detail for the first time. The examples are the effects of optical link distortions on QAM transmission.

References

[1] Weisser, S., et al., "Dry-Etched Short-Cavity Ridge-Waveguide MQW Lasers Suitable for Monolithic Integration with Direct Modulation Bandwidth Up to 33 GHz and Low Drive Currents," *Proc. ECOC,* 1994, pp. 973–976.

[2] Van der Ziel, A., *Noise,* Upper Saddle River, NJ: Prentice Hall, 1954.

[3] Chang, K., *Handbook of Microwave and Optical Components,* Vol. 2, New York: Wiley, 1990.

[4] Yariv, A., *Optical Electronics,* New York: Wiley, 1991.

[5] Agrawal, G. P., and N. K. Dutta, *Semiconductor Lasers,* 2nd ed., New York: Van Nostrand, 1995.

[6] Cox, C., F. Betts, and L. Johnson, "An Analytic and Experimental Comparison of Direct and External Modulation in Analog Fiber-Optic Links," *IEEE Trans. on Microwave Theory and Techniques,* Vol. MTT-38, No. 5, 1990, pp. 501–509.

[/] Cox, C., E. J. Ackerman, and F. Betts, "Relationship Between Gain and Noise Figure in an Optical Analog Link," *IEEE MTT-S Int. Microwave Symp. Digest,* Vol. 3, San Francisco, CA, June 1996, pp. 1551–1554.

[8] Simons, R., *Optical Control of Microwave Devices,* Norwood, MA: Artech House, 1990.

[9] Darioush, A., et al., "Interfaces for High-Speed Fiber Optic Links," *IEEE Trans. on Microwave Theory and Techniques,* Vol. 39, No. 12, Dec. 1991, pp. 2031–2044.

[10] Frigyes, I., et al., "Noise and Loss Characteristics of Microwave Direct Modulated Optical Links," *Proc. 27th European Microwave Conference,* Jerusalem, Israel, Sept. 1997, pp. 309–317.

[11] Hilt, A., "Optical Transmission and Processing of Microwave Signals," Ph.D. Thesis, Institut National Polytechnique de Grenoble, France, 1999.

[12]   Tamir, T., *Guided-Wave Optoelectronics,* Berlin: Springer-Verlag, 1990.

[13]   Wake, D., "50 GHz Edge-Coupled Pin Photodetector," *Electron. Lett.,* Vol. 27, No. 12, June 6, 1991, pp. 1073–1075.

[14]   Wey, Y. G., et al., "110 GHz GaInAs/InP Double-Heterostructure Pin Photodetectors," *J. Lightwave Technology,* Vol. 13, No. 7, July 1995, pp. 1490–1499.

[15]   Abrie, P. L. D., *The Design of Impedance Matching Networks,* Dedham, MA: Artech House, 1985.

[16]   Agrawal, G. P., *Fiber-Optic Communication Systems,* 2nd ed., New York: Wiley, 1997.

[17]   Frigyes, I., et al., "Loss and Noise Characteristics of Direct Modulated Microwave Fiber Optic Links," *Journal on Wireless Communications,* Vol. 14, No. 2, Aug. 2000, pp. 199–214.

[18]   Bjarklev, A., *Optical Fiber Amplifiers: Design and System Applications,* Norwood, MA: Artech House, 1993.

[19]   Olesen, H., and G. Jacobsen, "A Theoretical and Experimental Analysis of Laser Fields and Power Spectra," *IEEE J. Quantum Electronics,* Vol. 18, No. 12, Dec. 1982, pp. 2069–2080.

[20]   Ackerman, E. J., et al., "Input Impedance Conditions for Minimizing the Noise Figure of an Analog Optical Link," *1997 IEEE Microwave Theory and Techniques Digest,* Denver, CO, June 1997, Paper No. TU3E-6.

[21]   Alferness, R. C., "Waveguide Electro-Optic Modulators," *IEEE Trans. on Microwave Theory and Techniques,* Vol. 30, No. 8, 1982, pp. 1121–1137.

[22]   Frankel, M. Y., and R. D. Esman, "Optical Single-Sideband Suppressed-Carrier Modulator for Wideband Signal Processing," *J. Lightwave Technology,* Vol. 16, No. 5, May 1998, pp. 859–863.

[23]   Joindot, I., and M. Joindot, *Les Télécommunications par Fibres Optiques,* Paris: Dunod, 1996.

[24]   Noguchi, K., et al., "Millimeter Wave Ti:LiNbO$_3$ Optical Modulators," *J. Lightwave Technology,* Vol. 16, No. 4, April 1998, pp. 615–619.

[25]   Hilt, A., "Microwave Harmonic Generation in Fiber-Optical Links," presented at MIKON 2000, Wroclaw, Poland, May 2000.

[26]   Gradshteyn, R., *Table of Integrals, Series and Products,* 5th ed., New York: Academic Press, 1994.

[27]   Smith, G. H., and D. Novak, "Broad-Band Millimeter-Wave Fiber-Wireless Transmission System Using Electrical and Optical SSB Modulation," *IEEE Phot. Techn. Lett.,* Vol. 10, No. 1, Jan. 1998, pp. 141–143.

[28]   Izutsu, M., S. Shikama, and T. Sueta, "Integrated Optical SSB Modulator/Frequency Shifter," *IEEE J. Quantum Electronics,* Vol. QE-17, No. 11, Nov. 1981, pp. 2225–2227.

[29]   Heinzelman, R., et al., "Optical Add-Drop Multiplexing of 60-GHz Millimeterwave Signals in a WDM Radio-on-Fiber Ring," presented at IEEE-OSA OFC 2000, Baltimore, MD, March 2000.

[30] Gliese, U., S. Nłrskov, and T. N. Nielsen, "Chromatic Dispersion in Fiber-Optic Microwave and Millimeter-Wave Links," *IEEE Trans. on Microwave Theory and Techniques,* Vol. 44, No. 10, Oct. 1996, pp. 1716–1724.

[31] Berceli, T., et al., "Millimeter-Wave Generation by Optical Means," presented at MIKON 2000, Wroclaw, Poland, May 2000.

[32] Wake, D., C. R. Lima, and P. A. Davies, "Optical Generation of Millimeter-Wave Signals for Fiber-Radio Systems Using a Dual-Mode DFB Semiconductor Laser," *IEEE Trans. on Microwave Theory and Techniques,* Vol. 43, No. 9, Sept. 1995, pp. 2270–2276.

[33] O'Reilly, J. J., et al., "Optical Generation of Very Narrow Linewidth Millimeter Signals," *Electron. Lett.,* Vol. 28, No. 25, 1992, pp. 2309–2311.

[34] Ramos, R. T., and A. J. Seeds, "Fast Heterodyne Optical Phase-Locked Loop Using Double Quantum Well Laser Diodes," *Electron. Lett.,* Vol. 28, No. 1, 1992, pp. 82–83.

[35] Brown, M., and L. Grosskopf, "Optical Feeding of Base Stations in Millimeter-Wave Mobile Communications," *ECOC '98,* Madrid, Spain, 1998, pp. 665–666.

[36] Schmuck, H., "Effect of Polarization Mode Dispersion in Fiber-Optic Millimeter-Wave Systems," *Electron. Lett.,* Vol. 18, 1994, pp. 1503–1504.

[37] Galtarossa, A., et al., "Polarisation Mode Dispersion in Long Single-Mode Fiber Links: A Review," *Fiber and Integrated Optics,* Vol. 13, 1999, pp. 215–229.

[38] Hofstetter, R., H. Schmuck, and R. Heidman, "Dispersion Effects in Optical Millimeterwave Systems Using Self Heterodyne Method for Transport and Generation," *IEEE Trans. on Microwave Theory and Techniques,* Vol. 43, No. 9, Sept. 1995, pp. 2263–2269.

[39] Frigyes, I., Z. Szabó, and P. Ványai, *Digital Microwave Transmission,* Amsterdam: Elsevier, 1989.

[40] Kolner, B. H., and D. W. Dolfi, "Intermodulation Distortion and Compression in an Integrated Electrooptic Modulator," *Applied Optics,* Vol. 26, No. 17, Sept. 1, 1987, pp. 3676–3680.

[41] Agrawal, G. P., *Nonlinear Fiber Optics,* San Diego, CA: Academic Press, 1995.

2

Subcarrier Optical Fiber Transmission Systems
Mohsen Kavehrad

2.1 Introduction

During the last decade, fiber optic analog transmission of microwave signals has received considerable attention in many applications. This has happened because optical fiber provides an excellent transmission medium for information distribution networks. The use of subcarrier multiplexing (SCM) transmission using an optical carrier instead of the traditionally used supercarrier over optical fibers is very attractive. This technology has found widespread application because of its simplicity and cost-effectiveness.

However, in SCM systems, the available bandwidth of optical fiber is generally limited by the processing speed of electronics. However, in the field of digital microwave radio, multistate modulation techniques such as 16-, 64-, and 256-QAM have been rapidly developed for both their power and spectral efficiency. The QAM technique combines two quadrature components of a continuous-wave tone and transmits the amplitude and phase of the carrier signal. It has been proven very useful for band-limited applications.

Error correction coding techniques, such as block, convolutional, and trellis, have advanced, further enhancing the noise immunity of multistate modulation schemes. Thus, the type of modulation mentioned plus coding techniques can be very good candidates for SCM applications.

Technological progress, mainly in the production of wide-modulation-bandwidth lasers, has also greatly contributed to increased interest in SCM systems. Optical analog microwave transmission started as a short-term, low-cost implementation of wideband optical networks. Possible applications for this solution are numerous. Video distribution is a strong motivation for the anticipated widespread penetration of optical fiber in the local loop. Moreover, fiber optic transmission is one of the main alternatives being considered for distribution of digital high-definition TV (HDTV).

This chapter, for the most part, describes and presents results on single-mode fiber optic transmission of microwave QAM signals. More specifically, we present fiber optic transmission of 64- and 256-QAM signals at 10-, 50-, and 90-Mbps transmission rates. A detailed analysis of direct modulation was presented in Chapter 1.

Two important methods of improving system performance are discussed and demonstrated: laser intensity noise minimization and error correction coding using a block code and a convolutional code. Applicability of this transmission technique to distribution of digital TV services is assessed with some encouraging results. Theoretical and some experimental bit-error-rate (BER) performance results are described for two types of 256-QAM signals. It is feasible to use 256-QAM signals as subcarriers in fiber distribution networks.

Semiconductor lasers, which are used as optical sources in SCM systems, exhibit nonlinear characteristics. Nonlinearity degrades the performance of SCM transmission systems. For nonlinearity compensation, predistortion can be incorporated prior to the laser. One characteristic of this nonlinear block is that it compensates that of the laser source. Performance of a radio subcarrier multiplexed system with and without predistortion is evaluated by harmonic and intermodulation distortion analyses. The nonlinear distortion level decreases significantly when predistortion is employed prior to laser intensity modulation. This enhancement results in the system requirements in terms of carrier-to-noise ratio being met for FM, four-level QAM, and HDTV radio subcarriers.

2.2 Fiber Optic Transmission of Microwave 64-QAM Signals

In recent years, fiber optic analog transmission of microwave signals has received considerable attention. This has happened for several reasons, among them being the high capacity of fiber as a transmission medium, the availability of relatively low-cost microwave components, and the simplicity of the SCM approach compared to, for example, the time-division multiplexing (TDM) technique. Technological advances, mainly in the production of wide-modulation-bandwidth laser diodes [1], have also contributed to increased interest in SCM systems.

Optical analog microwave transmission started as a short-term, low-cost implementation of broadband optical networks. Possible applications were and still are numerous: analog video distribution [2–4], transmission of digitally modulated RF carriers [5–7], transmission of a mixture of digital and analog signals [8], multiple access networks [9], and so forth. There is little doubt that video distribution is a strong motivation for the anticipated widespread penetration of optical fiber in the local loop. Moreover, fiber optic transmission is one of the main alternatives in sight for distribution of digital HDTV. A paper by Kanno and Ito [10] describes a subcarrier multiplexing proposal for the distribution of digital HDTV using a multilevel quadrature amplitude modulation (M-QAM) format.

In this section, we would like, in general, to demonstrate the practicality of long-distance fiber optic transmission of a microwave digitally modulated signal, using a 64-QAM format. M-QAM has been widely accepted for voiceband modems and terrestrial microwave communications. Hence, by using it here, a simple interface between radio and fiber systems becomes possible. Another great advantage of M-QAM modulation is its combined power and bandwidth efficiency. Although the fiber band is very large, in SCM systems, the available bandwidth is a limited resource due to the relatively slow processing speed of electronics, so it must be used as efficiently as possible. Of course, wavelength-division multiplexing (WDM) can be combined with SCM [11] to gain access to wider bands. However, this is a separate issue beyond the scope and purpose of this chapter.

The rest of Section 2.1 is organized as follows. In Section 2.2.1, a description of an experimental setup is presented. In Section 2.2.2, a theoretical signal-to-noise evaluation and BER analyses are presented. The effects of thermal noise, intrinsic laser intensity noise, and reflection-induced laser noise on SCM systems performance are discussed, together with methods for improving this performance. Measurement results are presented and interpreted in Section 2.2.3.

2.2.1 Experimental Setup

A block diagram of an experimental test setup is shown in Figure 2.1. RF signals from two 90-Mbps 64-QAM radio transmitter units, after some

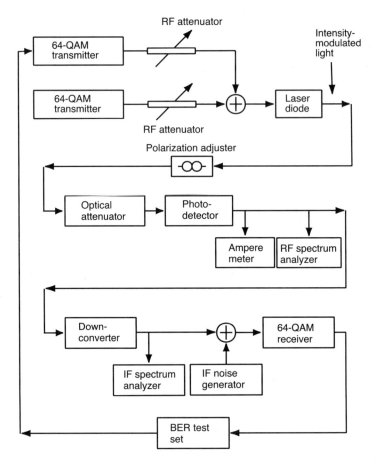

Figure 2.1 Experimental system setup in laboratory.

attenuation is applied, are combined and used to intensity modulate a laser diode. A polarization controller is used to reduce the reflection-induced laser intensity noise. An optical attenuator is used to simulate possible optical path loss.

At the receiver end, the optical signal is detected and downconverted to an intermediate frequency signal, where white Gaussian noise is added to it from a noise generator. By adding a controlled amount of Gaussian noise, it is possible to obtain experimental data about the average BER as a function of bit energy per noise spectral density.

After the intermediate frequency (IF) stage, a 90-Mbps 64-QAM receiver recovers digital information and sends it to a BER test set. The latter test set also generates a pseudorandom digital sequence provided to the transmitter.

This radio system was equipped with an error-correcting code using a shortened rate, 18/19 self-orthogonal convolutional code similar to the one described in [12]. The effective coding rate is $R_c = 12/13$. As will be demonstrated in the following sections, error correction capability greatly enhances the performance of such a system. The experimental setup permits us to measure error probability before error correction decoding. This is done by using a cyclic redundancy check error-detecting code embedded in the transmitted bit stream for frame synchronization purpose. Also, this experimental setup permits us to measure error probability after the decoding, using a BER test set.

The parameters for the 90-Mbps 64-QAM digital radio system are as follows: Carrier frequency is 4 GHz; overall bit rate is 95.96 Mbps; RF bandwidth is 20 MHz; intermediate frequency is 70 MHz; symbol rate is 15.99 megasymbols per second; spectral efficiency is 4.8 bits per second per hertz of bandwidth; and receiver noise figure (ratio of output-to-input S/N) is 4.5 dB.

An ORTEL laser diode used in this experiment is a multilongitudinal mode diode with a central wavelength of about 1.3 λm and a full-width at half-maximum line width of about 2 nm. Its 3-dB bandwidth is at 6 GHz, and its relaxation oscillation frequency is near this point. The photodetector used in the experiment has a responsivity of 0.6 A/W and a 50Ω load resistor.

A polarization controller reduces the reflection-induced noise by making the state of polarization of reflected light orthogonal to that of the emitted light. In a practical system, polarization direction varies along the fiber length. Also, there may be multiple reflections. Therefore, use of a polarization controller may not be appropriate. An optical isolator is a better alternative.

2.2.2 S/N and Average BER Analysis

There are several sources of noise in this system: receiver thermal noise, Gaussian noise added at the receiver intermediate frequency stage, laser intensity noise (intrinsic and reflection induced), dark current noise, and shot noise. Dark current noise is negligible, compared to other ones; thus, it can be neglected. For now, we concentrate on the case of using a single subcarrier for transmission, so that the effects of harmonic distortion, intermodulation, and cross-modulation distortion can be set aside.

Laser intensity noise has long been recognized as a major detrimental factor in analog fiber optic transmission. A very comprehensive analysis of the effects of laser diode intensity noise on system performance is presented in [13]. A detailed theoretical and experimental study of characteristics of intensity noise itself can be found in [14, 15]. A quantitative measure of laser intensity noise is RIN defined in [7] and [13] as:

$$\text{RIN} = \frac{\Delta \overline{P}^2}{\overline{P}_o^2} \qquad \text{dB/Hz} \qquad (2.1)$$

In (2.1), $\overline{P}_o^2$ is the time-averaged laser light intensity and $\Delta \overline{P}^2$ is the mean-square intensity fluctuation spectral density of the output light. The RIN spectrum exhibits a maximum near the laser relaxation oscillation frequency, but within a relatively narrow frequency band (such as a modulated subcarrier band), it can be assumed flat. For a laser diode biased well above threshold, intensity noise can be assumed to have a Gaussian probability distribution [14, 15].

The optical power generated by a laser diode is linearly proportional to the input electric driving current. Optical modulation index is defined as

$$m = \frac{\Delta I}{I_b - I_{th}} \qquad (2.2)$$

where ΔI is the variation of electric driving current around a bias point in modulating a laser light source, I_b is the laser bias current, and I_{th} is the laser threshold current beyond which the light source starts to emit coherent light. To avoid overmodulation of a light source, we make sure that $0 \leq m \leq 1$.

The symbol error rate on in-phase or quadrature-phase channel of a M-QAM signal is expressed by [16] as:

$$P_s(e) = \left(1 - \frac{1}{\sqrt{M}}\right) \text{erfc}\left(\sqrt{\gamma_s \frac{3}{2(M-1)}}\right) \qquad (2.3)$$

where γ_s is the average S/N defined for QAM symbols, and

$$\text{erfc}(x) = \frac{2}{\sqrt{\pi}} \int_x^\infty e^{-t^2} dt \qquad (2.4)$$

Equations (2.1) through (2.4) can be used to evaluate average BER analytically as a function of S/N or the received optical power, knowing that detected photocurrent is expressed by

$$I_D = R_0 P_r \qquad (2.5)$$

In (2.5) R_0 is the responsivity of the photodiode in amperes per watt and P_r is the average received optical power.

Carrier-to-noise power ratio (C/N) at the photodiode output can be defined as:

$$\text{C/N} = \frac{\frac{1}{2}(m \cdot I_D)^2}{\left(\dfrac{4kTF}{R_L} + 2eI_D + \text{RIN} \cdot I_D^2 + N_0 \right) \Delta f} \qquad (2.6)$$

In (2.6), Δf is the double-sided receiver intermediate frequency bandwidth, k is Boltzmann's constant (1.38×10^{-23} J/K), T is the absolute temperature (Kelvin), e is an electron charge (1.602×10^{-19} Coulomb), F is the electronic receiver amplifier noise figure, and R_L is the photodiode load resistor with a 50Ω nominal value. The effect of thermal noise added at IF is shown by the $N_0 \Delta f$ term in the denominator of (2.6), where N_0 is the equivalent spectral density height of the added Gaussian noise.

We present part of our results in terms of bit energy-to-noise density E_b/N_0 in the following sections. We may relate this to γ_s in (2.3) by

$$\gamma_s = \frac{E_b}{N_0} \times \frac{R_b}{\Delta f} \qquad (2.7)$$

where R_b is the transmission bit rate.

Benefits of error correction techniques in fiber optic systems have long been recognized and exploited [17, 18]. Forward error correction (FEC) coding is a particularly valuable approach to compensate for power-dependent BER impairment [19]. It can also be used to reduce error rate floors and/or relax stringent technological requirements otherwise imposed on optical and electronic components.

Performance of the type of error correction code used in this experiment is described in detail in [12]. Self-orthogonal codes are a class of convolutional codes that are rather simple to implement, and they can simply be decoded with majority-logic decoding. The particular code used in this experiment can correct up to 2-bit errors within a constraint length of $N_c = 949$ bits. What is most needed in the analysis at hand is a relationship between input and output BER for the decoder, and this can be found in [12] expressed as:

$$P_b \leq \frac{12}{18} \times \frac{1}{N_c R_c} \sum_{i=3}^{N_c} \binom{N_c}{i} p^i (1-p)^{N_c-i} \qquad (2.8)$$

where P_b is the bit error probability after decoding and p is the BER at the decoder input. The upper bound in (2.8) is a good approximation for values of p less than 10^{-4}. This is derived under an assumption that a Poisson distribution of errors occurs at the decoder input. Conversely, if the distribution departs from being Poisson significantly, the input/output BER performance of the decoder will change and (2.8) may no longer be accurate enough.

2.2.3 Measurement Results

The optical modulation index m was 0.3 throughout all of these experiments. Our measurements did not extend below 10^{-8} BER due to limitations of our test equipment. The first measurement is on the average BER on one channel of QAM signal versus E_b/N_0, without any attempt to minimize the intensity noise. Received average photocurrent is kept to a constant value of 0.6 mA. The value of E_b/N_0 is controlled by inserting an appropriate amount of white Gaussian noise at the receiver intermediate frequency. Results obtained are shown in Figure 2.2.

The ideal performance of a 64-QAM signal, that is, with only a Gaussian noise source, has been plotted in Figure 2.2 as a reference. A large penalty due to intensity noise is clearly seen on the error rate curve at the decoder input. This is measured by a cyclic redundancy check code performing error detection for frame synchronization. Error correction decoding improves the performance of this system greatly, bringing it very close to that of an ideal system, down to an error probability equal to about 10^{-6}.

Figure 2.3 shows the results of a test where Gaussian noise source is removed and the optical signal is gradually attenuated. No intensity noise minimization is applied. In this case, RIN is worse than in the previous measurement. This is due to the presence of an optical attenuator, which introduces more connectors and, consequently, more reflecting surfaces. One can clearly see the occurrence of an error floor for large values of received optical power. Again, performance at the decoder output is substantially better.

In the next experiment, intensity noise is minimized by using a polarization controller. Similar measurements of BER versus E_b/N_0 are taken, both before and after error correction decoding. Results are shown in Figure 2.4. Clearly, the situation is much better than in the previous two experiments.

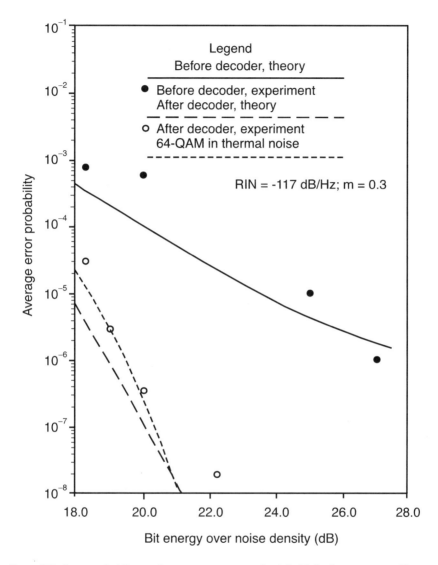

Figure 2.2 Error probability performance versus required E_b/N_0 in the presence of intensity noise, before and after error correction decoding of the 64-QAM signal.

The curve closest to the reference one represents a case of measured back-to-back performance for the radio; that is, when the optics are replaced by an RF attenuator such that the same nominal power is received by the radio receiver. There is about a 1-dB modem degradation at an error probability of 10^{-8}. In the same experiment, another radio is added for simultaneous transmission. The power budget of the system will be lower, of course, but the

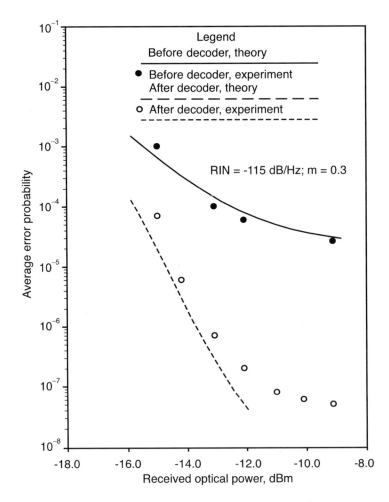

Figure 2.3 Average error probability versus received optical power in the presence of intensity noise.

important fact is that there is no sign of BER degradation from interference or distortion, at least for two 64-QAM signals. The guard band between the two signals is equal to the receiver bandwidth of 20 MHz and appears to be adequate.

Results of the fourth experiment are shown in Figure 2.5. Average BER versus received average optical power is measured for the case of minimized intensity noise. Down to around a BER of 10^{-6}, there is no error floor, even on the error rate curve measured at the decoder input. However, this floor will still occur at higher optical power levels, since the intrinsic intensity noise of the laser is still present.

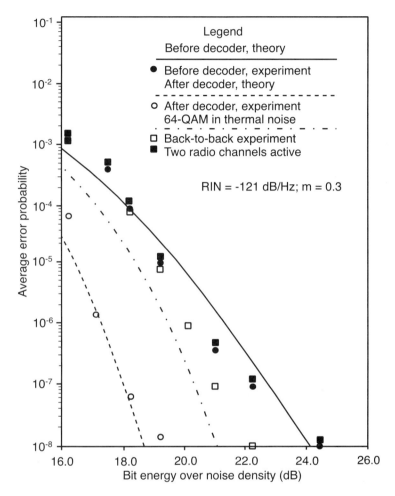

Figure 2.4 Error probability performance versus required E_b/N_0 with minimized intensity noise, before and after the error correction decoding of the 64-QAM signal, for one and two active modulated carriers.

We note that the values for RIN are quite high, even when reflections are minimized. The reason for this is the relatively high microwave input power transferring low-frequency noise into the transmission band [13]. The intrinsic RIN of the free-running laser, as specified by the manufacturer, is typically about −129 dB/Hz. This number can become larger by up to 10 dB in the case of a high RF power at the laser input. In our case, 8 dB (1 mW) of RF power was needed to achieve an $m = 0.3$. To avoid distorting the QAM signal constellation corner points, RF power must be reduced to below 3 dB (1 mW), for which $m \leq 0.17$.

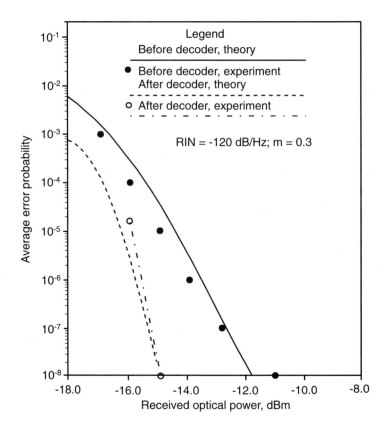

Figure 2.5 Average error probability versus received optical power with minimized intensity noise.

These experimental results are generally in good agreement with theoretical predictions. The occasional larger differences are due to system imperfections not accounted for. One should also keep in mind that the decoder input error rate is not measured by the BER test set and, therefore, cannot be very accurate. Also, (2.8) is an approximate expression that produces accurate results only under the conditions specified in Section 2.2.2.

Figure 2.6 shows the received RF spectrum and signal constellation when two channels are transmitted, simultaneously. RIN was minimized in this case. The optical power was kept constant at a 0-dB (1-mW) level. Figures 2.6(b)–(d) correspond to different levels of added Gaussian noise. Figure 2.6(d) corresponds to a BER $= 4 \times 10^{-4}$. In Section 2.3, we focus our attention on a higher level of modulation, namely, 256-QAM, and consider again error correction coding.

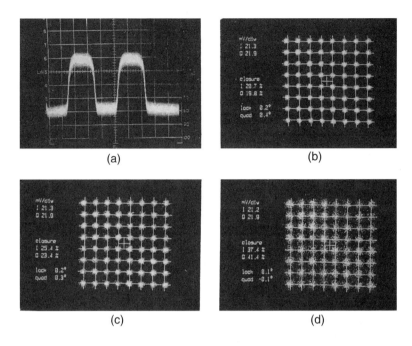

Figure 2.6 Received two-channel signal with a minimized intensity noise. (a) RF spectrum with two radio signals received. (b) Reference received signal constellation for error-free received bit streams. (c) Reference received signal constellation for error-free received bit streams with an increased eye closure compared to case (b). (d) Reference received signal constellation for an error probability of 4×10^{-4}.

2.3 The 256-QAM Subcarrier Transmission Using Coding and Optical Intensity Modulation

As stated earlier, optical fiber can provide a good transmission medium for information distribution networks. SCM transmission over optical fibers is very simple and has found widespread applications because of its simplicity and cost-effectiveness [20–22]. However, in SCM systems, the available bandwidth of optical fiber is limited by the processing speed of electronics. However, multistate modulation techniques such as 16-, 64-, and 256-QAM have been rapidly developed to increase spectral efficiency in various applications [23, 24]. Error-correcting codes such as BCH and Trellis codes have further advanced, enhancing the noise immunity of multistate modulation schemes [25, 26]. BCH codes are cyclic error-correcting codes that allow multiple error correction.

The source encoder for a BCH code maps its input information blocks to longer coded blocks by adding redundant parity bits to the end of information blocks before modulation. Radio equipment using LSI technology for BCH codes has been installed in Japan [24]. For similar applications, SCM systems using the QAM signal format have also been reported [27, 28]. The main purpose of this section is to show the feasibility of 256-QAM subcarrier transmission by using optical intensity modulation. A detailed analysis on the effect of direct modulation on RIN is given in Chapter 1.

2.3.1 Theory

In general, to allow avalanche photodetection, the C/N of a subcarrier signal at the output of photodetector is represented as:

$$\text{C/N} = \frac{\frac{1}{2}(mMR_0P_r)^2}{\left[\dfrac{4kTF}{R_L} + 2e\{(R_0P_r + I_{dm})M^{2+x} + I_{du}\} + \text{RIN}(MR_0P_r)^2\right]\Delta f} \tag{2.9}$$

where m is the optical modulation index, M is the signal current avalanche gain, R_0 is the photodiode responsivity, P_r is the average received optical power, k is Boltzmann's constant, T is absolute temperature (Kelvin), F is the noise figure of the receiver front-end electronic amplifier, R_L is the load resistor of the photodetector (Ω), e is an electron charge, I_{dm} is the multiplied portion of primary dark current, I_{du} is the unmultiplied portion of the dark current, x is excess noise due to avalanche multiplication, RIN stands for relative intensity noise of a laser diode, and Δf is the double-sided receiver filter IF bandwidth.

For the C/N in (2.9), the average BER of uncoded 256-QAM signals using Gray coded symbols can be expressed as:

$$P_e = \frac{15}{65}\text{erfc}\left(\sqrt{\frac{[\text{C/N}]}{170}}\right) \tag{2.10}$$

A BCH code [25] is a powerful random error-correcting cyclic code. In this section, we consider using BCH codes, and we calculate the BER after decoding by an approximate expression with numerical results close to experimentally measured data in terms of BER versus C/N performance.

For an i error-correcting BCH code of constraint length $L = 2^8 - 1$ provided that $p < \dfrac{(2i + 1)}{L}$, the average BER, P_b, after decoding is approximately:

$$P_b = \frac{1}{4}\left[5F\left(\frac{p}{8}\right) + F\left(\frac{p}{4}\right) + F\left(\frac{p}{2}\right) + F(p)\right] \tag{2.11}$$

with

$$F(p) = \binom{L}{i}p^i(1 - p)^{(L-i)}\frac{2i - 1}{L} \tag{2.12}$$

where p is the BER corresponding to the least significant bit [29] before decoding, and is expressed as:

$$p = \frac{1}{2}\,\mathrm{erfc}\left(\sqrt{\frac{C/N}{170}}\right)$$

Note that p is not the same as the uncoded bit error ratio in (2.10). Furthermore, the required C/N to achieve a BER of 10^{-9} is about 35.5 dB for a 256-QAM signal before decoding, as obtained from (2.10). Single- and double-error-correcting BCH codes exhibit a coding gain of approximately 2.3 and 3.5 dB, respectively, at a BER equal to 10^{-9}. The latter values are obtained using (2.4).

2.4 Experimental Results

The experimental setup in the laboratory [30, 31] and a 256-QAM signal constellation are shown in Figure 2.7. The optical power generated by a laser diode is linearly proportional to the input driving current. Therefore, the optical modulation index m can also be defined by the electric average input power P_{in} as

$$m = \frac{\sqrt{\dfrac{2P_{in}}{R}}}{I_b - I_{th}} \tag{2.13}$$

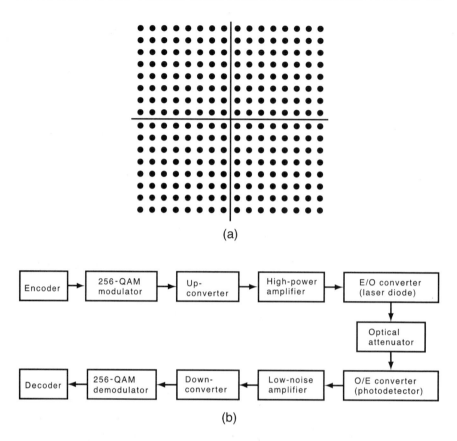

(a)

(b)

Figure 2.7 Experimental setup using 256-QAM signals with FEC: (a) 256-QAM signal constellation and (b) experimental setup.

where I_b is the bias current, I_{th} is the threshold current, and R is the input impedance of the laser diode. The modulation index m is equal to 0.5 in the case of $P_{in} = +10$ dBm.

Analytical and experimental results are compared for two kinds of 256-QAM signals. In one case, the symbol rate is 1.25 MHz, and we use a double-error-correcting BCH code (code rate $R = 239/255$) with a double-sided receiver bandwidth $\Delta f = 2$ MHz. The baseband filtering is Nyquist type, using a roll-off factor equal to 0.5. These parameters correspond to an uncoded bit rate of 10 Mbps. In the other case, the symbol rate is 6.25 MHz, and we use a single-error-correcting BCH code (code rate $R = 247/255$) with a receiver double-sided bandwidth of $\Delta f = 10$ MHz. Again, the Nyquist baseband filtering has a roll-off factor of 0.5, and these parameters correspond to a transmission bit rate of 50 Mbps.

A p-i-n photodiode is used in the experiment. The responsivity is 0.6 A/W, and the avalanche gain M is set to unity in (2.9). A load resistor $R_L = 50\Omega$ and a noise figure of $F = 3$ dB are parameters in the experiment. The results in the form of BER versus average received optical power are shown in Figure 2.8. The theoretical performance for the two types of 256-QAM implementations are plotted as a reference.

Use of the BCH codes improves the transmission performance substantially, compared to the uncoded performance. The double-error-correcting BCH code with a coding gain of 4 dB at a BER of 10^{-10} improves the optical received power by 2 dBm. The BER performance is sensitive to receiver bandwidth. Here, because lesser noise is seen by the subcarrier, the BER improves for smaller receive band. Intermodulation distortion, however, is increased when the total bandwidth is fixed and multiple channels are employed. A system designer should be careful in this case.

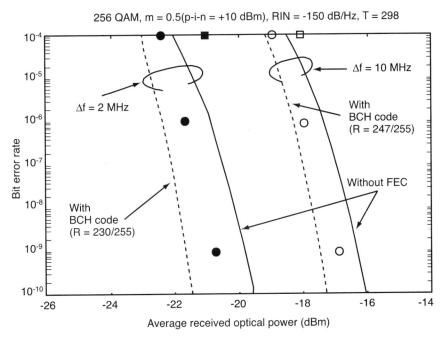

Figure 2.8 256-QAM BER performance versus average received optical power with or without FEC for two values of Δf. Experimental data: $\Delta f = 10$ MHz, without FEC; $\Delta f = 10$ MHz, with FEC (BCH with $R = 247/255$); $\Delta f = 2$ MHz, without FEC; and $\Delta f = 2$ MHz, with FEC (BCH with $R = 239/255$).

2.4.1 Laser Nonlinearity Compensation for Radio Subcarrier Multiplexed Fiber Optic Transmission Systems

A block diagram for an end-to-end subcarrier multiplexing system is shown in Figure 2.9. Modulated carriers are combined to form an electric input signal of a laser. The output optical signal reaches the receiver via an optical fiber. At the receiver a photodetector converts the optical power to an electrical signal. The electrical signal is demultiplexed by bandpass filters (listed as BPF in the figure) and demodulated.

A semiconductor laser is a nonlinear device. Figure 2.10 shows the input/output characteristic of a laser diode. Applying a large signal to the system causes nonlinear distortion. In CATV transmission, several subcarriers are multiplexed to modulate a single laser diode light. Total current that forms the input current to a laser is a phasor sum of amplitude and phase of these subcarriers. This current can form a large signal at the laser input.

The first source of nonlinearity in semiconductor lasers is threshold current, an inherent property of semiconductor lasers, and it causes population inversion in energy states in lasing mode. The second source is due to saturation, which depends on the value of total input current. The total value depends on the number of subcarriers and the individual modulation indices. Linearization of laser characteristic yields a lower nonlinear distortion. Consequently, a larger C/N is achieved. This can be done by introducing a predistortion block placed prior to the laser diode. The predistortion characteristic compensates for that of the laser diode. Hence, the overall system exhibits less nonlinearity than the laser diode itself. In other words, the nonlinearity of the laser is compensated.

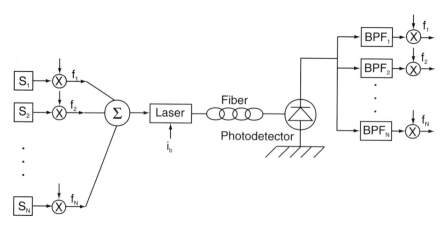

Figure 2.9 Block diagram of an end-to-end subcarrier multiplexing system.

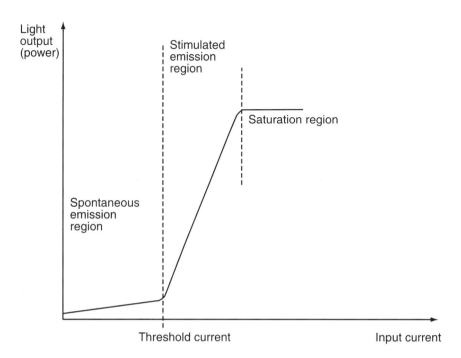

Figure 2.10 Input/output characteristic of a laser diode.

To linearize the characteristic in a laser active region, we need a model. Because laser nonlinearity has memory, a Volterra series expansion has been shown to be an appropriate model for the system [32]. Semiconductor lasers exhibit a weak nonlinearity when laser driving current value is well above the threshold value. Therefore, Volterra transfer functions of low order are employed in the modeling. In this case, up to a third-order transfer function modeling is adequate [32, 33].

Performance of a subcarrier multiplexed system with and without predistortion is evaluated by harmonic and intermodulation distortion analyses. Both analyses show a significant reduction in nonlinear noise terms with predistortion. We apply the proposed system on several modulation schemes that can possibly be utilized in CATV transmission. Without the predistortion block, the requirements on C/N are not met for some of these systems. The reason is, of course, that nonlinear distortion degrades system performance. We show that with predistortion, these requirements are met due to reduction of nonlinear noise.

Next we present harmonic and intermodulation distortions analyses in Section 2.4.2. In Section 2.4.3 we present system performance evaluations. Numerical results are presented in Section 2.5.

2.4.2 Harmonic and Intermodulation Distortion Analyses

The dynamic response of a laser diode is expressed by rate equations, which are two coupled equations relating the density of the injected electrons to the density of emitted photons [34]. For a detailed explanation of symbols and equations presented in this section, see [35].

$$\frac{dN}{dt} = \frac{I_0}{V} - \frac{N}{\tau_s} - g(N - N_0)(1 - \epsilon S)S \qquad (2.14a)$$

$$\frac{dS}{dt} = \Gamma g(N - N_0)(1 - \epsilon S)S - \frac{S}{\tau_{ph}} + \Gamma \beta \frac{N}{\tau_s} \qquad (2.14b)$$

Descriptions and typical numerical values of the parameters in the rate equations are presented in Table 2.1. We used this table, with values corresponding to an Ortel SL-620 laser diode, in our simulation software programs.

Before proceeding with the Volterra series expansion of rate equations, the coupled rate equations are first combined into one equation expressing the input current in terms of the output photons' density:

$$\frac{I_0}{V} = \frac{N_0}{\tau_s} + \left[\frac{1}{\tau_s} + g(1 - \epsilon S)S\right] \frac{\dfrac{dS}{dt} + \dfrac{S}{\tau_{ph}} - \dfrac{\Gamma \beta N_0}{\tau_s}}{\Gamma g(1 - \epsilon S)S + \dfrac{\Gamma \beta}{\tau_s}} \qquad (2.15)$$

$$+ \frac{d}{dt} \left\{ \left[\frac{dS}{dt} + \frac{S}{\tau_{ph}} - \frac{\Gamma \beta N_0}{\tau_s} \right] \Big/ \left[\Gamma g(1 - \epsilon S)S + \frac{\Gamma \beta}{\tau_s} \right] \right\}$$

Volterra transfer functions for this equation can be derived by expansion of the nonlinear term into a Taylor series [33]. Transfer functions of the block diagram shown in Figure 2.11 model the predistortion block and laser diode in terms of a Volterra series. Because a laser is a weakly nonlinear system, Volterra transfer functions up to the third order are adequate for description of its behavior [32]. We used higher order Volterra transfer functions for modeling the predistortion block, but results would be almost the same if we modeled by up to a third-order Volterra operator. The reason is that higher order operators contribute to the generation of large harmonic

Table 2.1

Laser Parameters

Parameter	Description	Values	Units
I_{th}	Threshold current	21	mA
V	Volume of active region times an electron charge	1.44E − 35	m^3-coulomb
τ_{ph}	Photon lifetime	2	ps
τ_s	Electron lifetime	3.72	ns
N_0	Transparent carrier density	4.6E24	m^{-3}
g	Optical gain coefficient	1E − 12	$s^{-1} m^{-3}$
Γ	Optical confinement factor	0.646	—
β	Spontaneous emission factor	0.001	—
ϵ	Gain compression factor	2.6E − 23	m^3

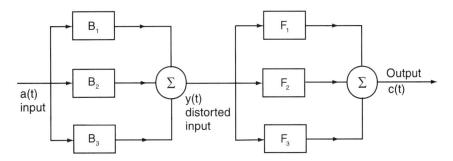

Figure 2.11 Block diagram representation of laser diode and its inverse system modeled in a Volterra series.

components at the output, which are filtered out by the system itself and, therefore, their contribution is negligible.

2.4.2.1 Harmonic Distortion Analysis

The performance of a cascade of a predistorter and a laser in terms of harmonic distortion can be evaluated by applying a single frequency tone at the input. Because the cascade blocks are the inverse of one another, the overall system performs almost as a linear system and the level of harmonic distortion at the output is very small. This argument excludes saturation and clipping regions. To proceed with the analysis, a frequency domain approach is used. Consider $y(t)$ as the output of operator, which represents a pth-order Volterra system. The frequency domain description of this output is

$$Y_p(\omega) = \frac{1}{(2\pi)^{p-1}} \int\limits_{-\infty}^{+\infty} \int\limits_{-\infty}^{+\infty} \cdots$$

$$\int\limits_{-\infty}^{+\infty} H_p(\omega - \omega_1 - \omega_2 - \ldots - \omega_p, \omega_1, \omega_2, \ldots, \omega_p)$$

$$\times A(\omega - \omega_1 - \omega_2 - \ldots - \omega_p)A(\omega_1)A(\omega_2) \ldots$$
$$A(\omega_p)d\omega_1, d\omega_2, \ldots, d\omega_p \qquad (2.16)$$

Recalling Figure 2.11, a frequency domain description of output $y(t)$ is:

$$Y(\omega) = B_1(\omega)A(\omega) + \frac{1}{2\pi} \int\limits_{-\infty}^{+\infty} B_2(\omega - \omega_1, \omega_1)A(\omega - \omega_1)A(\omega_1)d\omega_1$$

$$+ \frac{1}{(2\pi)^2} \int\limits_{-\infty}^{+\infty} \int\limits_{-\infty}^{+\infty} B_3(\omega - \omega_1 - \omega_2, \omega_1, \omega_2)A(\omega - \omega_1 - \omega_2)$$

$$A(\omega_1)A(\omega_2)d\omega_1 d\omega_2 \qquad (2.17)$$

If $a(t)$ is $e^{j\omega_0 t}$, then by the fact that $A(\omega) = \delta(\omega - \omega_0)$, we get:

$$Y(\omega) = B_1(\omega_0)\delta(\omega - \omega_0)$$

$$+ \frac{1}{2\pi} \int\limits_{-\infty}^{+\infty} B_2(\omega - \omega_1, \omega_1)\delta(\omega - \omega_1 - \omega_0)\delta(\omega_1 - \omega_0)d\omega_1$$

$$+ \frac{1}{(2\pi)^2} \int\limits_{-\infty}^{+\infty} \int\limits_{-\infty}^{+\infty} B_3(\omega - \omega_1 - \omega_2, \omega_1, \omega_2)\delta(\omega - \omega_1 - \omega_2 - \omega_0)$$

$$\times \delta(\omega_1 - \omega_0)\delta(\omega_2 - \omega_0)d\omega_1 d\omega_2 \qquad (2.18)$$

and

$$Y(\omega) = B_1(\omega_0)\delta(\omega - \omega_0)$$

$$+ \frac{1}{2\pi} B_2(\omega_0, \omega_0)\delta(\omega - 2\omega_0) \qquad (2.19)$$

$$+ \frac{1}{(2\pi)^2} B_3(\omega_0, \omega_0, \omega_0)\delta(\omega - 3\omega_0)$$

Now, $y(t)$ is an input to the second system with Volterra transfer functions denoted by F, so in (2.17) $A(\omega)$ and $B(\omega)$ are substituted by $Y(\omega)$ and $F(\omega)$, respectively. The frequency-domain description of the final output is $C(\omega)$, which consists of terms as $\delta(\omega - n\omega_0)$ with $n = 1, 2, \ldots,$ 9. Expressions for B and F transfer functions can be found in [35].

The ratio of coefficients of harmonics to fundamental frequency is a measure of linearization effectiveness. When this ratio is small enough, the effect of nonlinearity is small. Sample results of our numerical analysis are shown in Figure 2.12. As can be seen from the graphs, the level of harmonic distortion is significantly decreased when a predistortion block is employed prior to laser diode.

2.4.2.2 Intermodulation Distortion Analysis

In practice, measurement of intermodulation noise in a multiplexed system is performed by noise loading the system with band-limited white noise, thus making this load resemble that of peak hours. Figure 2.13 shows the frequency spectrum of the input signal in this case. Subcarrier frequencies are chosen to fall between the lowest subcarrier f_l, and the highest subcarrier f_h. If we assume that the transmitted power of each subcarrier is P_i, $S_a(f) = \frac{NP_i}{2(df)}$. Here, df is the available bandwidth for transmission or the bandwidth that subcarriers occupy. The transmitted power of each subcarrier is $I_m^2/2$ per 1Ω resistance where I_m is the peak of the current amplitude. Earlier in the chapter, we introduced the modulation index as $m = \frac{\Delta I}{I_b - I_{th}}$. Using this definition, I_m is expressed as:

$$I_m = \frac{m(I_b - I_{th})}{\rho} \qquad (2.20)$$

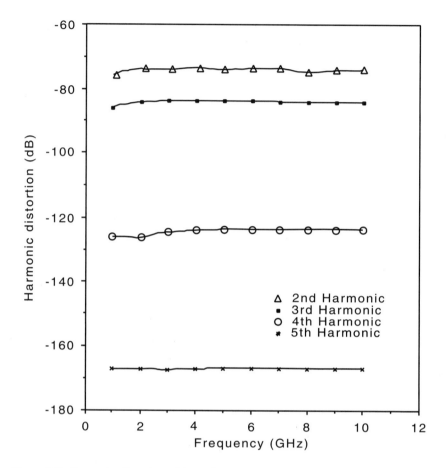

Figure 2.12 Harmonic distortion with predistortion.

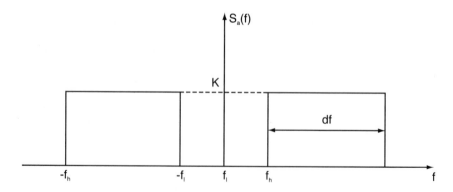

Figure 2.13 Frequency spectrum of subcarriers.

where ρ is the slope of the light current characteristic of the semiconductor laser that is well approximated by $|F_1(0)|$. Since $P_i = \dfrac{I_m^2}{2}$, the frequency spectrum of the input is

$$S_a(\omega) = \frac{Nm^2(I_b - I_{th})^2}{4\rho(df)} \qquad (2.21)$$

This is the value of constant K in Figure 2.13.

At the output, the intermodulation noise is measured in a narrow band (referred to as a frequency notch) from which the noise has been excluded. The first step is to find the frequency spectrum of the second- and third-order intermodulation noise at the output. The autocorrelation of output signal when the system is represented by Volterra kernels of first, second, and third orders is derived directly by finding the expectation of $y(t) \cdot y(t - \tau)$. The resulting power spectrum for the first system is [36]:

$$S_y(\omega) = \langle a(t)^2 \rangle \delta(\omega) + S_a(\omega) \left| B_1(\omega) + \frac{1}{2} \int_{-\infty}^{+\infty} S_a(\lambda) B_3(\omega, \lambda, -\lambda) d\lambda \right|^2$$

$$+ \frac{1}{2!} \int_{-\infty}^{+\infty} S_a(\lambda) S_a(\omega - \lambda) |B_2(\lambda, \omega - \lambda)|^2 d\lambda \qquad (2.22)$$

$$+ \frac{1}{3!} \int_{-\infty}^{+\infty} \int S_a(\lambda) S_a(\gamma) S_a(\omega - \lambda - \gamma) |B_3(\lambda, \gamma, \omega - \lambda - \gamma)|^2 d\lambda \, d\gamma$$

In a similar way, the power spectrum of the final output is derived as

$$S_c(\omega) = \langle c(t)^2 \rangle \delta(\omega)$$

$$+ S_y(\omega) \left| F_1(\omega) + \frac{1}{2} \int_{-\infty}^{+\infty} F_3(\omega, \lambda, -\lambda) S_y(\omega - \lambda) d\lambda \right|^2 \qquad (2.23)$$

$$+ \frac{1}{2!} \int_{-\infty}^{+\infty} S_y(\lambda) S_y(\omega - \lambda) |F_2(\lambda, \omega - \lambda)|^2 d\lambda$$

$$+ \frac{1}{3!} \int_{-\infty}^{+\infty} \int_{-\infty}^{+\infty} S_y(\lambda) S_y(\gamma) S_y(\omega - \lambda - \gamma) |F_3(\lambda, \gamma, \omega - \lambda - \gamma)|^2 d\lambda \, d\gamma$$

The δ function terms of $S_y(\omega)$ and $S_c(\omega)$ are zero frequency components and are removed at both outputs by highpass filtering. However, major sources of distortion are third and fourth terms, which correspond to second- and third-order intermodulation distortion, respectively.

Equation (2.22) can be written in a way that makes the analysis simpler. Recalling expressions B_2 and B_3 from [35] and substituting them in (2.22) results in

$$S_y(\omega) = \langle a(t)^2 \rangle \delta(\omega)$$

$$+ S_a(\omega) \left| B_1(\omega) + \frac{1}{2}(6r - 2g\omega^2 + 2j\sigma\omega) \int_{-\infty}^{+\infty} S_a(\lambda)d\lambda \right|^2$$

$$+ \frac{1}{2!}[(-2l + n\omega^2)^2 + m^2\omega^2] \int_{-\infty}^{+\infty} S_a(\lambda)S_a(\omega - \lambda)d\lambda \qquad (2.24)$$

$$+ \frac{1}{3!}[(6r - 2g\omega^2)^2 + 4\sigma^2\omega^2] \int_{-\infty}^{+\infty}\int_{-\infty}^{+\infty} S_a(\lambda)S_a(\gamma)S_a(\omega - \lambda - \gamma)d\lambda\, d\gamma$$

If we call the fundamental component S_{y1} and the second- and third-order intermodulation products S_{y2} and S_{y3}, then:

$$S_{y1}(\omega) = S_a(\omega) \left| B_1(\omega) + \frac{1}{2}(6r - 2g\omega^2 + 2j\sigma\omega) \int_{-\infty}^{+\infty} S_a(\lambda)d\lambda \right|^2$$

$$(2.25)$$

$$S_{y2}(\omega) \propto \int_{-\infty}^{+\infty} S_a(\lambda)S_a(\omega - \lambda)d\lambda = S_a(\omega) * S_a(\omega) \qquad (2.26)$$

$$S_{y3}(\omega) \propto \int_{-\infty}^{+\infty} S_a(\lambda)S_a(\gamma)S_a(\omega - \lambda - \gamma)d\lambda\, d\gamma = S_a(\omega) * S_a(\omega) * S_a(\omega)$$

$$(2.27)$$

So the second- and third-order intermodulation terms can be expressed by single and double convolution of the fundamental terms, respectively.

The intermodulation products of the laser without a predistortion (inverse) block are shown in Figures 2.14 and 2.15. The cascade system operates linearly. Therefore, the intermodulation distortion is very small compared to the fundamental carrier component, once the laser is linearized. The results in this case where the inverse predistortion block has been used prior to the laser are shown in Figures 2.16 and 2.17.

As the figures indicate, second-order intermodulation distortion is at least 70 dB lower than the case for the system with no predistortion. For the third-order intermodulation distortion (IM3) this amount is about 20 dB. We computed the integral of the power spectrum by a numerical integration rule.

We obtained the results corresponding to different values of the optical modulation index. The modulation index considered in this part of the

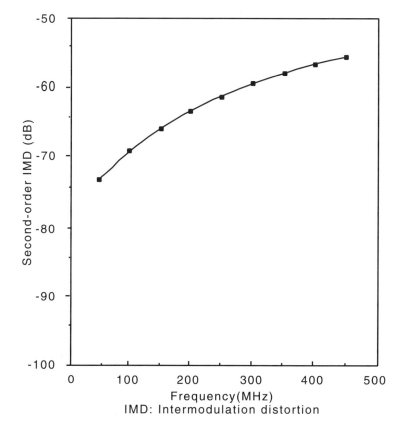

Figure 2.14 Second-order intermodulation distortion variations versus frequency without predistortion.

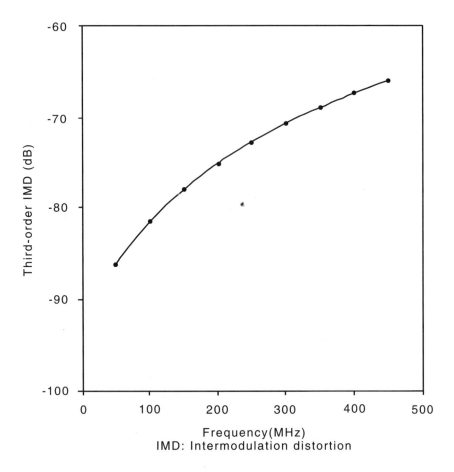

Figure 2.15 Third-order intermodulation distortion variations versus frequency without predistortion.

analysis and in our computer programs is the optical modulation index per carrier. As shown in Figures 2.18 and 2.19, increasing the optical modulation index causes increased intermodulation distortion, as expected. This is because the effective amplitude of the applied signal is smaller. The reason is that, with a smaller modulation index, a smaller portion of the laser characteristic is utilized, that is, the range is less nonlinear. Figure 2.20 shows variation in the intermodulation distortion level as a function of different bias current values when a predistortion block is used prior to the laser. As can be seen from the corresponding curve, nonlinear effects depend on laser bias current. The optimum bias current is at the midpoint between the threshold current and the saturation. The optical modulation index remains a constant 0.04 per carrier in this case. The total modulation index is [37]:

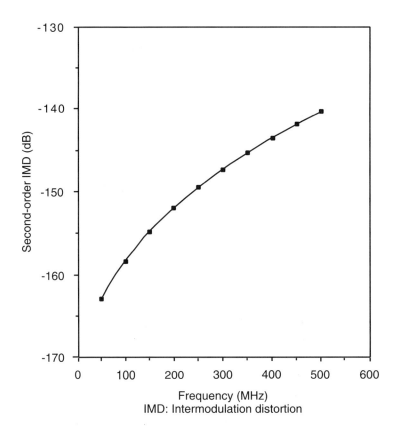

Figure 2.16 Second-order intermodulation distortion variations at the output with predistortion.

$$m = \sqrt{\sum_{i=1}^{N} m_i^2} \qquad (2.28)$$

where N is the number of subcarriers and m_i is the optical modulation index per carrier. Typical values for the total modulation index are in the range of 0.3–0.6 [37]. Therefore, if we assume equal modulation indices for all the subcarriers as expressed by:

$$m_i = \frac{m}{\sqrt{N}} \qquad (2.29)$$

this would result in an m_i in the range of 0.03–0.07 for 75 subcarriers ($N = 75$). The standard desired subcarrier number in CATV transmission is $N = 75$.

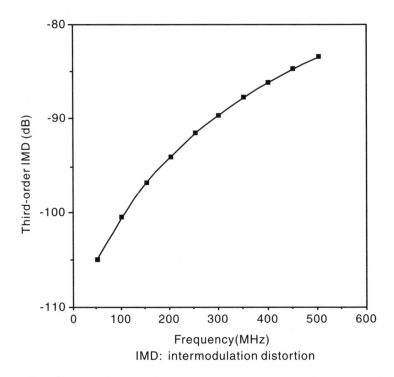

IMD: intermodulation distortion

Figure 2.17 Third-order intermodulation distortion variations at the output with pre-distortion.

2.4.3 System Performance Evaluation

A baseband signal modulates an electrical subcarrier prior to intensity modulation of a light source. In CATV transmission systems a subcarrier can be modulated in any of the appropriate methods such as AM-VSB, FM, four-level QAM, and HDTV. Nonlinearities in the optical source include distortions, intermodulation products, and RIN in the laser. Primary concerns in a receiver are its sensitivity and bandwidth. Major receiver noise sources are quantum noise in photodetection and thermal amplifier noises in the receiver front end. In performance analysis of analog systems, as we saw earlier in Section 2.3, the ratio of the carrier power to the noise power at the output of the optical receiver termed the C/N is expressed as:

$$C/N = \frac{\text{Carrier power}}{\text{Source noise} + \text{Receiver noise}} \qquad (2.30)$$

For multiplexing systems, intermodulation noise is considered as well. The signal power C at the output of the photodetector is

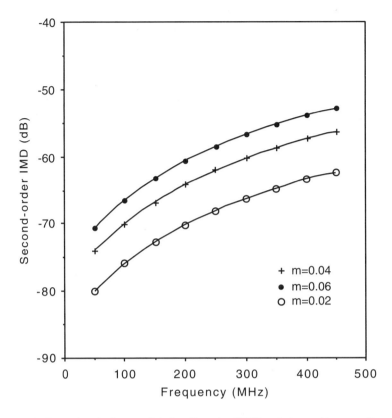

Figure 2.18 Second-order intermodulation distortion (IMD) variations without predistortion for different modulation index values, *m*.

$$C = \frac{1}{2}(mI_D)^2 \tag{2.31}$$

where in (2.31) I_D is the total receiver photocurrent and *m* is the optical modulation index per subcarrier [38]. Noise sources in a fiber optic link that are of concern in C/N calculations consist of shot noise and thermal noise at the receiver and RIN and nonlinear distortions introduced at the source laser.

For different systems with different modulation type, a C/N of a specified value is required for an acceptable signal quality at the receiver [39]. In analog communications applications, a bias point is first identified on the source characteristic curve. The analog signal is then transmitted after it modulates a light carrier. The simplest form of optical source modulation is intensity modulation. Intensity modulation of optical output from the

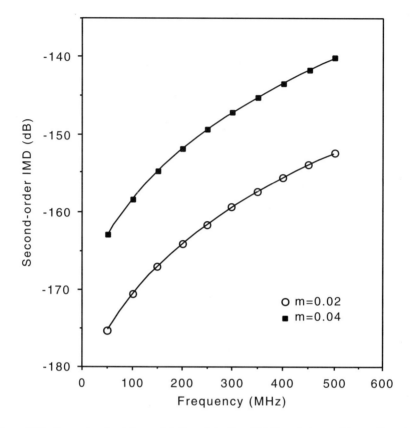

Figure 2.19 Second-order intermodulation distortion (IMD) variations with predistortion for different modulation index values, *m*.

source is accomplished by varying the bias current around the bias point in proportion to the RF signal level.

A microwave RF signal is a modulated form of the original baseband signal. The baseband signal is translated into an electrical subcarrier prior to intensity modulation of the source. This can be done by using amplitude-vestigial-sideband modulation, frequency modulation (FM), or any other type of appropriate modulation technique.

Regardless of the type of modulation technique used, nonlinearity of the optical source must be considered carefully. These include harmonic distortions, intermodulation products, and the RIN of a laser. For each system, using a given modulation method, a specified level of C/N is required to meet an acceptable signal quality at the receiver. Noise terms that limit the C/N include thermal noise from the receiver preamplifier, light signal

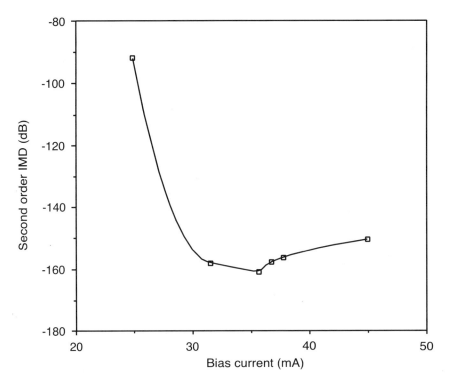

Figure 2.20 Second-order intermodulation distortion (IMD) variations versus laser bias current.

shot noise, and the RIN of light source due to reflections. Hence, using the fact that the specified noise terms have independent sources, the total system C/N is

$$C/N_{total} = \frac{(mI_D)^2}{2\Delta f \left[I_n^2 + I_D^2 \left(RIN + \frac{2e}{I_D} \right) \right]} \qquad (2.32)$$

The expression for C/N$_{total}$ can be converted to

$$\frac{1}{C/N_{total}} = \{ I_n^2 + I_D^2 \, RIN + 2eI_D \} \frac{2\Delta f}{(mI_D)^2}$$

or

$$\frac{1}{C/N_{total}} = \frac{1}{C/N_{shot-noise}} + \frac{1}{C/N_{RIN}} + \frac{1}{C/N_{thermal-noise}} \quad (2.33)$$

Any other carrier-to-noise allocation, such as that for clipping or inter-modulation distortion, can be added to the above expression in the same manner in order to find the total system C/N.

For example, if the ratio of the carrier to second-order intermodulation products is $(C/N)_{IMD2}$, then a $\frac{1}{(C/N)_{IMD2}}$ term is added to the right-hand side of (2.33) and similarly for the third-order products a $\frac{1}{(C/N)_{IMD3}}$ term is added to the sum. According to (2.24), a second-order intermodulation product in the frequency domain has a power spectrum equal to

$$\frac{1}{2!}[(-2l + n\omega^2)^2 + m^2\omega^2] \int\limits_{-\infty}^{+\infty} S_a(\lambda)S_a(\omega - \lambda)d\lambda \quad (2.34)$$

For each value of ω this integral is computed by numerical methods with enough accuracy. The frequency power spectrum part corresponding to the fundamental component is the former term

$$\left| S_a(\omega) \left[B_1(\omega) + \frac{1}{2}(6r - 2g\omega^2 + 2j\sigma\omega) \int\limits_{-\infty}^{+\infty} S_a(\lambda)d\lambda \right] \right|^2 \quad (2.35)$$

Therefore, the ratio of the second-order intermodulation distortion to the carrier or fundamental part is the ratio of the above two terms. Hence, we can, for example, substitute this ratio for $(C/N)_{IMD2}$ in the C/N expression.

2.5 Numerical Results

The amount of C/N and whether it meets the required C/N or not determine the number of subcarriers that can be transmitted in a given bandwidth.

Another parameter that determines this number is the optical modulation index per carrier. A large number of subcarriers are like a large number of small signal inputs that combine to form a large input signal. In some cases even zero-level clipping occurs in the optical output because it cannot be negative. If the number of subcarriers is limited because of a small optical modulation index, the nonlinearity effects are reduced [40].

For each system with a specified modulation format, the system designer must allow for a certain C/N in order to receive signals having the desired quality. The required C/N has been obtained by considering the transmitted power, various system losses such as coupling loss due to fibers, receiver photodetector responsivity, and the system noise margin. For example, for an amplitude-vestigal-sideband system, the utilized bandwidth for transmission is 50–500 MHz. The noise equivalent bandwidth according to [41] is 4 MHz with a channel spacing of 6 MHz; 75 subcarriers are accommodated in the specified bandwidth. The required C/N is 40 dB and the system margin is 12 dB. Similar parameters are defined for other systems with modulation schemes such as FM, four-level QAM, and HDTV. C/Ns with no predistortion have been obtained by computer simulation for all video signal formats considered here [42].

For each system, a dominant noise term exists that plays a major role in setting the total noise level. This dominant noise is not necessarily the nonlinear distortion noise. For systems dominated by nonlinear distortion, the use of predistortion shows an enhancement in the per-carrier total C/N. To be compatible with the present NTSC/TV transmission standard, assume it is desirable to have an amplitude-vestigal-sideband system. For 75 subcarriers, the per-carrier required C/N is 40 dB, which is not met according to our evaluations, as shown in Table 2.2. For the amplitude-vestigal-sideband system, dominant noise is caused by zero-level clipping; hence, predistortion does not affect the total C/N. Reducing the number of subcarriers to 20 causes a decrease in the nonlinear distortion noise level and hence the required C/N is met. This can be observed in Table 2.3. For an FM scheme, as shown in Table 2.4, with an optical modulation index of 0.02, before using predistortion, the C/N value is less than the requirement. With predistortion,

Table 2.2
C/N Results for Amplitude-Vestigal-Sideband Modulation Scheme

Modulation Scheme	OMI* *m*	C/N Required per Carrier (dB)	C/N Without Predistortion (dB)	C/N with Predistortion (dB)
AM-VSB	0.02	40	28.71	31.72
	0.04	40	35.68	37.687
	0.06	40	30	34
	0.08	40	20	22.96

*OMI = optical modulation index. Number of subcarriers = 75.

Table 2.3
C/N Results for Amplitude-Vestigal-Sideband Modulation Scheme

Modulation Scheme	OMI* m	C/N Required per Carrier (dB)	C/N Without Predistortion (dB)	C/N with Predistortion (dB)
AM-VSB	0.04	40	36.9	37.72
	0.06	40	39	41.12[†]
	0.08	40	40.16	43.74

*OMI = optical modulation index. Number of subcarriers = 20.
[†]The C/N requirement is met by using the predistortion block prior to the laser.

Table 2.4
C/N Results for FM Scheme

Modulation Scheme	OMI* m	C/N Required per Carrier (dB)	C/N Without Predistortion (dB)	C/N with Predistortion (dB)
FM	0.01	16.5	12	12.6
	0.02	16.5	11.75	18[†]
	0.03	16.5	7.7	19.49[†]

*OMI = optical modulation index. Number of subcarriers = 75.
[†]The C/N requirement is met by using the predistortion block prior to the laser.

a C/N that is even higher than that required by the system is achievable. Therefore, predistortion enables us to use an FM scheme with a modulation index of 0.02 with 75 subcarriers. We have considered the effect of using a predistortion block for the cases of four-level QAM and HDTV (Zenith). The results are shown in Tables 2.5 and 2.6.

Table 2.5
C/N Results for Four-Level QAM Scheme

Modulation Scheme	OMI* m	C/N Required per Carrier (dB)	C/N Without Predistortion (dB)	C/N with Predistortion (dB)
Four-level QAM	0.02	15.6	12.1	12.42
	0.04	15.6	15.98	16.87
	0.06	15.6	15	18.3[†]
	0.08	15.6	15.2	16.5[†]

*OMI = optical modulation index. Number of subcarriers = 75.
[†]The C/N requirement is met by using the predistortion block prior to the laser.

Table 2.6
C/N Results for HDTV (Zenith) Modulation Scheme

Modulation Scheme	OMI* m	C/N Required per Carrier (dB)	C/N Without Predistortion (dB)	C/N with Predistortion (dB)
HDTV	0.02	18	12.4	12.47
	0.04	18	17.8	18.48[†]
	0.06	18	17	21.567[†]
	0.08	18	17.6	18.68[†]

*OMI = optical modulation index. Number of subcarriers = 75.
[†]The C/N requirement is met by using the predistortion block prior to the laser.

2.6 Conclusions

From the results presented in this chapter, it is evident that the fiber optic transmission of a 64-QAM microwave signal is feasible and practical. Two important ways of improving the system performance were discussed and demonstrated: laser intensity noise minimization and the use of error correction coding. The transmission technique investigated is suitable for many different types of digital information. It could be particularly attractive as a cost-effective approach for the distribution of digital video services.

We have also presented results on 256-QAM subcarrier transmission by using optical intensity modulation and BCH coding. Good agreement exists between the theoretical and experimental results for two types of 256-QAM signals. We also showed that application of BCH codes is very effective for this type of fiber optic transmission.

In an application, video signals modulate equally spaced subcarriers. The sum of the subcarriers is applied as an input to the transmit laser. Laser diodes exhibit nonlinear characteristics that degrade the system performance in terms of C/N and capacity.

We presented a technique to compensate for the nonlinearity in laser characteristics by using a predistortion block prior to the laser diode. The characteristics of this block invert those of the laser, so the overall system has a lower nonlinear noise level. We evaluated the performance of the linearized system by harmonic distortion and intermodulation distortion analyses.

As seen from the intermodulation distortion versus bias current curves shown in the various figures, there is an optimum bias current for which the amount of intermodulation distortion noise is low. This is at the midpoint between threshold current and the saturation current. Dependence of nonlinear noise on the optical modulation index is another factor. Larger modulation indices cause more nonlinear noise. The modulation index for system operation is determined by the number of channels and the system power budget. We considered an optical modulation index of 0.04 per carrier for a 75-subcarrier system. This yields an effective optical modulation index of about 0.32 for the total modulating current.

We used the proposed system, including the predistortion block, with different modulation schemes such as amplitude-vestigal-sideband, FM, four-level QAM, and HDTV (Zenith). Cancellation of nonlinear distortion greatly enhanced the system performance for cases where nonlinear noise is dominant or considerable compared to other sources of noise in the system. For example, for an FM system, as shown in Table 2.4, the C/N is improved when using predistortion. Here, nonlinear noise is comparable to other noise levels in the system. In the case of amplitude-modulated signals and for four-level QAM, which contains envelope zero crossings corresponding to 180° phase jumps of the modulated carrier, clipping noise dominates. This noise is not reduced by the predistorter; hence, the predistortion does not greatly improve the overall system performance for signals with abrupt temporal variations in the modulation envelope. However, it works efficiently for FM-type signals.

References

[1]   Olshansky, R., et al., "InGaAsP Buried Hetero-Structure Laser with 22-GHz Bandwidth and High Modulation Efficiency," *Electron. Lett.,* Vol. 23, July 1987, pp. 839–841.

[2]   Olshansky, R., and V. A. Lanzisera, "60-Channel FM Video Subcarrier Multiplexed Optical Communication System," *Electron. Lett.,* Vol. 23, Oct. 1987, pp. 1196–1198.

[3]   Way, W. I., et al., "90-Channel FM Video Transmission to 2048 Terminals Using Two Inline Traveling-Wave Laser Amplifiers in a 1300 nm Subcarrier Multiplexed Optical System," *Proc. IEEE 14th Europe. Conf. Opt. Commun.,* Sept. 1988, pp. 37–40.

[4]   Way, W. I., et al., "Applications of Traveling-Wave Laser Amplifiers in Subcarrier Multiplexed Lightwave Systems," *Proc. IEEE Int. Conf. Commun.,* 1987, pp. 987–995.

[5]   Bowers, J. E., "Optical Transmission Using PSK-Modulated Subcarriers at Frequencies to 16 GHz," *Electron. Lett.,* Vol. 22, Oct. 1986, pp. 1119–1121.

[6]   Bowers, J. E., et al., "Direct Fiber Optic Transmission of Entire Microwave Satellite Antenna Signals," *Electron. Lett.,* Vol. 23, 1987, pp. 185–187.

[7]   Way, W. I., "Fiber Optic Transmission of Microwave 8-Phase PSK and 16-Ary Quadrature-Amplitude-Modulated Signals at the 1.3 μm Wavelength Region," *J. Lightwave Technol.,* Vol. LT-6, Feb. 1988, pp. 273–280.

[8]   Olshansky, R., V. Lanzisera, and P. Hill, "Design and Performance of Wide-Band Subcarrier Multiplexed Lightwave Systems," *Proc. IEEE 14th Europe. Conf. Opt. Commun.,* Sept. 1988, pp. 143–146.

[9]   Darcie, T. E., "Subcarrier Multiplexing for Multiple-Access Lightwave Networks," *J. Lightwave Technol.,* Vol. LT-5, Aug. 1987, pp. 1103–1110.

[10] Kanno, N., and K. Ito, "Fiber Optic Subcarrier Multiplexing Transport for Broadband Subscriber Distribution Network," *Proc. Int. Conf. Commun.,* 1989, pp. 996–1003.

[11] Liew, S. C., and K. W. Cheung, "A Broadband Optical Local Network Based on Multiple Wavelengths and Multiple RF Carriers," *IEEE Proc. Int. Conf. Commun.,* 1989, pp. 996–1003.

[12] Kavehrad, M., "Convolutional Coding for High-Speed Microwave Radio Communications," *AT&T Tech. J.,* Vol. 64, Sept. 1985, pp. 1625–1637.

[13] Sato, K., "Intensity Noise of Semiconductor Laser Diodes in Fiber Optic Analog Video Transmission," *IEEE J. Quantum Electronics,* Vol. QE-19, Sept. 1983, pp. 1380–1391.

[14] Yamamoto, Y., "AM and FM Quantum Noise in Semiconductor Lasers—Part 1: Theoretical Analysis," *IEEE J. Quantum Electronics,* Vol. QE-19, Jan. 1983, pp. 34–46.

[15] Yamamoto, Y., S. Saito, and T. Mukai, "AM and FM Quantum Noise in Semiconductor Lasers—Part 2: Comparison of Theoretical and Experimental Results for AlGaAs Lasers," *IEEE J. Quantum Electronics,* Vol. QE-19, Jan. 1983, pp. 47–58.

[16] Proakis, J. G., *Digital Communications,* New York: McGraw-Hill, 1983, Chap. 4.

[17] Yoshikai, N., K. Katagiri, and T. Ito, "mB1C Code and Its Performance in an Optical Communication System," *IEEE Trans. on Communication,* Vol. COM-32, Feb. 1984, pp. 163–168.

[18] Kawanishi, S., et al., "DmB1M Code and Its Performance in a Very High-Speed Optical Transmission System," *IEEE Trans. on Communication,* Vol. COM-36, Aug. 1988, pp. 951–956.

[19] Grover, W. D., "Forward Error Correction in Dispersion-Limited Lightwave Systems," *J. Lightwave Technol.,* Vol. LT-6, May 1988, pp. 643–654.

[20] Olshansky, R., V. A. Lanzisera, and P. M. Hill, "Subcarrier Multiplexed Lightwave Systems for Broadband Distribution," *J. Lightwave Technol.,* Vol. LT-7, 1989, pp. 1328–1340.

[21] Way, W. I., "Subcarrier Multiplexed Lightwave System Design Consideration for Subscriber Loop Application," *J. Lightwave Technol.,* Vol. LT-7, 1989, pp. 1806–1818.

[22] Yoneda, E., "Subcarrier Multiplexed Optical Fiber Transmission Systems," *JECEJ Tech. Rep.,* 1990, pp. 15–22, MW89146 (in Japanese).

[23] Yamamoto, H., "Advanced 16-QAM Techniques for Digital Microwave Radio," *IEEE Communications Magazine,* Vol. 19, 1981, pp. 36–45.

[24] Kasai, H., T. Murase, and H. Ueda, "Synchronous Digital Transmission Systems Based on CCN7 SDH Standard," *IEEE Communications Magazine,* Vol. 28, 1990, pp. 50–59.

[25] Lin, S., and D. J. Costello, *Error Control Coding,* Upper Saddle River, NJ: Prentice Hall, 1983.

[26] Ungerboeck, G., "Trellis-Coded Modulation with Redundant Signal Sets—Part 2: State of the Art," *IEEE Communications Magazine,* Vol. 25, 1987, pp. 12–21.

[27] Kavehrad, M., and E. Savov, "Fiber Optic Transmission of Microwave 64-QAM Signals," *IEEE J. Selected Areas in Communication,* Vol. 8, No. 7, Sept. 1990, pp. 1320–1326.

[28] Kanno, N., and K. Ito, "Fiber Optic Subcarrier Multiplexing Video Transport Employing Multilevel QAM," *IEEE J. Selected Areas in Communication,* Vol. 9, 1990, pp. 1313–1319.

[29] Saito, Y., S. Komaki, and M. Murotani, "Feasibility Consideration of High-Level QAM Multicarrier System," *IEEE Proc. Int. Conf. Commun.,* 1989, pp. 665–671.

[30] Ohtsuka, H., Y. Nakamura, and O. Kagami, "Fiber Optic Transmission of Microwave 256-QAM Signals Using Radio Countermeasures," presented at the MICE Nat. Conv., Japan, Oct. 1990.

[31] Ohtsuka, H., et al., "256-QAM Subcarrier Transmission Using Coding and Optical Intensity Modulation in Distribution Networks," *IEEE/LEOS Photon. Technol. Lett.,* Vol. 3, April 1991, pp. 381–383.

[32] Czylwik, A., "Nonlinear System Modeling of Semiconductor Lasers Based on Volterra Series," *J. of Optical Communications,* No. 3, July 1986, pp. 104–114.

[33] Biswas, T. K., and W. F. McGee, "Analytical Predictions of the Intermodulation Noise of Semiconductor Laser Diode," presented at Canadian Conference on Electrical and Computer Engineering, Ottawa, Ontario, Canada, Sept. 1990.

[34] Kressel, H., and J. K. Bulter, *Semiconductor Lasers and Heterojunction LED's,* New York: Academic Press, 1977.

[35] Tayebi, N., and M. Kavehrad, "Laser Nonlinearity Compensation for Radio Subcarrier Multiplexed Fiber Optic Transmission Systems," *IEICE Trans. on Communications (Japan),* Vol. E76-B, No. 9, Sept. 1993, pp. 1103–1114.

[36] Bedrosian, E., and S. O. Rice, "The Output Properties of Volterra Systems (Nonlinear Systems with Memory) Driven by Harmonic and Gaussian Inputs," *Proc. IEEE,* Vol. 59, No. 12, Dec. 1971.

[37] Keiser, G., *Optical Fiber Communications,* New York: McGraw-Hill, 1983.

[38] Darcie, T. E., and G. E. Bodeep, "Lightwave Multichannel Analog AM Video Distribution Systems," *Proc. ICC,* 1989, pp. 32.4.1–32.4.4.

[39] Senior, J. M., *Optical Fiber Communications,* London: Prentice-Hall International, 1985.

[40] Saleh, A. A. M., "Fundamental Limit on Number of Channels in Subcarrier-Multiplexed Lightwave CATV System," *Electron. Lett.,* Vol. 25, No. 12, June 1989, pp. 776–777.

[41] Darcie, T. E., et al., "Lightwave System Using Microwave Subcarrier Multiplexing," *Electron. Lett.,* Vol. 22, 1986, pp. 774–775.

[42] Neusy, P., and W. F. McGee, "Effects of Laser Nonlinearities on TV Distribution Using Subcarrier Multiplexing," *Proc. Canadian Conference on Electrical and Computer Engineering,* Montreal, Sept. 1989, pp. 883–836.

3

Low-Cost Fiber Optic Links for Cellular Remote Antenna Feeding
Stephan Hunziker

3.1 Introduction

Remote antenna feeding, particularly in cellular telephone systems, has become a very attractive field of application for fiber optic links. Frequency-independent low fiber loss, high bandwidth, and increasing availability of low-cost optoelectronic and RF integrated components are the major drivers for fiber optic RF solutions. Almost all existing remote antenna feeding concepts use coaxial cable or radio links.[1] Replacing such well-tried and reliable technology through a new technology is only justified if the new technology offers a significant cost reduction or decisive technical advantages. It is the aim of this chapter to give an overview and to point out advantages and problems related to fiber optic links for cellular remote antenna feeding applications. Chapters 1 and 2 contain a comprehensive analysis for direct modulation and its effect on relative intensity noise.

The chapter starts with a system-level analysis and considerations of the requirements for cellular base station remote antenna links. We then show that these requirements can be fulfilled with direct intensity-modulated analog fiber optic links. A comparison of conventional coaxial links and

---

1. Some use CATV-grade fiber optic links with expensive optically isolated distributed feedback laser diodes.

fiber optic links considering technical as well as economical aspects shows that the use of the latter is justified even for in-house applications.

One cost-effective link concept is bidirectional transmission over one fiber. The most important parameter determining the quality of a fiber optic link is its dynamic range, which is determined by noise and nonlinearities. Nonlinearities and noise performance of laser diodes and their influence on link performance are discussed. Laser diode intermodulation including external optical feedback is analyzed using the Volterra theory of nonlinear systems. A subsection treats kink nonlinearities. It turns out that links with low-cost, multilongitudinal-mode, Fabry-Perot multiple-quantum-well laser diodes offer high dynamic range at low cost for link spans of up to about 10 km. If the dynamic range and link length requirements are somewhat relaxed, light emitting diodes and vertical cavity surface emitting lasers can also be used. For link spans above 10 km, distributed feedback laser diodes are required.

A section is devoted to fiber-induced effects. These include increased intermodulation and noise of multilongitudinal-mode Fabry-Perot laser links at long fiber spans and bandwidth limitations caused by fiber dispersion. Multicarrier intermodulation and carrier allocation optimization in subcarrier multiplexing systems are discussed, as are driver and receiver design issues.

The last part of this chapter is devoted to alternative concepts for remote antenna feeding fiber optic links and interesting new components. Plastic fibers are a promising low-cost technology for future short-haul fiber optic links. Downconversion of the RF signals leads to relaxed bandwidth requirements in narrowband links. Hybrid fiber/coaxial solutions are recommended as optimum feeding concepts in distributed antenna systems. An up-to-date literature survey is provided.

3.2 System-Level Analysis and Design of Fiber Optic Remote Antenna Feeding Links

Modern cellular telephone systems like the European Groupe Spéciale Mobile (GSM) system are SCM^2 and time-division multiple access (TDMA) systems. Therefore, intermodulation is a key issue.

Some useful system-level definitions for fiber optic SCM systems are given in this section. A new and simple method demonstrating the importance of usually neglected fifth-order nonlinearities of modern RF devices in link

---

2. In SCM systems many modulated carriers are transmitted over one link [1].

design is presented. Dynamic range and output power considerations for cellular remote antenna feeding links are made.

3.2.1 Link Dynamic Range Considerations

Like all parts of a communication system, a fiber optic link suffers from essential limitations: noise and nonlinearity (bandwidth may also be an issue in broadband fiber optic links). These two limitations and the combination of both, the dynamic range, are now discussed. Useful information on fiber optic link design and dynamic range considerations can also be found in [2, 3].

3.2.1.1 Weakly Nonlinear Systems and Intercept Points

For simplicity the weakly nonlinear system is assumed to be static or, in other words, it is operated well below its upper cutoff frequency. This means that the nonlinearity can be approximated by a Taylor series. We therefore start our analysis by using a fifth-order polynomial for the relation between input and output voltage of the amplifier v_{in} and v_{out} [4]:

$$v_{out}(v_{in}) = a_1 v_{in} + a_2 v_{in}^2 + a_3 v_{in}^3 + a_4 v_{in}^4 + a_5 v_{in}^5 \tag{3.1}$$

Equation (3.1) can calculate signal, intermodulation, and harmonic products at the amplifier output from the so-called one- and two-tone tests with single- and two-tone input signals, respectively:

$$v_{in,\,1\text{-tone}}(t) = A\cos(\omega_1 t) = 0.5A(e^{j\omega_1 t} + e^{-j\omega_1 t}) \tag{3.2}$$

$$\begin{aligned} v_{in,\,2\text{-tone}}(t) &= A\cos(\omega_1 t) + A\cos(\omega_2 t) \\ &= 0.5A(e^{j\omega_1 t} + e^{-j\omega_1 t}) + 0.5A(e^{j\omega_2 t} + e^{-j\omega_2 t}) \end{aligned} \tag{3.3}$$

and the multinomial theorem [5] with $p = 1$ or 2 for one- and two-tone tests, respectively:

$$(x_1 + x_2 + x_3 + \ldots + x_p)^n = \sum_{\substack{n_j \geq 0, \\ \Sigma_{i=1}^{p} n_j = n}} \frac{n!}{n_1! n_2! n_3! \ldots n_p!} x_1^{n_1} x_2^{n_2} x_3^{n_3} \ldots x_p^{n_p} \tag{3.4}$$

The one-tone test is used to find the gain (G) and the 1-dB compression point (P_{1dB}) of the device. P_{1dB} is the value of the output power for which the output power level is 1 dB below the output power level of an ideally linear device with the same small signal gain.

The output voltage of the one-tone test at ω_1 is:

$$v_{\text{out}}^{\omega_1}(t) = \left(\frac{1}{2} a_1 A + \frac{3}{8} a_3 A^3 + \frac{5}{16} a_5 A^5 \right) \cos(\omega_1 t) \qquad (3.5)$$

In an amplifier with compression, the A^3 and A^5 terms are responsible for the compression.

The output of a device under the two-tone test at the intermodulation angular frequency $2\omega_1 - \omega_2$ is:

$$v_{\text{out}}^{2\omega_1 - \omega_2}(t) = \left(\frac{3}{4} a_3 A^3 + \frac{25}{8} a_5 A^5 \right) \cos[(2\omega_1 - \omega_2)t] \qquad (3.6)$$

At low input levels the $2\omega_2 - \omega_1$ and $2\omega_1 - \omega_2$ products are caused by the third-order nonlinearity of the device. But generally intermodulation at $2\omega_1 - \omega_2$ as well as the signal (gain compression) are influenced by odd-order nonlinearities because they are combinations of positive and negative odd harmonics of input frequencies ($m\omega_1 + n\omega_2$, where m, n are positive or negative integers and $|n| + |m| = 1, 3, 5, \ldots$). The term containing a_1 is due to the gain and the a_3 and a_5 terms are the third- and fifth-order gain compression. It is often stated that $2\omega_1 - \omega_2$ products are third-order products. This is only true for small signal excitation. At medium and strong input signals, higher odd orders also contribute to this product. Taking into account that (3.3) consists of exponentials with $\pm\omega_1$ and $\pm\omega_2$ in their exponents, one can easily find the spectrum generated by an nth-order term in (3.1) by adding the angular frequencies $\pm\omega_1$ and $\pm\omega_2$ because nth power means just adding up combinations of the various exponents (angular frequencies). So the fifth-order term of the nonlinearity generates among others products at $+\omega_1 + \omega_1 + \omega_1 - \omega_2 - \omega_2 = 3\omega_1 - 2\omega_2$ but also at $+\omega_1 + \omega_1 + \omega_2 - \omega_2 - \omega_2 = 2\omega_1 - \omega_2$ and so on.

A simple means to characterize the intermodulation behavior at small signal operation of a weakly nonlinear system is to use the second- and third-order intercept points (IP$_2$ and IP$_3$). Due to the fact that the nth-order intermodulation curve and the signal curve at low levels have steepness n and 1, respectively, in double log coordinates there is an intercept point between the two. The nth-order intercept point (IP$_n$) is as follows, where variables in decibels are denoted with "dB" in their index:

$$IP_n = \frac{CIM_{n,dB}}{n-1} + C_{dBm} \tag{3.7}$$

with the carrier power level C_{dBm} and the carrier-to-nth-order intermodulation ratio $CIM_{n,dB}$. At the output the IP is called output intercept point (OIP); at the input it is called input intercept point (IIP).

Figure 3.1 illustrates the definition of the intercept points for second- and third-order intermodulation. The signal curve steepness is 1 at low input levels, whereas the pure second- and third-order curves of intermodulation products of angular frequencies $\omega_1 + \omega_2$ and $2\omega_1 - \omega_2$ have a steepness of 2 and 3, respectively. The dashed curves are asymptotes of the measured intermodulation (straight).

In most small signal and narrowband (less than 1 octave) systems, only third-order intermodulation (IM_3) has to be considered. Then the following relations can be derived from above:

$$OIP_{3,dBm} = \frac{CIM_{3,out,dB}}{2} + C_{out,dBm} \tag{3.8}$$

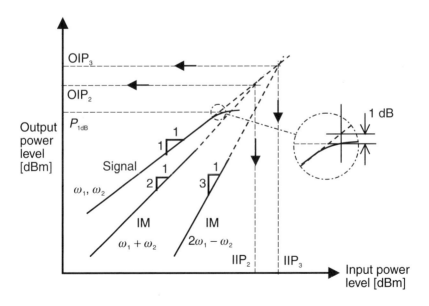

Figure 3.1 Output and input second- and third-order intercept point and 1-dB compression point definitions for a weakly nonlinear system (OIP_2, OIP_3, IIP_2, IIP_3, and P_{1dB}).

$$IM_{3,out,dBm} = 3C_{out,dBm} - 2OIP_{3,dBm} \qquad (3.9)$$

$$CIM_{3,out,dBm} = 2(OIP_{3,dBm} - C_{out,dBm}) \qquad (3.10)$$

The total OIP_n of a cascade of m blocks with gains G_i and nth-order output intercept points $OIP_{n,i}$ is found by adding the cumulative intermodulation power:

$$(OIP_{n,tot})^{\frac{1-n}{2}} = (OIP_{n,m})^{\frac{1-n}{2}} + (G_m OIP_{n,m-1})^{\frac{1-n}{2}}$$
$$+ (G_m G_{m-1} OIP_{n,m-2})^{\frac{1-n}{2}} + (G_m \ldots G_2 OIP_{n,1})^{\frac{1-n}{2}}$$
$$(3.11)$$

Table 3.1 lists the output signal components of weakly nonlinear system[3] excited by three input tones at angular frequencies ω_1, ω_2, and ω_3 and amplitudes A_1, A_2, and A_3, respectively.

3.2.1.2 Noise

The spectral noise power density $N_{dBm/Hz}$ in a narrowband system can usually be considered constant over the system bandwidth (BW). The noise power N_{dBm} in dBm is

$$N_{dBm} = N_{dBm/Hz} + 10 \log(BW \text{ [Hz]}) \qquad (3.12)$$

The C/N, C/N_{dB}, in decibels for a carrier with power C_{dBm} in dBm is then:

$$C/N_{dB} = C_{dBm} - N_{dBm} \qquad (3.13)$$

The noise figure F_{dB} is defined as [6]

$$F_{dB} = 10 \log\left(\frac{C/N_{input}}{C/N_{output}}\right) \qquad (3.14)$$

The noise power for an amplifier $N_{amp,dBm}$ with gain G and noise figure F is

$$N_{amp,dBm} = 10 \log[(F - 1)k_B TG + N_{in} G] + 30 \text{ dB} \qquad (3.15)$$

---

3. The systems exhibit nonlinearities up to third order.

Table 3.1
Signal Components at the Output of a Weakly Nonlinear System Excited by Three
Input Tones at Angular Frequencies ω_1, ω_2, and ω_3 and
Amplitudes A_1, A_2, and A_3, Respectively

Order	Component	Physical Significance
1	$A_1 A_i \cos(\omega_i t)$ $i = 1 \ldots 3$	Three first-order components; result of linear gain
2	$1/2\, a_2 A_i^2$ $i = 1 \ldots 3$	Three dc components; rectification effect of second-order nonlinearity
2	$a_2 A_i A_j \cos[(\omega_i \pm \omega_j)t]$ $i, j = 1 \ldots 3; i \neq j$	Three second-order intermodulation products; new frequencies
2	$1/2\, a_2 A_i^2 \cos(2\omega_i t)$ $i = 1 \ldots 3$	Three second-order harmonics at double frequency; new frequencies
3	$1/4\, a_3 A_i^3 \cos(3\omega_i t)$ $i = 1 \ldots 3$	Three third-order harmonics at triple frequency; new frequencies
3	$3/2\, a_3 A_i A_j A_k \cos[(\omega_i \pm \omega_j \pm \omega_k)t]$ $i, j, k = 1 \ldots 3; i \neq j \neq k$	Four third-order intermodulation components ("triple beat"); new frequencies
3	$3/4\, a_3 A_i^2 A_j \cos[(\omega_i \pm \omega_j)t]$ $i, j = 1 \ldots 3; i \neq j$	Four third-order intermodulation products (beat); new frequencies
3	$3/4\, a_3 A_i^3 \cos(\omega_i t)$ $i = 1 \ldots 3$	Three third-order products, carrier frequency (self-compression/expansion)
3	$3/2\, a_3 A_i A_j^2 \cos(\omega_i t)$ $i, j = 1 \ldots 3; i \neq j$	Three third-order products, carrier frequency (cross compression/expansion)

with temperature T, Boltzmann's constant $k_B = 1.38 \times 10^{-23}$ J/K and input noise power N_{in}, whereas for an attenuator with loss L, one finds

$$N_{att, dBm} = 10 \log\left[\left(1 - \frac{1}{L}\right)k_B T + \frac{N_{in}}{L}\right] + 30 \text{ dB} \qquad (3.16)$$

$$F_{att, dB} = 10 \log(L) \qquad (3.17)$$

It is often useful to consider an equivalent input noise power level EIN_{dB} instead of the noise power at the output or at any other point in the link. It is a measure for the amount of noise that has to be fed into the input of an equivalent ideal noise-free link to generate the amount of noise that appears at the output of the actual noisy link:

$$EIN_{dBm} = N_{out,dBm} - G_{dB} \tag{3.18}$$

For a matched source with purely resistive source impedance at temperature $T = 293K$ the link noise figure is then:

$$F_{dB} = 10 \log \left(\frac{C/N_{input}}{C/N_{output}} \right)$$

$$= C_{in,dBm} - (-173.9 \text{ dBm/Hz}) \tag{3.19}$$

$$- [C_{in,dBm} + G_{dB} - (EIN_{link,dBm/Hz})]$$

$$= 173.9 \text{ dBm/Hz} + EIN_{link,dBm/Hz}$$

Friis's formula gives the total noise figure of a cascade of m blocks with gains G_n and noise figures F_i [7]:

$$F_{tot} = F_1 + \frac{F_2 - 1}{G_1} + \frac{F_3 - 1}{G_1 G_2} + \frac{F_4 - 1}{G_1 G_2 G_3} + \ldots + \frac{F_m - 1}{G_1 \ldots G_{m-1}} \tag{3.20}$$

3.2.1.3 Spurious Free Dynamic Range and Sensitivity

When the signal power is increased just up to a point where the intermodulation products reach the noise floor for a certain noise bandwidth, noise and intermodulation power are equal. For stronger signal levels, the usable signal range is limited by the intermodulation products, whereas for weaker signals it is limited by the noise. Due to the fact that the third-order intermodulation products increase by $3 \times x$ dB when the signal power is increased by x dB, the useful signal range will shrink if the intermodulation products surpass the noise floor [determined by carrier-to-intermodulation (CIM) ratios and no longer by C/N].

The maximum signal power that gives the maximum dynamic range is reached at the point where noise and intermodulation are equally large. The difference between signal and noise floor/intermodulation product is called spurious free dynamic range (SFDR) as defined in Figure 3.2. To calculate SFDR of a narrowband system we need noise and third-order intermodulation parameters.

Equating noise and third-order intermodulation, we find

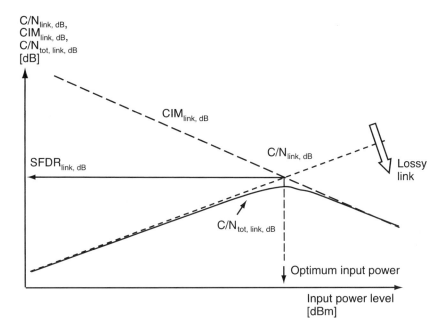

Figure 3.2 SFDR of a link. If the link loss increases, the C/N_{link} curve moves in the direction of the broad arrow. Then the $C/N_{tot,link}$ equals C/N_{link} at the operating input power, which was optimum in the former case.

$$N_{out,dBm} = EIN_{dBm/Hz} + 10 \log(BW) + G_{dB} \qquad (3.21)$$

$$= IM_{3,out,dBm} = 3C_{out,dBm} - 2OIP_{3,dBm}$$

Subtracting $3N_{out,dBm}$ and adding $2OIP_{3,dBm}$ one gets

$$2OIP_{3,dBm} - 2N_{out,dBm} = 3C_{out,dBm} - 3N_{out,dBm} = 3C/N_{out,dB} \Big|_{N=IM_3}$$

$$= 3SFDR_{dB} \qquad (3.22)$$

$$SFDR_{dB} = \frac{2}{3}(OIP_{3,dBm} - N_{out,dBm}) \qquad (3.23)$$

$$= \frac{2}{3}[OIP_{3,dBm} - EIN_{dBm/Hz} - 10 \log(BW) - G_{dB}]$$

The sensitivity S of a link defines the minimum signal level for which a certain minimum C/N_{out} at the link output is maintained:

$$S_{dBm} = EIN_{dBm/Hz} + 10\log(BW) + C/N_{out,dB} \qquad (3.24)$$

3.2.2 Intermodulation of Power Amplifiers Used in Remote Antenna Feeding Links

In narrowband communication systems such as cellular radio systems the nonlinearity is characterized essentially by the $2\omega_1 - \omega_2$ or $2\omega_2 - \omega_1$ intermodulation product, respectively. These intermodulation products are usually called third-order intermodulation products and can be calculated by means of the third-order intercept point[4] IP_3 (see Section 3.2.1). In general, mainly when it comes to power amplifiers, this is not fully true because the amplifier nonlinearity is not purely third order. For low signal amplitudes the $2\omega_1 - \omega_2$ or $2\omega_2 - \omega_1$ intermodulation is essentially third order. As the input signal of the amplifier exceeds a certain critical power, however, fifth-order or even higher order components contribute to those intermodulation products. The resulting P_{in}-intermodulation curve does not have a constant steepness of 3 as expected. Just above the critical input power, the steepness exceeds 3 caused by the amplifier's fifth-order nonlinearity.

3.2.2.1 Critical Power

The amount of intermodulation in an RF amplifier for weak input signals can be calculated if IP_3 is given. In doing so, we assume a steepness of 3 of the intermodulation curve. As already mentioned in Section 3.2.1 this estimation is no longer valid if the signal exceeds a certain limit. For the RF designer it is essential to know this limit. It is particularly useful for high dynamic range amplifiers in SCM systems with relatively low output powers where the intermodulation is low but only for signals that are small enough.

We now derive a formula to determine this limit. Using this formula we can calculate the maximum output power below which the $2\omega_1 - \omega_2$ or $2\omega_2 - \omega_1$ intermodulation is a pure third-order intermodulation, and below which we can use IP_3 to estimate the intermodulation. The model can also be used to estimate the influence of the fifth-order nonlinearity on these intermodulation products just above the limit. The limit is determined by two amplifier parameters available from the data sheet: the third-order output intercept point (OIP$_3$) and the output 1-dB compression point (P_{1dB}).

---

4. Output third-order intermodulation products are characterized by the third-order output intercept point (OIP$_3$), whereas input third-order intermodulation products are characterized by the third-order input intercept point (IIP$_3$).

The OIP_3 validity limit for the output power per carrier is a very important design parameter for low-cost high dynamic range links in multi-carrier systems. It is common practice to design the circuit by specifying its gain and OIP_3 and to stay well below P_{1dB} with the output signal. For a given OIP_3 the OIP_3 validity limit depends on P_{1dB}. The difference between P_{1dB} and maximum usable output power is often underestimated for modern low-cost amplifiers.

3.2.2.2 Importance of the Difference $OIP_3 - P_{1dB}$

The amplifier gain using (3.5), omitting the a_3 and a_5 terms; OIP_3 using (3.6), omitting the a_5 term and its P_{1dB}; and using (3.5) can be expressed by the a coefficients defined in Section 3.2.1.1. The equation for P_{1dB} is:

$$20 \log \left(\frac{a_1 A_{P_{1dB}} + \frac{3}{4} a_3 A_{P_{1dB}}^3 + \frac{5}{8} a_5 A_{P_{1dB}}^5}{a_1 A_{P_{1dB}}} \right) \overset{!}{=} -1 \text{ dB} \qquad (3.25)$$

with the input amplitude $A_{P_{1dB}}$ at the 1-dB compression point. We find that, for a 50Ω impedance environment,

$$a_5 = -\frac{\frac{3}{4} a_3 A_{P_{1dB}}^3 + a_1 A_{P_{1dB}} \left(1 - 10^{\frac{-1dB}{20}} \right)}{\frac{5}{8} A_{P_{1dB}}^5} \qquad (3.26)$$

$$A_{P_{1dB}} = \frac{\sqrt{10^{\frac{P_{1dB,out} + 1}{10}} \cdot 50\Omega \cdot 1 \text{ mW}}}{a_1} \cdot \sqrt{2} \qquad (3.27)$$

From the two-tone test we get (3.3) and OIP_3 by equating signal and IM_3:

$$|a_1|A = \frac{3}{4}|a_3|A^3 \Rightarrow OIP_{3,dBm} = 10 \log \left[\frac{(|a_1|A)^2}{50\Omega \cdot 1 \text{ mW} \cdot 2} \right] \qquad (3.28)$$

$$= 10 \log \left[\frac{2|a_1|^3}{50\Omega \cdot 1 \text{ mW} \cdot 3|a_3|} \right]$$

$$a_3 = -\frac{2}{3} a_1^3 \frac{10^{-OIP_3/10}}{50\Omega \cdot 1 \text{ mW}} \tag{3.29}$$

$$a_1 = 10^{gain/20} \tag{3.30}$$

For devices showing compression—as most amplifiers do at not too high frequencies—a_3 and a_5 must be negative.

Comparing the two parts of (3.6), we define that if the fifth-order part is smaller than the third-order part by x dB, we approximately reach the limit of pure third-order intermodulation. We call the output power at this critical point P_{crit} [4]:

$$10 \log\left[\frac{\left(\frac{3}{4}|a_3|A_{P_{crit,out}}^3\right)^2}{50\Omega \cdot 1 \text{ mW} \cdot 2}\right] = 10 \log\left[\frac{\left(\frac{25}{8}|a_5|A_{P_{crit,out}}^5\right)^2}{50\Omega \cdot 1 \text{ mW} \cdot 2}\right] + x \rightarrow A_{P_{crit,out}} \tag{3.31}$$

$$P_{crit,out,dBm} = 10 \log\left[\frac{\left(a_1 A_{P_{crit,out}}\right)^2}{50\Omega \cdot 1 \text{ mW} \cdot 2}\right]$$

$$= 2(P_{1dB,out} + 1) - OIP_3 - \frac{x}{2} - 4.594 \text{ dB} \tag{3.32}$$

$$- 10 \log(0.1088 - 0.75 k_1 k_2)$$

$$k_1 = \frac{10^{(P_{1dB,out}+1)/10}}{10}, \quad k_2 = \frac{40}{3} \cdot 10^{-OIP_3/10}$$

Figure 3.3 shows a plot of (3.32) for $x = 10$ dB, which is fundamental because it is valid for every amplifier independent of its gain. This curve can be approximated by a plane [last term of (3.32) omitted] for $OIP_3 - P_{1dB} > 15$ dB. It can be seen from Figure 3.3 that there is a fundamental lower limit for the difference $OIP_3 - P_{1dB}$ at approximately 10 dB. If the difference is lower, there is no more compression but expansion. For a given OIP_3, the limit of its applicability is higher for a higher P_{1dB}. Modern small signal field effect transistor or hetero-bipolar transistor RF amplifiers are very linear well below saturation and therefore have a high OIP_3 but a low P_{1dB} [8]. For that reason they are suited for high dynamic range applications

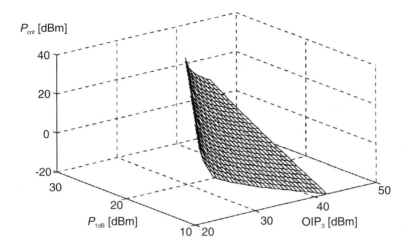

Figure 3.3 Dependence of an amplifier's critical output power P_{crit} on third-order output intercept and 1-dB compression points OIP_3 and P_{1dB}, respectively. (*Source:* [4]. © 1996 IEE.)

only if the signal level is low enough or if the number of carriers in an SCM link is low.

If higher output levels are needed or many carriers have to be transmitted, an amplifier with at least the same OIP_3 but a higher P_{1dB} has to be chosen. Common bipolar amplifiers usually have a difference $OIP_3 - P_{1dB}$ of about 11 dB, whereas modern hetero-bipolar transistor monolithic microwave integrated circuit amplifiers can have 17 dB or more. If the difference is small, the influence of the fifth-order nonlinearity is small, too. We should also mention that a high $OIP_3 - P_{1dB}$ difference can be very desirable for linear amplifiers in very low power applications such as mobile phones and in other applications. It combines very low intermodulation at low output power with high efficiency.

Figure 3.4 shows the measured and calculated intermodulation end signals at the output of an amplifier. Calculated (from data sheet parameters) and measured useful ranges of the OIP_3 definition are practically the same. The difference is mainly caused by the fact that the measured gain is about 1 dB smaller than the data sheet typical value. The prediction of the limit of pure third-order intermodulation by (3.32) is quite good. If guaranteed data sheet values are used, a worst-case estimate can be made. The formula is a good means to estimate the performance of an amplifier with given data sheet parameters.

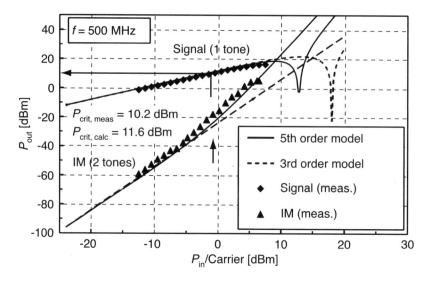

Figure 3.4 Measured and modeled signal (f_1) and intermodulation (IM) ($2f_1 - f_2$) output power of an RF amplifier integrated circuit. (*Source:* [4]. © 1996 IEE.)

These considerations are important not only from a technical but also from an economical point of view. For a given OIP_3 the costs of an amplifier depend on P_{1dB}. The formulas are valid for frequencies well below the amplifier's cutoff frequency. For higher frequencies a more sophisticated model has to be developed using, for instance, the Volterra theory of nonlinear systems (see Section 3.4).

3.2.3 Dynamic Range Requirements for a Cellular Remote Antenna Feeding Link

The dynamic range requirements for a remote antenna link depend on many points, such as the cellular standard, geographic aspects, and traffic. As an example, a GSM link (Figure 3.5) is now considered that gives the reader a general idea for this and for other standards. Requirements defined by network providers are usually derived from field studies. This example is based on discussions with network providers and component suppliers.

We make these assumptions:

- Noise and intermodulation are caused mainly by the fiber optic part (the laser diode). This assumption is made because in practice it is usually possible to improve amplifiers and passive components for much lower costs than the laser diode. Therefore, the SFDR value

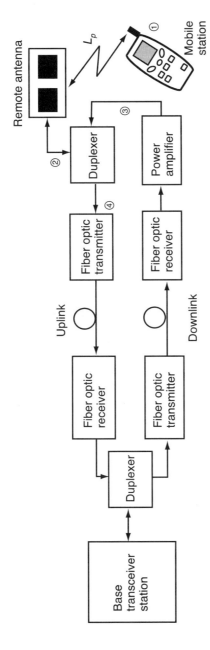

Figure 3.5 Bidirectional GSM remote antenna feeding link for a microcell repeater. Uplink: 890–915 MHz, downlink: 935–960 MHz. Dynamic range requirements: high for uplink, low for downlink.

for the laser diode determines its suitability for a fiber optic remote antenna link. The actual power levels are matched by means of amplifiers and variable attenuators.

- Power balance is assumed. This means that with increasing path loss between mobile stations and remote antennas, uplink and downlink connections will be interrupted simultaneously.

- Uplink and downlink frequency bands do not overlap. Even intermodulation and signal bands do not overlap in many cases.

- The noise bandwidth for a GSM channel is BW = 200 kHz.

The following downlink requirements are necessary:

1. Any interference, crosstalk, and noise (spurious) at antenna input, $N_{sp,dBm}$, must be less than −36 dBm.
2. The path loss from a mobile station (MS) to a remote antenna (1 to 2) must be in the range of L_p = 37–102 dB.
3. C/N, $C/N_{min,dB}$, must be greater than 10 dB.
4. Sensitivity, $S_{MS,dB}$, must be equal to −92 dBm (worst case).

From requirements 1, 2, and 4, the minimum OIP_3 at downlink (DL) output (2) is found with (3.21):

$$C_{out,DL,dBm} = S_{MS,dBm} + L_{p,max,dB} = 10 \text{ dBm} \qquad (3.33)$$

$$OIP_{3,DL,dBm} = \frac{3C_{out,DL,dBm} - N_{sp,dBm}}{2} = 33 \text{ dBm} \qquad (3.34)$$

If the critical power in the sense of (3.32) is set to $C_{out,DL,dBm}$, the minimum P_{1dB} for the downlink output is found (x = 10 dB)[5]:

$$P_{1dB,DL,dBm} \approx 20 \text{ dBm} \qquad (3.35)$$

The duplexer loss is of the order of 2 dB. Therefore, $OIP_{3,PA,dBm}$ and $P_{1dB,PA,dBm}$ at the power amplifier (PA) output (3) should be some decibels higher:

---

5. For about four GSM carriers; for a higher number of carriers, P_{1dB} should be increased (see Section 3.7).

$$\text{OIP}_{3,\text{DL,dBm}} = \frac{3C_{\text{out,DL,dBm}} - N_{sp,\text{dBm}}}{2} = 35 \text{ dBm} \qquad (3.36)$$

$$P_{1\text{dB,DL,dBm}} \approx 22.23 \text{ dBm} \qquad (3.37)$$

Downlink noise is critical. The requirement of $N_{sp,\text{dBm}} < -36$ dBm must be fulfilled, and the duplexer isolation from downlink to uplink has to be high enough to avoid an uplink (UL) sensitivity (S_{UL}) decrease due to crosstalk of noise from downlink to uplink (3 to 4). The latter is analyzed later in this section.

From (3.23) it follows that

$$\text{SFDR}_{\text{DL,dBm}} = \frac{2}{3} [\text{OIP}_{3,\text{DL,dBm}} - N_{sp,\text{dBm}}] = 46 \text{ dB} \qquad (3.38)$$

This shows that the SFDR requirement for the downlink is not very high.

The most important downlink requirements are the power amplifier OIP_3 and $P_{1\text{dB}}$ as well as the duplexer isolation at the antenna input.

The following uplink requirement is necessary:

5. Apart from the downlink requirements, an additional requirement for the mobile station (MS) output power has to be considered[6]: $C_{\text{MS,max,dBm}} = 27$ dBm (0.5W).

From requirements 1, 2, and 5, the minimum IIP_3 at uplink input (4) is found with (3.9):

$$C_{\text{in,UL,dBm}} = C_{\text{MS,max,dBm}} - L_{p,\text{min,dBm}} = -10 \text{ dBm} \qquad (3.39)$$

$$\text{IIP}_{3,\text{UL,dBm}} = \frac{3C_{\text{in,UL,dBm}} - N_{sp,\text{dBm}}}{2} = 3 \text{ dBm} \qquad (3.40)$$

and from requirements 2 and 3 with (3.24)

$$S_{\text{UL,dBm}} = C_{\text{MS,max,dBm}} - L_{p,\text{max,dB}} = -85 \text{ dBm} \qquad (3.41)$$

$$\text{EIN}_{\text{UL,dBm/Hz}} = S_{\text{UL,dBm}} - C/N_{\text{min,dB}} - 10\log(\text{BW}) = -148 \text{ dBm/Hz} \qquad (3.42)$$

---

6. This is due to the low mobile station power consumption requirement.

$$\mathrm{SFDR_{UL,dB}} = \frac{2}{3} \left[\mathrm{IIP_{3,UL,dBm}} - \mathrm{EIN_{dBm/Hz}} - 10 \log(\mathrm{BW}) \right]$$

$$= 65.3 \text{ dB} \qquad\qquad (3.43)$$

The SFDR requirements for cellular remote antenna feeding links are summarized in Table 3.2. In special cases the SFDR requirements may be some decibels higher (up to ≈ 75 dB).

3.2.4 Analysis of a Generic Analog Fiber Optic Link

Figure 3.6 shows a generic fiber optic link like that used for remote antenna feeding [2, 3]. We now analyze the link loss, noise, and dynamic range of this generic link.

Table 3.2
SFDR Requirements for Cellular Remote Antenna Feeding (BW = 200 kHz)

	SFDR Downlink (dB)	SFDR Uplink (dB)
Indoor	> 46	≈ 50–60
Outdoor	> 46	> 65–70

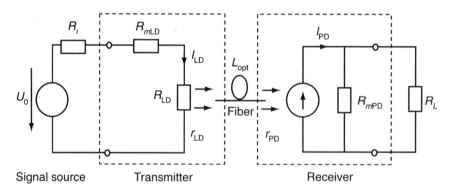

Figure 3.6 Generic fiber optic link. r_{LD}, r_{PD}: laser diode and photodiode responsivities; R_{mLD}, R_{mPD}: laser diode and photodiode matching resistors; R_{LD}: laser diode input resistance; R_i, R_L: source and load resistance; U_0: source voltage. For resistive matching $R_{mPD} + R_{LD} = 50\Omega$ and $R_{mPD} = R_L = 50\Omega$.

3.2.4.1 Link Loss

At several interfaces of the link losses are introduced:

- *Source laser diode:* Depends on matching network insertion loss. The resistive matching shown in Figure 3.6 introduces an approximate 9- to 13-dB loss due to the low laser diode impedance $R_{\text{LD}} \approx 5\Omega$.
- *Laser diode conversion efficiency:* Depends on laser diode type.
- *Laser diode–fiber coupling:* Depends on pigtail and coupling optics.
- *Fiber and optical connector losses:* Depend on wavelength and connector cleanliness.
- *Fiber-photodiode coupling:* Depends on pigtail.
- *Photodiode quantum efficiency:* Depends on photodiode type.
- *Photodiode responsivity:* Depends on wavelength.

A laser diode (LD) with threshold current I_{th} and operating current I_{LD} generates an optical power of $P_{\text{opt}} \propto I_{\text{LD}} - I_{th}$. This optical power is converted into an electrical current again by the photodiode (PD): $I_{\text{PD}} \propto P_{\text{opt}}$, which causes an electrical power in the load of the receiver: $P_L \propto I_{\text{PD}}^2 \propto P_{\text{opt}}^2 \propto (I_{\text{LD}} - I_{th})^2$. Therefore, an optical power loss L between laser-diode output and photodiode input leads to an electrical power loss of L^2 between link input and output.

The lower limit for the link loss L_0 of the generic link is:

$$L_0 = \left. \frac{P_{Qav}}{P_L} \right|_{\substack{R_{\text{LD}} + R_{m\text{LD}} = R_j \\ = R_L = R_{m\text{PD}} = 50\Omega}} = \frac{I_{\text{LD}}^2 \cdot 50\Omega}{\frac{1}{4} I_{\text{PD}}^2 \cdot 50\Omega} = \frac{I_{\text{LD}}^2}{\frac{1}{4}(I_{\text{LD}} \, r_{\text{LD}} \, r_{\text{PD}})^2}$$

$$= \frac{4}{(r_{\text{LD}} \, r_{\text{PD}})^2} \tag{3.44}$$

with the available source power P_{Qav}. Including optical losses L_{opt} from fibers, connectors, splices, isolators, and so on, the link loss is:

$$L_{\text{link}} = L_0 L_{\text{opt}}^2 G_{\text{TX}}^{-1} G_{\text{RX}}^{-1} = \frac{4}{(r_{\text{LD}} \, r_{\text{PD}})^2} \frac{L_{\text{opt}}^2}{G_{\text{TX}} G_{\text{RX}}} \tag{3.45}$$

where G_{TX} and G_{RX} are the additional matching gains if reactive instead of resistive transmitter and receiver matching networks are used (see Section 3.9).

3.2.4.2 Link Noise

In a fiber optic link noise is primarily caused by the laser diode laser relative intensity noise (RIN) and the detection process (photodiode shot noise). The laser RIN is defined as the relative optical intensity fluctuation or the relative photocurrent fluctuation with the shot noise subtracted [9, 10]:

$$\text{RIN}_{\text{laser}} = \frac{\langle \Delta P_{\text{opt}}^2 \rangle}{\langle P_{\text{opt}} \rangle^2} \cdot \frac{1}{\text{BW}} = \frac{\langle \Delta I_{\text{PD}}^2 \rangle - 2e\,\text{BW}\langle I_{\text{PD}} \rangle}{\langle I_{\text{PD}} \rangle^2} \cdot \frac{1}{\text{BW}} \quad [1/\text{Hz}]$$

$$(3.46)$$

The shot noise itself, which is due to the quantum nature of the photodetection process, is not included in the shot noise. The relative shot noise is often called RIN_Q:

$$\text{RIN}_Q = \frac{2e\langle I_{\text{PD}} \rangle}{\langle I_{\text{PD}} \rangle^2} = \frac{3.2 \cdot 10^{-16}\ \text{mA/Hz}}{\langle I_{\text{PD}} \rangle} \hat{=} -155\ \text{dB/Hz} - 10\log(I_{\text{LD}}\ [\text{mA}])$$

$$(3.47)$$

This curve is shown in Figure 3.7.

The contribution of the laser RIN to the link equivalent input noise (EIN) power at the laser-diode input is

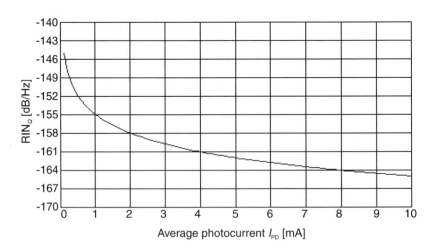

Figure 3.7 Relative intensity noise due to shot noise (RIN_Q) for given average photocurrent $\langle I_{\text{PD}} \rangle$.

$$\text{EIN}_{\text{RIN}} = \text{RIN} \left(\frac{P_{\text{opt}}}{r_{\text{LD}}} \right)^2 (R_{\text{LD}} + R_{m\text{LD}}) \frac{1}{G_{\text{TX}}} \qquad (3.48)$$

with the laser's average optical power P_{opt}. The shot noise contribution is

$$\text{EIN}_{sh} = \left(\frac{R_{m\text{PD}}}{R_{m\text{PD}} + R_L} \right)^2 2er_{\text{PD}}(R_{\text{LD}} + R_{m\text{LD}})L_0 L_{\text{opt}} P_{\text{opt}} \frac{1}{G_{\text{TX}}} \qquad (3.49)$$

and the receiver thermal noise contribution is

$$\text{EIN}_{th} = k_B TL_0 L_{\text{opt}}^2 \left(\frac{R_{\text{LD}} + R_{m\text{LD}}}{R_L} \right) \frac{1}{G_{\text{TX}}} \qquad (3.50)$$

In a practical link with low-intensity laser diodes, the resistor $R_{m\text{PD}}$ may generate too much noise. Therefore, the photodiode should directly feed a transimpedance amplifier input without a matching resistor $R_{m\text{PD}}$. The receiver noise is then characterized by the receiver thermal input spectral noise current density $I_{n,th}$ with $R_{m\text{PD}} \to \infty$ and $G_{\text{RX}} = 4$ or $G_{\text{RX,dB}} = 6$ dB, respectively (see Section 3.9). We find for EIN_{th}:

$$\text{EIN}_{th} = \frac{L_0 (R_{\text{LD}} + R_{m\text{LD}})}{4 G_{\text{TX}}} \cdot (I_{n,th} L_{\text{opt}})^2 \qquad (3.51)$$

The total equivalent input noise power is the sum of the three contributions:

$$\text{EIN}_{\text{link}} = \text{EIN}_{\text{RIN}} + \text{EIN}_{sh} + \text{EIN}_{th} \qquad (3.52)$$

Figure 3.8 shows the contributions of the three terms on the total equivalent input noise for a typical link example. Fiber loss is approximately 0.3 dB/km (1,310 nm) and 0.15 dB/km (1,550 nm); connector loss is less than 0.5 dB/connector.

For a remote antenna link having a length of less than 5 km, the link loss is usually below 3 dB.

3.2.4.3 Dynamic Range

From Sections 3.2.3 and 3.2.4, two new expressions for SFDR can be derived using EIN and RIN definitions:

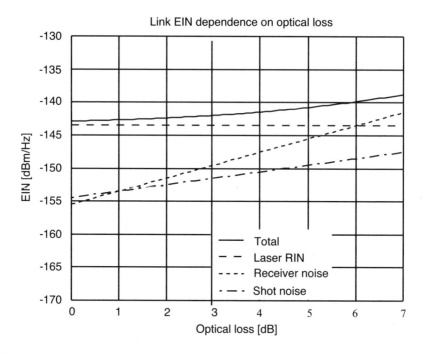

Figure 3.8 Noise contributions at the input of an optical receiver in a typical fiber optic link depending on optical loss.

$$\text{SFDR}_{dB} = \frac{2}{3}(\text{IIP}_{3,dB} - \text{EIN}_{dB}) \tag{3.53}$$

$$\text{SFDR}_{dB} = \frac{2}{3} \cdot \tag{3.54}$$

$$\left\{ \text{IIP}_{3,dB} - \text{RIN}_{\text{laser},dB/Hz} - 10\log\left[\left(\frac{P_{LD}}{r_{LD}}\right)^2 \frac{R_{LD} + R_{mLD}}{1\ \text{mW} \cdot G_{TX}}\right] - 10\log(\text{BW}) \right\}$$

To estimate the achievable SFDR for a particular laser diode, the curves given in Figures 3.9 and 3.10 can be used. A laser diode's IIP$_3$, RIN, r_{LD}, and nominal optical output power as well as a photodiode's r_{PD} are either available from the data sheets or can be measured rather easily. For high link loss the SFDR values of Figure 3.9 can only be achieved with low-noise receivers (see Section 3.9).

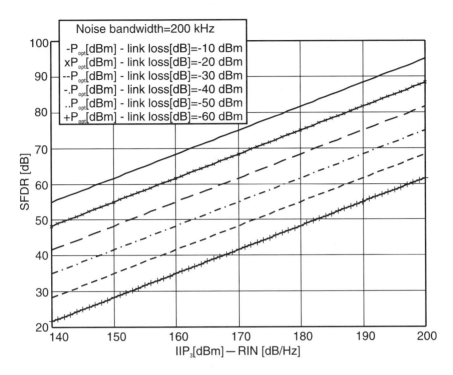

Figure 3.9 Maximum achievable SFDR for a fiber optic link realized using a laser diode with known IIP$_3$ for 50Ω resistive input matching and relative intensity noise (RIN = RIN$_Q$ + RIN$_{laser}$) values. The curves are parameterized for various values of the difference between laser diode optical output power and link loss L_0 for a 0-m fiber span. The photodiode responsivity is set to r_{PD} = 0.85 A/W.

3.3 Comparison Between Coaxial and Fiber Optic Remote Antenna Feeding Links

The use of fiber optic links for remote antenna feeding is justified only if there are technical and mainly economical advantages. Therefore, the feasibility and performance of low-cost bidirectional single-fiber links for analog RF fiber optic antenna feeding are now studied. We consider two options:

1. Bidirectional transmission over two fibers. In this scheme one fiber is used for the uplink and the other one for the downlink.

2. Bidirectional transmission over one single fiber. Wavelength-division multiplexing (WDM) allows transmission of uplink (λ_{opt} = 1,310 nm) and downlink (λ_{opt} = 1,550 nm) signals over one single standard fiber in this scheme. Designs using low-cost

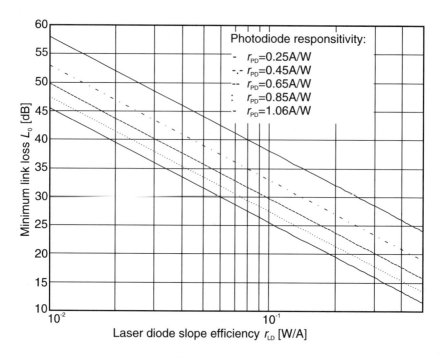

Figure 3.10 Minimum link loss of a fiber optic link with given photodiode responsivity and laser diode slope efficiency.

multiple-quantum-well Fabry-Perot laser diodes in bidirectional transceiver modules or in separate receiver/transmitter modules are compared with coaxial cable links.

3.3.1 Introduction

Low-cost multiple-quantum-well Fabry-Perot lasers are well suited as high dynamic range transmitters [11, 12] and have become an interesting alternative to coaxial cable or solutions with expensive optically isolated distributed feedback lasers mainly in cases where the link span is below some kilometers and the modulation frequency is below some gigahertz. In the case of remote base station antennas, the link span is usually in the range of some 100m up to about 2 or 3 km, and the modulation frequency is of the order of some 100 MHz up to about 2 GHz. Therefore, fiber dispersion is no issue for single-mode standard fibers at 1,310-nm laser wavelength even if multilongitudinal-mode laser diodes are used.

The classical solution relies on two separate fibers: one for the uplink and one for the downlink. There are two important reasons why it could be highly desirable to spare one of the two fibers:

1. At fiber spans approaching the kilometer range, fiber costs become significant.
2. In many cases as few installed fibers as possible should be used for a single application and it would often be advantageous to save fibers for other applications (dark fibers) at the price of slightly increased transmitter and receiver complexity.

Figure 3.11 shows the two topologies (single- and two-fiber bidirectional links).

3.3.2 Transceivers for Remote Antenna Feeding Links

Cellular base station antennas are operated in a receiving and a transmitting mode. Therefore, the feeding link has to be bidirectional. The separation between uplink and downlink is done by means of an antenna duplexer. Transceivers with separate laser diode transmitters and photodiode receivers

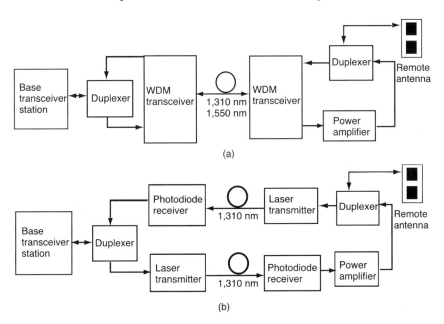

Figure 3.11 (a) Single-fiber and (b) two-fiber bidirectional remote antenna feeding links.

are usually the most convenient solution [Figure 3.11(a)], but bidirectional modules [13], already known from SDH/Sonet and hybrid fiber-coaxial return path applications, can also be used as "full-duplex" transceivers [Figure 3.11(b)].

A bidirectional module consists of a laser diode, a photodiode, and a WDM beam splitter. Bidirectional modules are used as a pair where one module usually emits at 1,310 nm and the other module emits at 1,550 nm (some bidirectional modules emit at only one wavelength as well). For the separation and isolation of the two wavelengths, a WDM beam splitter is used. Figure 3.12 shows a cross section of such a bidirectional module. The dynamic range requirements for the downlink are less demanding than for the uplink as derived in Section 3.2.3. The dispersion induced by the 1,550-nm source over some kilometers of standard fibers can be tolerated.

3.3.3 Experimental Results for Practical Links

For the measurements, bidirectional modules, laser diodes, and photodiodes were mounted on the edge of a Duroid 5880 substrate. With this mounting it is possible to keep the highly inductive connecting pins very short. To achieve simple matching to 50Ω, a 47Ω resistor was put in front of the laser diode assuming a laser diode resistance of approximately 3Ω. Output matching consists of a 51Ω resistor in parallel with the photodiode. The following devices have been measured [13]:

- Infineon SBM81314 (1,310-nm) and SBH51414 (1,550-nm) bidirectional modules;

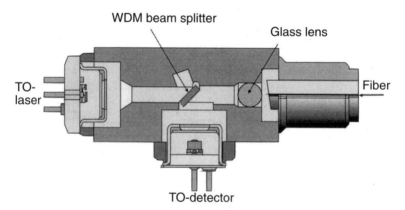

Figure 3.12 Cross section of a bidirectional module; TO: transistor outline package. (Courtesy of Infineon Technologies AG.)

- MRV MRTXLSPS010 (1,310-nm) and MRTXLSPS5010 (1,550-nm) bidirectional modules;
- Hitachi HL1326CF (1,310-nm), Photon PT3543-12-3-1FC (1,550-nm), and Fujitsu FID3Z1KX modules.

The MRV devices are equipped with subscriber connector/physical contact connectors. All other devices have fiber connector/angled physical contact connectors. The Hitachi and the Photon devices are coaxially packaged laser diodes. In combination with the Fujitsu photodiode, they form a simple unidirectional optical link. If the optical power at the receiver is of the order of at least 1 mW, the receiver thermal noise is not significant if a low-noise transimpedance or high-impedance receiver is used [14]. Noise is mainly caused by the laser-diode RIN and shot noise in that case.

Another important parameter of an optical link is the link gain. To measure the link gain (negative link loss) up to 3 GHz, a network analyzer was used. Figure 3.13 shows the S parameter S_{21} of a bidirectional module. In this case the MRV MRTXSPS010 module with the 1,310-nm laser diode (biased at 10 mA) was used as the transmitter. The measurement results of all bidirectional modules are summarized in Table 3.3.

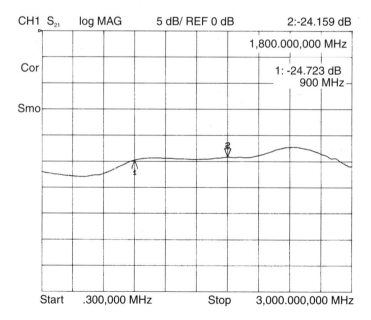

Figure 3.13 Link gain up to 3 GHz with MRV MRTXLSP010 as transmitter (biased at 10 mA) and MRV MRTXLSP5010 as receiver. (*Source:* [15]. © 2000 IOS Press.)

Table 3.3
S_{21} (Link Gain) of Bidirectional Modules Between 900 and 1,800 MHz
as a Function of Laser Diode Bias Current

Transmitter	Infineon (1,550 nm) SBH51414	Infineon (1,310 nm) SBM81314	MRV (1,550 nm) MRTXLSPS5010	MRV (1,310 nm) MRTXLSPS010
Receiver	Infineon (1,310 nm) SBM81314	Infineon (1,550 nm) SBH51414	MRV (1,310 nm) MRTXLSPS010	MRV (1,550 nm) MRTXLSPS5010
Laser Diode Current [mA]		S_{21} 0.9–1.8 GHz [dB]		
10	—	—	—	-24.4 ± 0.3
15	-31.8 ± 1.5	—	—	-25.7 ± 0.3
20	-29.0 ± 1.3	-29.5 ± 1.2	-31.1 ± 0.9	-26.3 ± 0.6
25	-30.1 ± 0.7	-30.8 ± 0.6	-32.1 ± 1.1	-26.5 ± 0.7
30	-30.7 ± 0.4	-31.4 ± 0.3	-32.1 ± 1.2	-27.0 ± 0.8

From Steiner et al. [15].

It is important to note that switching the laser in the receiving module on does not affect the SFDR of any of the tested transmitting modules. This means that the isolation of the two wavelengths is very good. Measured crosstalk is below 41 dB at 900 MHz/48 dB at 1.8 GHz (MRV), and 55/61 dB (Infineon). This crosstalk has been identified as entirely electrical. The isolation between downlink intermodulation and downlink signal is guaranteed with these values. The SFDR values at a noise bandwidth of 200 kHz of all the measured links at 900 MHz and 1.8 GHz are summarized in Table 3.4.

Referring to the requirements derived in Section 3.2.3, this means that WDM links with bidirectional WDM transceiver modules presented above can be used for indoor as well as outdoor remote antenna feeding in the 900-MHz, 1.8-GHz, and 2-GHz ranges.

The 1,550-nm link, which shows slightly worse performance than the 1,310-nm link, is used as the downlink (lower SFDR requirement), whereas the 1,310-nm link is used for the uplink (higher SFDR requirement). At the upper link span limit of about 5 km, the downlink SFDR can be degraded by several decibels due to dispersion. Table 3.4 reveals that a fair amount of the SFDR margin is left even in this case.

It is well known that semiconductor laser diodes suffer from severe second-order distortion (see Section 3.4). In the case of operating a link simultaneously at 900 MHz and 1.8 GHz, designers must be able to guarantee

Table 3.4
Measured SFDR of Bidirectional and Unidirectional Links at 0.9 and 1.8 GHz
(BW = 200 kHz)

Device, Bias Current	SFDR [dB] 0.9 GHz	SFDR [dB] 1.8 GHz
Infineon SBM81314 (1,310 nm), 40 mA	74	69
MRV MRTXSPS010 (1,310 nm), 31 mA	76	73
Hitachi HL1326CF (1,310 nm), 32 mA	76	71
Infineon SBH51414 (1,550 nm), 40 mA	67	64
MRV MRTXLSPS5010 (1,550 nm), 31 mA	76	71
Photon PT3543-12-3-1FC (1,550 nm), 32 mA	78	71

From Steiner et al. [15].

that the second harmonics/intermodulation of the 900-MHz band will not affect the 1.8-GHz band. It would also be possible to downconvert the upper band in the range of 100–400 MHz before transmission. It has to be pointed out that all of these results have been achieved with multiple-quantum-well Fabry-Perot laser diodes. No distributed feedback laser diodes have been used for cost reasons. It is very important to mention that good performance with low-cost laser diodes can usually only be achieved if angled connectors are used. Then expensive optical isolation and cooling are not necessary.

3.3.4 Comparison of Fiber Optic and Coaxial Cable Link

Coaxial cable and fiber optic links are now compared with respect to hardware costs (transmitters, receivers, repeaters, and cables). Coaxial copper cable links need repeater amplifiers. Moreover, they need remote power supply. Particularly if many repeaters are required, this approach will additionally cause reliability and maintenance problems. Figure 3.14 shows the relative costs for two different coaxial and four single-mode fiber optic bidirectional links (uplink and downlink).

All components are low cost as required for microcellular repeaters. Installation costs are not considered here. For a distance exceeding 100m, a fiber optic link is clearly more cost effective than coaxial links [11, 12, 15]. At first sight it seems that single-fiber links with bidirectional transceiver modules are only cost-effective compared to two fiber links for link lengths above 1 km. But the difference between the two is quite small. Therefore, it will be advantageous in many cases to use bidirectional transceiver modules to spare one dark fiber for slightly higher cost. If separate transmitters

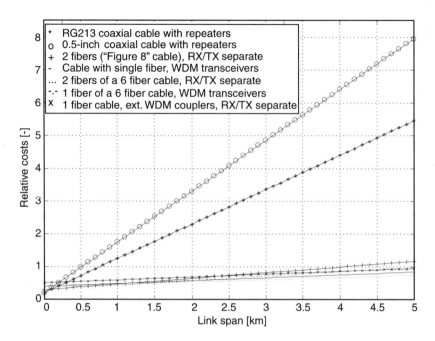

Figure 3.14 Comparison of relative hardware costs for fiber optic and coaxial indoor/
outdoor remote antenna feeding links. The following link types are compared:
two coaxial cable links; fiber optic links with one single fiber and WDM
transceiver modules; fiber optic links with one single fiber; separate RX and
TX modules, and external WDM couplers; and two fiber links with separate
RX and TX modules. Curves for single-fiber, duplex, and six-fiber cables as
well as RG213 and 0.5-inch coaxial cables are shown. Installation costs are
not considered. (*Source:* [14]. © 2000 IOS Press.)

and receivers together with external WDM couplers are used, bidirectional
transmission over one single fiber with even higher SFDR is possible.

External WDM couplers are rather expensive. So the use of external
couplers is only recommended for long outdoor links and an exceptionally
high number of carriers (SFDR > 75 dB) where fibers have to be saved. In
fiber optic links it is generally cost-effective to use multifiber cable as duplex
or even 6- or 12-fiber cables, because usually more than one application is
installed in a ducting or tube.

The main cost driver for significant link spans is the cable. Installation
costs are roughly proportional to the cable cross section area. If they were
also considered, optical fiber solutions would be even more advantageous.
Coaxial cables have higher cross sections and higher weight. Fibers can be
installed very easily compared to coaxial cables (e.g., consider methods using

compressed air such as the "blown fiber" technique by British Telecom or Blolite by BICC [16]).

Good multiple-quantum-well laser diodes in coaxial packages with optical output power above 1 mW and photodiodes for the 1,310-nm wavelength are available for less than $50 in high quantities.

3.4 Laser and Photodiode Nonlinearities and Their Influence on Link Performance

The main source of intermodulation in a directly modulated fiber optic link is the laser diode. Laser-diode nonlinearities are analyzed using Volterra and Fourier series. Photodiode nonlinearities are discussed here also, and measurements and predictions are compared.

Most nonlinearities arising in fiber optic link components, particularly laser diodes, can be divided into four main categories [14]:

1. *Static weak nonlinearity:* A component that shows frequency-independent intermodulation or harmonic distortions has a "static" nonlinearity. The transfer characteristic of the component can then in general be described by a set of nonlinear algebraic functions. These functions can often be expanded into Taylor series. For low-input signals only some low-order terms of the Taylor series have to be considered to compute the distortion products. The nonlinearity is then called "weak" because the lower the input signal amplitude is, the higher the S/N. Higher-order terms of the Taylor expansion have to be considered only for high-input signal amplitudes. A static component reacts instantaneously upon excitation at its input; no relaxation processes can be observed. Therefore, it is referred to as "memoryless." The amplitudes of the generated harmonic or intermodulation products do not depend on the frequencies of the input signals but on their amplitudes.

2. *Static hard nonlinearity:* If the functions describing the transfer characteristic of a static nonlinearity cannot be expanded into a Taylor series because they contain discontinuities or discontinuous derivatives, the nonlinearity is called "hard." The S/N can then be low even for low-input amplitudes and even increase with increasing input amplitude. The nonlinearity is called hard even if the amplitude of generated intermodulation or harmonic distortion is very small because high orders of nonlinearity are involved even with small input signals.

3. *Dynamic weak nonlinearity:* The dynamic behavior of the component can be described by a set of nonlinear differential equations. The nonlinearities appearing within the components do not exhibit discontinuities or discontinuous derivatives and can therefore be expanded by Taylor series. The component contains energy-storing elements, a fact that is reflected by the differential equations. Differential equations lead to relaxation processes. Therefore, a component exhibiting a dynamic nonlinearity is considered to possess a "memory." The memory property of a system should be considered if the period of the highest excitation frequency is of the order of the system's lowest relaxation time constant. The amplitudes of the generated harmonic and intermodulation products generally depend not only on the amplitudes but also on the frequencies of the input signals.

4. *Dynamic hard nonlinearity:* If the nonlinearities appearing within a component described by a set of differential equations cannot be expanded into a Taylor series, the nonlinearity is called "hard." As in the case of the static hard nonlinearity, higher-order harmonics and intermodulation products are generated at the output of such a component even for small input signals.

Components or subsystems of communication systems often exhibit weak nonlinearities. Weakly nonlinear systems can be analyzed with Taylor expansions in the static case as shown in Section 3.2 or by the so-called Volterra series in the dynamic case. The Volterra theory combines two concepts that are very well known in communications engineering: the Taylor series (well known for the analysis of static weakly nonlinear systems) and the transfer function (well known for the analysis of linear dynamic systems). Because laser diodes show relaxation oscillations usually at some gigahertz and they are also operated not far below this region, the laser-diode intermodulation is now analyzed by means of the Volterra theory [17].

3.4.1 Analysis of a Laser Diode by Means of the Volterra Theory

3.4.1.1 Analysis of Weak Dynamic Nonlinearities Using the Volterra Series

The response $y(t)$ of a linear system for an input $x(t)$ is

$$y(t) = h(t) * x(t) = \int_{0}^{\infty} h(\tau)x(t - \tau)d\tau \qquad (3.55)$$

where $*$ denotes convolution.

If the nonlinear functions in the nonlinear differential equations describing a weakly nonlinear system with memory can be expanded in Taylor series, and if the resulting differential equation fulfills some quite general analytical properties, the response of the system can be described similarly by a so-called Volterra series [18–23]:

$$
\begin{aligned}
y(t) &= \int_0^\infty h_1(\tau)x(t - \tau)d\tau \\
&+ \int_0^\infty \int_0^\infty h_2(\tau_1, \tau_2)x(t - \tau_1)x(t - \tau_2)d\tau_1 d\tau_2 + \ldots \\
&+ \int_0^\infty \ldots \int_0^\infty h_N(\tau_1, \ldots, \tau_N)x(t - \tau_1) \ldots x(t - \tau_N)d\tau_1 \ldots d\tau_N + \ldots
\end{aligned}
\tag{3.56}
$$

with the ith-order impulse responses or Volterra kernels $h_i(\tau_1, \ldots, \tau_i)$.

If the system is static or the considered timescale is such that relaxation processes run off almost instantaneously, all impulse responses degenerate into Dirac delta functions, which leads to a Taylor expansion for $y(t)$. For very small input signal $x(t)$, all higher-order integrals vanish and only the first integral remains, which corresponds to the linear case (3.55).

In analog communication systems, the nonlinearity is usually weak enough such that higher-than-third-order terms in (3.56) can be neglected. For harmonic input signals one can then use the Fourier transforms $H_i(j\omega_1, \ldots, j\omega_i)$ of the Volterra kernels to find $y(t)$ [20, 21, 23]. The first- and third-order harmonics $y_{\omega 0}(t)$ and $y_{3\omega 0}(t)$, respectively, at the output for

$$
x(t) = \frac{A}{2}e^{j\omega_0 t} + \frac{A^*}{2}e^{-j\omega_0 t}
\tag{3.57}
$$

are then

$$
y_{\omega_0}(t) = \frac{A}{2}H_1(j\omega_0)e^{j\omega_0 t} + \frac{A^*}{2}H_1(-j\omega_0)e^{-j\omega_0 t}
$$

$$y_{3\omega_0}(t) = \left(\frac{A}{2}\right)^3 H_3(j\omega_0, j\omega_0, j\omega_0) e^{j3\omega_0 t}$$

$$+ \left(\frac{A^*}{2}\right)^3 H_3(-j\omega_0, -j\omega_0, -j\omega_0) e^{-j3\omega_0 t}$$

3.4.1.2 Nonlinear Laser Diode Model

A direct-modulated laser diode imposed by a noncoherent external feedback is a weakly nonlinear system for moderate input signals and is modeled by the rate equations for photon and electron densities $s(t)$ and $n(t)$, respectively [9, 24–26]:

$$\dot{n} = \frac{I}{eV} - g(n)(1 - \epsilon s)s - \frac{n}{\tau_e} \tag{3.58}$$

$$\dot{s} = \left[\Gamma g(n)(1 - \epsilon s) - \frac{1}{\tau_p}\right]s + \frac{\Gamma\beta}{\tau_e}n + \frac{\Gamma\mu s(t - \tau)}{\tau_p} \tag{3.59}$$

with $g(n) = B(n - n_t)$, B = gain coefficient, n_t = electron density at transparency, ϵ = gain saturation coefficient, τ_p = photon lifetime, τ_e = electron lifetime, Γ = confinement factor, β = spontaneous emission factor, e = electron charge, V = volume of active region, μ = external optical intensity reflection factor, and τ = delay of reflected light.

With operating point values (subscript 0) and signals (tilde)

$$I(t) = I_0 + \tilde{I}(t), \ n(t) = n_0 + \tilde{n}(t), \ s(t) = s_0 + \tilde{s}(t) \tag{3.60}$$

we find

$$\dot{\tilde{n}}(t) = k_s^1 \tilde{s}(t) + k_n^1 \tilde{n}(t) + k_{sn}^1 \tilde{s}(t)\tilde{n}(t) + k_{ss}^1 \tilde{s}^2(t) + k_{ssn}^1 \tilde{s}^2(t)\tilde{n}(t) + k_I^1 \tilde{I}(t) \tag{3.61}$$

$$\dot{\tilde{s}}(t) = k_s^2 \tilde{s}(t) + k_n^2 \tilde{n}(t) + k_{sn}^2 \tilde{s}(t)\tilde{n}(t) + k_{ss}^2 \tilde{s}^2(t) + k_{ssn}^2 \tilde{s}^2(t)\tilde{n}(t) \tag{3.62}$$

$$k_n^1 = Bs_0(\epsilon s_0 - 1) - \frac{1}{\tau_e}, \ k_s^1 = B(n_0 - n_t)(2\epsilon s_0 - 1), \ k_{sn}^1 = B(2\epsilon s_0 - 1)$$

$$k_{ss}^1 = B(n_0 - n_t)\epsilon, \ k_{ssn}^1 = B\epsilon, \ k_l^1 = \frac{1}{eV}, \ k_n^2 = \Gamma Bs_0(1 - \epsilon s_0) + \Gamma\frac{\beta}{\tau_e}$$

$$k_s^2 = \Gamma B(n_0 - n_t)(1 - 2\epsilon s_0) + \frac{\Gamma\mu T_\tau - 1}{\tau_p},$$

$$k_{sn}^2 = B\Gamma(1 - 2\epsilon s_0), \ k_{ss}^2 = \Gamma B(n_t - n_0)\epsilon$$

$$k_{ssn}^2 = -B\Gamma\epsilon, \ T_\tau : \text{shift operator } T_\tau s(t) = s(t - \tau)$$

Setting the derivatives in (3.58) and (3.59) to zero and neglecting gain saturation, the operating point values are found. The Fourier transforms of the Volterra kernels are found using the *growing exponential approach* [20]:

1. Let $\tilde{I}(t) = e^{\lambda_1 t} + e^{\lambda_2 t} + e^{\lambda_3 t}$ and

$$\tilde{n}(t) = \sum_m n_{m_1, m_2, \ldots, m_N}(\lambda_1, \lambda_2, \ldots, \lambda_N)e^{\sum_i m_i \lambda_i t} \quad (3.63)$$

$$\tilde{s}(t) = \sum_m s_{m_1, m_2, \ldots, m_N}(\lambda_1, \lambda_2, \ldots, \lambda_N)e^{\sum_i m_i \lambda_i t} \quad (3.64)$$

   where the m_i's are determined by the structure of the differential equation(s) and the number of harmonic input signals. Here three input tones are used since we are interested in three orders of kernels.

2. Replace $\tilde{I}(t)$, $\tilde{n}(t)$, and $\tilde{s}(t)$ in (3.61) and (3.62) by the expressions of entry 1.

3. Equate the coefficients of $e^{\lambda_1 t}$, $e^{(\lambda_1 + \lambda_2)t}$, and $e^{(\lambda_1 + \lambda_2 + \lambda_3)t}$ to provide the functions n_{100}, n_{110}, n_{111}, s_{100}, s_{110}, and s_{111}. The remaining coefficients are not used because the kernel transforms should be symmetric, which simplifies the analysis [20]. Let $\lambda_i = j\omega_i$.

4. The transfer functions for the photon (H) and electron (G) densities are found by using these relations:

$$H_1(j\omega) = s_{100}(j\omega)$$
$$G_1(j\omega) = n_{100}(j\omega)$$
$$H_2(j\omega_1, j\omega_2) = \frac{1}{2!}s_{110}(j\omega_1, j\omega_2)$$

$$G_2(j\omega_1, j\omega_2) = \frac{1}{2!} n_{110}(j\omega_1, j\omega_2)$$

$$H_3(j\omega_1, j\omega_2, j\omega_3) = \frac{1}{3!} s_{111}(j\omega_1, j\omega_2, j\omega_3)$$

$$H_1(j\omega) = \frac{k_n^2 k_l^1}{D(j\omega)} \quad G_1(j\omega) = \frac{j\omega_1 - k_s^2(j\omega_1)}{D(j\omega_1)} \tag{3.65}$$

The shift operator in k_s^2 is transformed to $e^{-j\omega t}$ in the frequency domain.

$$H_2(j\omega_1, j\omega_2) =$$

$$\frac{1}{2!} \frac{[k_{sn}^2 C_1(j\omega_1, j\omega_2) + k_{ss}^2 C_2(j\omega_1, j\omega_2)](j\omega_1 + j\omega_2 - k_n^1) + k_n^2[k_{sn}^1 C_1(j\omega_1, j\omega_2) + k_{ss}^1 C_2(j\omega_1, j\omega_2)]}{D(j\omega_1 + j\omega_2)}$$

$$\tag{3.66}$$

$$G_2(j\omega_1, j\omega_2) =$$

$$\frac{1}{2!} \frac{[k_{sn}^1 C_1(j\omega_1, j\omega_2) + k_{ss}^1 C_2(j\omega_1, j\omega_2)][j\omega_1 + j\omega_2 - k_{ss}^2(j\omega_1 + j\omega_2)] + k_s^1[k_{sn}^1 C_1(j\omega_1, j\omega_2) + k_{ss}^1 C_2(j\omega_1, j\omega_2)]}{D(j\omega_1 + j\omega_2)}$$

$$\tag{3.67}$$

$$H_3(j\omega_1, j\omega_2, j\omega_3) =$$

$$\frac{1}{3!} \frac{\begin{aligned}&[k_{sn}^2 C_3(j\omega_1, j\omega_2, j\omega_3) + k_{ss}^2 C_4(j\omega_1, j\omega_2, j\omega_3) \\ &+ k_{ssn}^2 C_5(j\omega_1, j\omega_2, j\omega_3)] \\ &\times (j\omega_1 + j\omega_2 + j\omega_3 - k_n^1) \\ &+ k_n^2[k_{sn}^1 C_3(j\omega_1, j\omega_2, j\omega_3) + k_{ss}^1 C_4(j\omega_1, j\omega_2, j\omega_3) \\ &+ k_{ssn}^1 C_5(j\omega_1, j\omega_2, j\omega_3)]\end{aligned}}{D(j\omega_1 + j\omega_2 + j\omega_3)}$$

$$\tag{3.68}$$

$$C_1(j\omega_1, j\omega_2) = H_1(j\omega_1)G_1(j\omega_2) + H_1(j\omega_2)G_1(j\omega_1) \tag{3.69}$$

$$C_2(j\omega_1, j\omega_2) = 2H_1(j\omega_1)H_1(j\omega_2) \tag{3.70}$$

$$
\begin{aligned}
C_3(j\omega_1, j\omega_2, j\omega_3) = 2! [& H_1(j\omega_1)G_2(j\omega_2, j\omega_3) \\
& + H_1(j\omega_2)G_2(j\omega_1, j\omega_3) \\
& + H_1(j\omega_3)G_2(j\omega_1, j\omega_2) \\
& + G_1(j\omega_1)H_2(j\omega_2, j\omega_3) \\
& + G_1(j\omega_2)H_2(j\omega_1, j\omega_3) \\
& + G_1(j\omega_3)H_2(j\omega_1, j\omega_2)]
\end{aligned} \tag{3.71}
$$

$$
\begin{aligned}
C_4(j\omega_1, j\omega_2, j\omega_3) = 2 \cdot 2! [& H_1(j\omega_1)H_2(j\omega_2, j\omega_3) \\
& + H_1(j\omega_2)H_2(j\omega_1, j\omega_3) \\
& + H_1(j\omega_3)H_2(j\omega_1, j\omega_2)]
\end{aligned} \tag{3.72}
$$

$$
\begin{aligned}
C_5(j\omega_1, j\omega_2, j\omega_3) = 2 [& G_1(j\omega_1)H_1(j\omega_2)H_1(j\omega_3) \\
& + G_1(j\omega_2)H_1(j\omega_1)H_1(j\omega_3) \\
& + G_1(j\omega_3)H_1(j\omega_1)H_1(j\omega_2)]
\end{aligned} \tag{3.73}
$$

$$
D(j\omega) = (j\omega)^2 - j\omega[k_s^2(j\omega) + k_n^1] - k_n^2 k_s^1 + k_s^2(j\omega)k_n^1 \tag{3.74}
$$

Now the two-tone second- and third-order intermodulation products and the carrier frequency component of the photon density as well as the CIM ratios are [23] (with the peak modulation current I_m):

$$
\mathrm{IM}_2 = \left| I_m^2 \cdot H_2(j\omega_1, j\omega_2) \right| \tag{3.75}
$$

$$
\mathrm{IM}_3 = \left| 0.75 \cdot I_m^3 \cdot H_3(j\omega_2, j\omega_2, -j\omega_1) \right| \tag{3.76}
$$

$$
C = \left| I_m \cdot H_1(j\omega_1) \right| \tag{3.77}
$$

$$
\left(\frac{\mathrm{IM}_2}{C} \right)_{\mathrm{dB}} = 20 \log \left| \frac{I_m \cdot H_2(j\omega_1, j\omega_2)}{H_1(j\omega_1)} \right| \tag{3.78}
$$

$$
\left(\frac{\mathrm{IM}_3}{C} \right)_{\mathrm{dB}} = 20 \log \left| \frac{0.75 \cdot I_m^2 \cdot H_3(j\omega_2, j\omega_2, -j\omega_1)}{H_1(j\omega_1)} \right| \tag{3.79}
$$

where the k coefficients of (3.61) and (3.62) were substituted by the laser-diode parameters.

The harmonic products are found by changing the arguments in (3.76) and (3.77): $j\omega_2$, $j\omega_2$, $-j\omega_1 \rightarrow j\omega_1$, $j\omega_1$, $j\omega_1$ for the third order and $j\omega_1$, $j\omega_2 \rightarrow j\omega_1$, $j\omega_1$ for the second order [23]. The factors $0.75I_m^3$ and $1.0I_m^2$ change to $0.25I_m^3$ and $0.5I_m^2$.

Figures 3.15 and 3.16 show the theoretical third-order intermodulation of a laser diode including external feedback.

The result is qualitatively reasonable. The strong intermodulation oscillations with respect to the frequency take effect on a measurement in such a way that the intermodulation is strongly time dependent. Small vibrations or movements of the optical fiber may influence τ or the polarization, and the effect is comparable to a small change of the frequency. Temperature may also be an issue.

The intermodulation change of some decibels for this order of μ can be observed experimentally. The intermodulation increase near the relaxation peaks and at low frequencies due to optical feedback is also observed experimentally. Here the feedback is assumed to be noncoherent. Therefore, the delay time τ has to be higher than the laser's coherence time. The influence of coherent external feedback on the purely dynamical intermodulation or

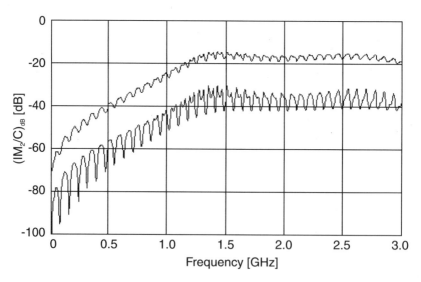

Figure 3.15 Calculated two-tone second-order intermodulation (IM) of a laser diode with noncoherent external optical feedback. Calculation assumes an optical modulation depth of 0.041 and external feedback of $\mu = 0.003$ (upper curve) and 0.01 (lower curve), respectively, $\tau = 0.3$ ns. The lower curve is downshifted by 20 dB. Frequency spacing: 0.2 MHz. $\beta = 3 \times 10^{-5}$ m^{3}s^{-1}; $\epsilon = 2.6 \times 10^{-23}$ m^{-3}. (Source: [17]. © 1995 SPIE.)

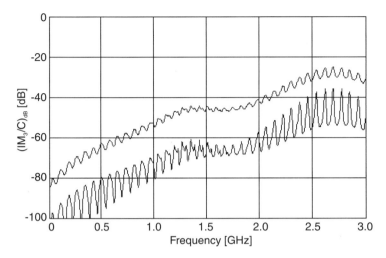

Figure 3.16 Calculated two-tone third-order intermodulation (IM) of a laser diode with non-coherent external optical feedback. Calculation assumes an optical modulation depth of 0.069 and external feedback of $\mu = 0.003$ (upper curve) and 0.01 (lower curve), respectively, $\tau = 0.3$ ns. The lower curve is downshifted by 20 dB. Frequency spacing: 0.2 MHz. $\beta = 3 \times 10^{-5}$ m^{3}s^{-1}; $\epsilon = 2.6 \times 10^{-23}$ m^{-3}. (*Source:* [17]. © 1995 SPIE.)

neglecting gain saturation has been reported in [27–29]. Other laser-diode Volterra analyses have also been published [30–33]. However, they did not consider gain saturation [32] or external optical feedback [31, 33]. This intermodulation analysis is not limited to signal pairs/triples of nearly the same frequency. The frequencies of the excitation tones are arbitrary.

3.4.1.3 Comparison with Measurements

Two-tone intermodulation measurements for a 1,310-nm metal clad ridge waveguide InGaAs/InP laser diode are now presented. The isolation between the two generators is essential and is provided by an RF amplifier-attenuator combination. The laser diode is packaged such that parasitics are negligible compared to the 47Ω input matching resistor. The light is coupled to an optical isolator to prevent backreflection.

Figure 3.17 shows the measured and calculated second-order intermodulation. The measured third-order intermodulation is compared to the Volterra analysis (3.79) and also to the perturbational approach of Wang et al. [34] in Figure 3.18.

The laser-diode parameters are:

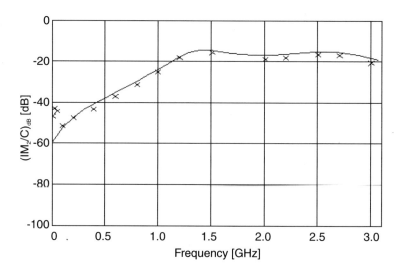

Figure 3.17 Comparison of measured and calculated second-order intermodulation (IM) of a laser diode. Straight line: Volterra theory; crosses: measurement. (*Source:* [17]. © 1995 SPIE.)

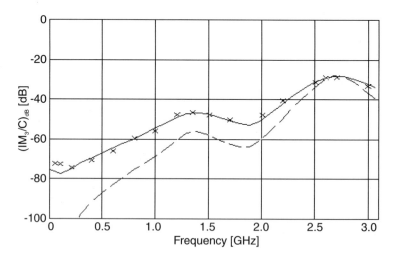

Figure 3.18 Comparison of measured and calculated third-order intermodulation (IM) of a laser diode. Straight line: Volterra theory; dashed line: perturbation theory of [34]; crosses: measurement. (*Source:* [17]. © 1995 SPIE.)

$$B = 2.75 \cdot 10^{-12} \text{ m}^3\text{s}^{-1}, \; n_t = 2.9 \cdot 10^{24} \text{ m}^{-3}, \; \tau_p = 1.1 \text{ ps}, \; \tau_e = 2.3 \text{ ns},$$

$$\Gamma = 0.3, \; \beta = 30 \cdot 10^{-5}, \; V = 120 \cdot 10^{-18} \text{ m}^3$$

The operating point values are: $I_0 = 40$ mA, $s_0 = 1.1 \times 10^{20}$ m^{-3}, and $n_0 = 4.0 \times 10^{24}$ m^{-3}.

The gain saturation parameter ϵ and the spontaneous emission coefficient β were tuned to fit the measured and calculated relaxation oscillation peak. β dampens the relaxation oscillation and enhances the distortion in the mean and low frequency range.

The relatively high static intermodulation at frequencies below 100 MHz is most probably due to the spatial hole burning effect. The gain saturation parameter ϵ is therefore split into a part caused by spectral hole burning and intraband relaxation (ϵ_{spec}) and a part caused by spatial hole burning ($\epsilon_{spat} = \epsilon_{spat0}/\sqrt{1 + (\omega \cdot \tau_{eff})^2}$). The roll-off time constant τ_{eff} is determined by the average photon density, the electron lifetime, the gain coefficient, and the uniformity of the electron density in the active region [35–37]. It is estimated to be a few nanoseconds. Here a value of 6.7 ns is assumed according to Kuo et al. [35, 36] and Vankwikelberge et al. [37]. Optimum values for the gain saturation parameters are $\epsilon_{spat0} = 4.0 \cdot 10^{-23}$ m^3 and $\epsilon_{spec} = 0.3 \cdot 10^{-23}$ m^3. In a distributed feedback laser the spatial hole burning contribution can be significantly higher than in a Fabry-Perot laser.

The curves show good agreement between theory and experimental data over a wide frequency range. The remaining excess static distortion may also be caused by the weak but not vanishing coherent optical feedback from the isolator input. Calculations show that the Wang formula is accurate for low gain saturation, low spontaneous emission, and frequencies near the relaxation oscillation.

3.4.2 Hard Nonlinearities

Apart from classical smooth second- and third-order intermodulation discussed above, higher order nonlinearities can play an important role in laser diodes even at low optical modulation index per carrier. A well-known phenomenon of that kind is clipping [38–42]. Clipping in subcarrier-multiplexing systems occurs at high carrier numbers, for example, in CATV applications, and is no longer considered here.

Another class of hard nonlinearities is *kinks*. In many cases measurements of $2f_1 - f_2$ intermodulation products of Fabry-Perot laser diodes show a classical third-order dependence on the signal only for particular (low) values of the operating current. Generally, the curves deviate from the pure third-order case even for small signals [43]. This behavior can be explained by introducing an additional nonlinearity, a *small kink*. Static small kinks are associated with longitudinal-mode jumps caused by temperature-dependent gain peak shifts [44].

Figure 3.19 shows the light power-current curve of a laser exhibiting small kinks during longitudinal-mode jumps. The kinks can only be seen if the curve is differentiated. It is well known that strong kinks can be caused by transversal-mode jumps, which is another phenomenon. At high modulation frequencies, the change of the gain spectrum leading to mode jumps is caused mainly by the gain saturation at higher operating currents rather than by the temperature effect.

To model these kinks at modulation frequencies of several hundred megahertz, we use an extensive multilongitudinal-mode rate equation model derived from the semiclassical laser theory [45–47]. Conventional multimode rate equation models [48, 49] do not exhibit kinks. The model has been implemented on a harmonic balance simulator (the Hewlett Packard microwave design system).

Multilongitudinal-mode rate equations have been studied to explain the occurrence of kinks in the laser diode's dynamic light-power current

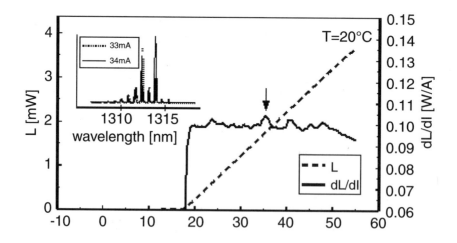

Figure 3.19 Measured light power (L) versus current (I) curve of a Fabry-Perot laser diode Mitsubishi FLD-45SDF (optically isolated) and its derivative when kinks occur. (*Source:* [43]. © 1996 IOS Press.)

curve. To model kinks associated with longitudinal-mode jumps, an accurate gain saturation model has to be implemented.[7] In a conventional model, even if it includes symmetric and asymmetric gain saturation, no kinks can be observed because the phase effects related to intermode beating are not considered. These effects, however, are crucial for the onset of small kinks as shown by Ogasawara and Ito [44] for low modulation frequencies where the temperature-dependent gain peak shift plays a major role.

A rate equation model developed by Yamada and Suematsu [47] and modified by Alalusi and Darling [50] has been used to model a laser diode:

$$\dot{n} = \frac{I}{eV} - \sum_i (G_{i,\text{lin}} - G_{i,\text{nonlin}})s_i - \frac{n}{\tau_e} \tag{3.80}$$

$$\dot{s}_i = \left[\Gamma \cdot (G_{i,\text{lin}} - G_{i,\text{nonlin}}) - \frac{1}{\tau_p}\right]s_i + \frac{\Gamma\beta}{\tau_e}n \tag{3.81}$$

$$G_{i,\text{nonlin}} = G_{i,\text{lin}}\sum_j\left(K_s\frac{1}{1+\delta_{ij}} + K_a\frac{1-\delta_{ij}}{\omega_{ij}}\right)\frac{2+\delta_{ij}}{\omega_{ij}^2 + \frac{1}{\tau_{\text{in}}^2}} \tag{3.82}$$

where n = electron density; s_i = photon density of mode i; $G_{i,\text{lin}}$ = linear gain of mode i; τ_e = electron lifetime; τ_p = photon lifetime; τ_{in} = intraband relaxation time; Γ = confinement factor; ω_{ij} = frequency spacing between modes i and j; and β = spontaneous emission factor.

The model of (3.82) is phenomenological, which means that the effects of various saturation mechanisms are summarized by symmetric and asymmetric gain saturation coefficients K_s and K_a, respectively. Nevertheless, intermode beating is modeled accurately through the inclusion of the Kronecker deltas δ_{ij}. The parameters, except for the coefficients K_s and K_a, are found from small signal measurements, measurements of the light power current curve, and the turn on delay. K_s and K_a are fitted such that the

---

7. The optical multilongitudinal-mode spectrum of an ideal Fabry-Perot laser diode has a regular distribution that is determined by the wavelength dependence of the optical gain (parabola). In a real laser diode this ideal spectral shape is modified by reflections inside and outside the laser diode's active region. By transformation of the reflections to the mirrors, a wavelength-dependent mirror reflectance occurs that modifies the spectral shape (see, e.g., Figure 3.19). Longitudinal-mode jumps can be forced by an optical spectrum with a strongly irregular shape because strong mode jumps are possible in that case [50–57].

kinks of the dynamic light power current curve occur at points that lead to the desired intermodulation behavior.

Figure 3.20 shows a comparison of simulated and measured intermodulation performance for a Hitachi Fabry-Perot laser diode. The kink is too weak to influence the $f_1 + f_2$ intermodulation. It is purely second order and dominated by the classical nonlinearity of the photon-electron interaction [17, 34, 49]. The kink has a strong effect on the $2f_1 - f_2$ intermodulation particularly at high operating currents where the third-order photon-electron nonlinearity decreases.

By changing the parameters in the model of (3.80) through (3.82) such that the number of longitudinal modes is increased, the kinks disappear. Due to the high number of modes, there are no hard mode jumps anymore and the change of the mode pattern is smooth. This observation is supported by measurements. Figure 3.21 shows the intermodulation curves for a high threshold current Lasertron multiple-quantum-well Fabry-Perot laser diode. Its intermodulation is smooth and the optical spectrum at the operating point is rather wide. This laser shows a low (multiple-quantum-well, high operating current) classical (many modes) intermodulation.

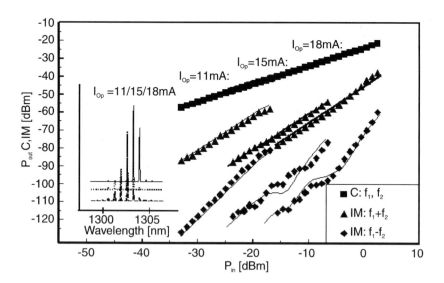

Figure 3.20   Measured (symbols) and calculated (lines) carrier (C) and intermodulation (IM) products at the output of the modulated Fabry-Perot laser diode Hitachi HL 1326CN (optically isolated). Modulation frequencies: f_1 = 944.9 MHz, f_2 = 945.1 MHz. Three operating currents are considered (threshold current: I_{th} = 8 mA). (*Source:* [43]. © 1996 IOS Press.)

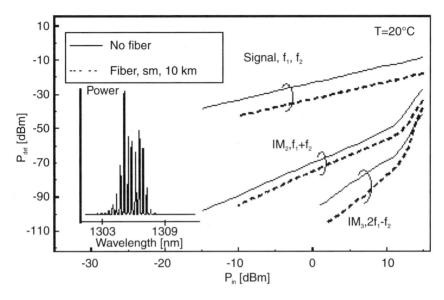

Figure 3.21   Measured signal and intermodulation (IM) products $IM_{2,3}$ at the output of a modulated Lasertron multiple-quantum-well Fabry-Perot laser diode (no optical isolator!) for 0- and 10-km fiber. SM: single mode. Modulation frequencies: f_1 = 944.9 MHz, f_2 = 945.1 MHz, where the operating point is 73 mA, 48 mA above threshold current. (*Source:* [43]. © 1996 IOS Press.)

Another well-known advantage of strongly multimodal lasers is their low RIN [10]. This laser diode has an RIN below −160 dB/Hz at 73 mA, which is very good. This result shows that laser diodes with spectra showing 5–10 longitudinal modes are well suited for high dynamic range low-cost fiber optic links but the application of these laser diodes is limited to short links (below 5–10 km) because, as a result of the wide spectrum, dispersion occurs in long fibers. Laser diodes exhibiting gain guiding, high spontaneous emission factor, and high threshold current show that type of spectra. The advantages of such laser diodes with relatively low coherence can be summarized as follows: low and smooth intermodulation (potentially suited for predistortion linearization [43]), low cost, reflection immunity, and low modal noise.

3.4.3   Photodiode Nonlinearities

In cellular fiber optic link applications photodiode nonlinearities can usually be neglected. They are an issue in cases where laser-diode power is very high (5–10 mW or higher) as is seen in, for example, fiber optic CATV links.

Photodiode nonlinearities are due to absorption saturation, electric field screening due to an external circuit effect, or a space charge in the intrinsic layer [58–61]. Technologies for high-power photodiodes have been developed mainly by fiber optic CATV optoelectronic device manufacturers.

3.5 Laser Diode Noise and Its Influence on Link Performance

3.5.1 Relative Intensity Noise

The laser-diode RIN has been defined in Section 3.2.4.2. Ideally it is mainly due to amplified spontaneous emission and relaxation oscillation.

Introducing Langevin noise sources into the single-mode rate equations [9, 10, 62, 63] (see Section 3.4.2.2), we find the following relation for the laser-diode RIN [10]:

$$
\text{RIN}_{\text{laser}}(\omega) = 2 \frac{\langle |\Delta s(\omega)|^2 \rangle}{\langle s \rangle^2} = 4 \frac{\beta n_0}{\tau_e} (\text{eV})^2 \frac{\left(\frac{1}{\tau_e} + \tau_p \omega_r^2\right)^2 + \omega^2}{\tau_p^2 \omega_r^4 s_0} |H(j\omega)|^2
$$

(3.83)

with the small signal transfer function $H(j\omega) = \Delta s(j\omega)/\Delta I(j\omega)$ where s and I are photon density and current, respectively.

Equation (3.83) shows the following:

- High spontaneous emission factor β increases the laser-diode RIN. This contribution will be higher in gain-guided laser diodes with broad spectra than in index-guided laser diodes with a low number of modes.

- The RIN curve essentially follows the squared small signal transfer function. Laser diodes with high relaxation oscillation frequency $f_r = \omega_r / 2\pi$.

- RIN is inversely proportional to the average photon density s_0.

For low RIN the laser diode should be operated at high intensity. This also increases f_r. High f_r not only decreases the RIN but also decreases intermodulation at high frequencies. Both increase the SFDR. This is applicable to single and multilongitudinal-mode laser diodes. In multilongitudinal-mode laser diodes, additional noise is generated as discussed next.

Mode Partition and Mode Hopping Noise.   Schimpe [64] analyzed the noise of a laser diode with a strong main mode and a weak side mode and found an additional noise source, called *mode partition noise* (MPN), due to the interaction between the strong main mode (photon density s_M) and the weak side mode (photon density s_S):

$$\text{RIN}_{\text{MPN}}(\omega) = 4 \frac{\frac{\langle s_S \rangle^3}{\langle s_M \rangle^2}}{1 + (\tau_S \omega)^2} \frac{\tau_p}{K_{\text{tot}} n_{sp}} \tag{3.84}$$

with the inversion factor $n_{sp} > 1$. The modal time constant τ_S for the side mode [10] is of the order of nanoseconds for Fabry-Perot laser diodes and picoseconds for distributed feedback laser diodes (side-mode suppression > 30 dB). The cutoff frequency of the mode partition noise is $1/(2\pi\tau_S) \approx$ 10–300 MHz. This shows that mode partition noise increases the RIN at frequencies in the range of some 10 MHz. Considering also gain saturation, Su et al. [48, 65, 66] showed that result (3.84) is masked by another effect: Low-frequency RIN is also affected by the character of the gain saturation. For symmetrical gain saturation the low-frequency RIN is very low, whereas for asymmetric gain saturation it is high.

RIN in the megahertz range can be further increased by mode hopping. Spontaneous emission can cause mode hopping [67]. This effect disappears in laser diodes with a high number of modes or, of course, in single-longitudinal-mode laser diodes. Strong gain saturation or specially designed cover layers can add hysteresis to the light power-current curve. This suppresses mode hopping noise effectively [68]. Another possibility is to operate the laser diode in a region where no mode hops occur at all.

Upconverted Low-Frequency Noise.   Although mode partition and mode hopping noise are in a frequency range that is not interesting for cellular remote antenna feeding links (except if the signal is downconverted for transmission), they have to be considered. Lau and Blauvelt [69] have found that low-frequency noise is upconverted by the laser-diode injection current modulation. It has been found that the RF noise increase due to low-frequency noise upconversion is significant for laser-diode input powers > 0 dBm (for resistively matched laser diodes) [15].

3.5.2 Influence of External Optical Reflections

Laser diodes generally suffer from external optical reflections. Coherent external reflections (mainly generated at the coupling optics) should be avoided. External reflectivity $R_{ext} < 10^{-5}$ (Fabry-Perot laser diodes) and 10^{-7} to 10^{-8} (distributed feedback laser diodes), respectively, should be guaranteed to avoid noise and intermodulation performance degradation [10, 70]. The use of angled connectors is usually sufficient for Fabry-Perot laser diodes. Distributed feedback laser diodes should be isolated optically.

3.6 Fiber Optic Microcell Repeater

Microcell repeaters are an important application of low-cost high dynamic range fiber optic links [11]. The aim of a microcell repeater is to improve the coverage of a base station in critical areas such as tunnels, streets, railway lines, and large buildings. It is often more convenient to improve coverage by adding remote antennas to a large base station than to use many small base stations, particularly in cases with inhomogeneous time-varying subscriber distribution within the cell. More information on microcellular systems and 3G is provided in Chapters 4 and 5.

3.6.1 Block Diagram of a Microcell Repeater

Figure 3.22 shows the block diagram of a low-cost fiber optic microcell repeater block diagram. Ideally, the link bandwidth should be large enough to cover all applications in a certain band (e.g., 0.8–1 GHz and/or 1.7–2.0 GHz).

The tuning of the variable attenuators and variable gain amplifiers for optimum input and output levels depends on the link loss. The link loss varies with the fiber length and the number of connectors and also depends on laser-diode and photodiode tolerances. The tuning can be done manually or by detecting the optical power in the receivers and then transmitting the control signals to the counter transceiver via a narrowband subchannel over the same fiber.

3.6.2 Choice of Low-Cost Devices

From the previous sections some criteria for the choice of optimum low-cost devices for cellular remote antenna feeding fiber optic links can be derived (Table 3.5). Today most Fabry-Perot laser diodes are multiple-

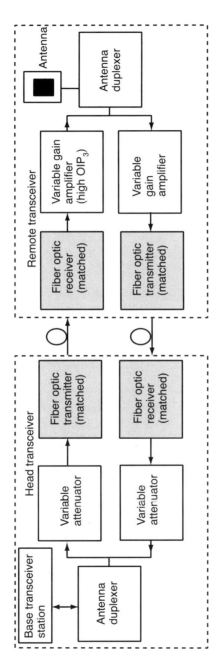

Figure 3.22 Low-cost fiber optic microcell repeater block diagram.

Table 3.5
Requirements for Devices Used in Fiber Optic Microcell Repeaters

Devices	Criteria
Laser diodes and driver	Pigtailed 1,310-nm multiple-quantum-well laser diode, 5–10 longitudinal modes, no isolator/cooling, matched/low-power amplifier
Photodiodes and receiver	Receptacle/pigtailed InGaAs low back reflection p-i-n photodiode (optical power = 1–5 mW), matched receiver (option) with GaAs FET receiver and monolithic microwave integrated circuit post amplifier
Low-noise amplifiers (variable gain)	GaAs monolithic microwave integrated circuit with high $OIP_3 - P_{1dB}$ difference
Power amplifier	GaAs monolithic microwave integrated circuit with low $OIP_3 - P_{1dB}$ difference
Fiber optic connectors	Angled (e.g., fiber connector/angled physical contact), single-mode standard fiber
Duplexers	Low insertion loss, high downlink/uplink isolation

quantum-well types. Due to the high-gain, increased carrier confinement multiple-quantum-well laser diodes are superior to bulk laser diodes. They show lower threshold currents, higher bandwidth, better linearity, better temperature stability, and lower back-reflection sensitivity.

Multiple-quantum-well laser diodes are less sensitive to reflections and temperature changes and have lower threshold currents and higher relaxation oscillation frequencies. This is particularly true for strained quantum well laser diodes. Driver and receiver design is discussed in Section 3.8.

3.6.3 Microcell Repeater Performance Evaluation

Link performance can be estimated using widely distributed software such as MATLAB or Microsoft Excel. A file or sheet that calculates signal, noise, intermodulation levels, noise figures, C/N and CIM ratios, SFDR, EIN, and input/output IP_3 of arbitrary links according to the theory of Section 3.2 containing attenuators (duplexers are modeled as attenuators, too), amplifiers, fiber optic transmitters, and receivers as building blocks. The blocks are fully characterized by data sheet parameters IP_3, P_{1dB}, and gain. Due to the narrowband character, implementation of frequency dependence is not necessary. Uplink and downlink consist of cascaded blocks.

Noise characterization: The input noise power of each block is amplified or attenuated and added incoherently (by adding powers, not voltages) to the noise power generated by the block itself according to its noise specification.

Intermodulation characterization: The blocks are specified by P_{1dB} and OIP_3. The input intermodulation of each block is amplified or attenuated and new intermodulation is generated in the block according to its P_{1dB} and OIP_3. The intermodulation products generated from the carriers in each nonlinear block are added incoherently at the link output. Fifth-order nonlinearities are considered only for the downlink power amplifier. The fifth-order nonlinearity increases the intermodulation at high power levels, which can then be expressed by an "equivalent" OIP_3 leading to the same carrier-to-intermodulation ratio.

Figures 3.23 through 3.25 show performance calculations for a fiber optic microcell repeater. The calculations were performed using the formulas derived in Section 3.2 for noise and intermodulation of the cascaded bidirectional link shown in Figure 3.22. Requirements for a GSM application were estimated in Section 3.2.3.

For 10-dBm output power per carrier, which is an expected maximum value for a GSM microcell application, the required OIP_3 value of 33 dBm can almost be achieved with the amplifier used in Figure 3.23. If necessary, an amplifier with higher P_{1dB} could be used to make sure that the highest possible levels can be processed without problems. Figure 3.24 shows the same calculation for $P_{1dB} = 27$ dBm.

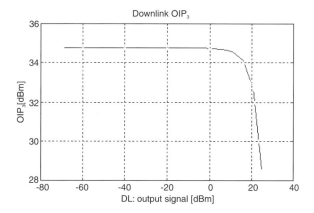

Figure 3.23 Dependence of downlink (DL) OIP_3 on output signal (fifth-order intermodulation influence) using a power amplifier with $P_{1dB} = 22$ dBm (as estimated in Section 3.2.3).

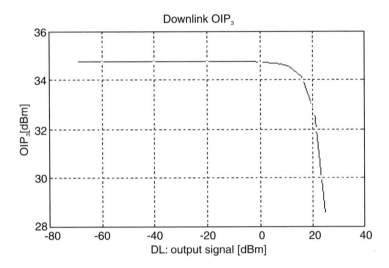

Figure 3.24 Dependence of downlink (DL) OIP_3 on output signal (fifth-order intermodulation influence) using a power amplifier with P_{1dB} = 27 dBm.

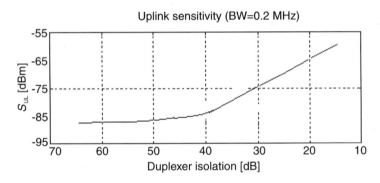

Figure 3.25 Uplink (UL) sensitivity dependence on duplexer isolation. A downlink isolation of >45 dB is required to meet the goal of $S_{UL,dBm}$ = −85 dBm. This is what can be expected from good low-cost ceramic resonator duplexers.

In Section 3.2.3, we stated that the antenna duplexer has to provide a certain isolation between downlink and uplink to avoid uplink sensitivity decrease due to noise crosstalk. An uplink sensitivity of at least $S_{UL,dBm}$ = −85 dBm should be guaranteed. Figure 3.25 shows the dependence of $S_{UL,dBm}$ on duplexer isolation. The measured microcell repeater uplink SFDR values using unisolated, uncooled multiple-quantum-well laser diodes (coaxial package, pigtail with fiber connector/angled physical contact connector) with 1- to 3-mW optical output power are 70 dB (three-mode laser diode, low I_{th}) to 78 dB (10-mode laser diode, high I_{th}) depending mainly on the laser diode threshold current and optical power. Laser diodes with high I_{th} are strongly multimode, reflection insensitive, and have low modal noise and low intermodulation.

3.7 Fiber and Multicarrier Effects

3.7.1 Fiber Loss

Fiber loss is very low. The loss of a standard single-mode fiber at 1,310 nm is around 0.3 dB/km [71] and therefore almost negligible for spans of less than 5 km. Furthermore, it is practically frequency independent.

3.7.2 Fiber Dispersion and Intermodulation

3.7.2.1 Introduction

As shown in previous sections, multimodal Fabry-Perot laser diodes are an alternative to distributed feedback laser diodes in low-cost analog fiber optic links. Modern multiple-quantum-well Fabry-Perot laser diodes are very linear (similar to distributed feedback laser diodes) and have relatively low intrinsic noise. However, if a multilongitudinal-mode laser diode is used as a transmitter in a link instead of a single-longitudinal-mode laser diode, a new source of nonlinearity is created by the fiber. It is well known that fiber dispersion can cause nonlinear distortions in fiber optic links with single-longitudinal-mode laser diodes such as distributed feedback laser diodes due to laser diode frequency chirping [72, 73].

With directly modulated laser diodes at 1,550 nm and standard single-mode fibers, the influence of frequency chirping is considerable due to the relatively high dispersion. At 1,310 nm it is usually negligible. With multilongitudinal-mode laser diodes as transmitters, another effect becomes important at high modulation frequencies. Every mode itself has a nonlinear light power-current curve although the total light power-current curve is very linear. This means that every mode itself generates strong intermodulation and harmonic products, whereas the total output, which is the sum of the intensities of all modes, is much less distorted. Due to the fact that the carrier reservoir is the same for all modes, there is a correlation between the modes and therefore a mutual compensation of the strong nonlinearities takes place [10]. If the intensity-modulated light of the multilongitudinal-mode laser diode is transmitted over a slightly dispersive fiber, the efficiency

of this compensation decreases mainly at high modulation frequencies due
to the different mode propagation speeds.

Let us consider a two-mode laser diode. If the difference of the propagation times equals half the period of the intermodulation product, the correlation between the two modes is negative and the compensation is completely
cancelled (see the comprehensive analysis in Chapter 2). The result is a
much higher nonlinearity of the link transfer characteristic. Not only the
intermodulation but also the signal amplitude changes with varying fiber
length. If the signal parts of the two modes just at the laser diode output
are in phase and have about the same amplitude ($k_1 \sim 1$), there is a certain
fiber length at which the phase difference between the two signals is 180°.

3.7.2.2   Theory and Modeling

A two-mode model is now used to analyze the effect of standard single-
mode fiber dispersion on intermodulation in a multilongitudinal-mode laser
link at 1,310 nm [74]. First-, second-, and third-order transfer characteristics
of laser diode modes 1 and 2 are modeled by first-, second-, and third-order
transfer functions according to the Volterra theory of nonlinear systems (see
Section 3.4). CIM ratios are calculated. The influence of fiber losses on the
CIM ratios is small and therefore neglected. The fiber dispersion is considered
by introducing delay time τ_1 for mode 1 and delay time τ_2 for mode 2,
respectively. The ith-order current-to-total light output transfer functions
are [20]:

$$H_i(\omega_1, \ldots, \omega_i) = H_i^1(\omega_1, \ldots, \omega_i)e^{j(\omega_1+\ldots+\omega_i)\tau_1} \quad (3.85)$$
$$+ H_i^2(\omega_1, \ldots, \omega_i)e^{j(\omega_1+\ldots+\omega_i)\tau_2}$$

where H_i^k are ith-order current-light transfer functions of mode k.

With (3.1) second- and third-order CIM ratios at angular frequencies
$\omega_1 + \omega_2$ and $2\omega_1 - \omega_2$, respectively, due to a two-tone input at ω_1, ω_2
with current amplitudes I_m are [17]:

$$\text{CIM}_2 = \left| \frac{H_1(\omega_1)}{I_m H_2(\omega_1, \omega_2)} \right| \quad (3.86)$$

$$= \left| \frac{H_1^1(\omega_1)}{I_m H_2^1(\omega_1, \omega_2)} \right| \cdot \left| \frac{e^{j\omega_1\tau_1} + \dfrac{H_1^2(\omega_1)}{H_1^1(\omega_1)} e^{j\omega_1\tau_2}}{e^{j(\omega_1+\omega_2)\tau_1} - e^{j(\omega_1+\omega_2)\tau_2}} \right|$$

$$\text{CIM}_3 = \left| \frac{H_1(\omega_1)}{0.75 I_m^2 H_3(\omega_1,\ \omega_1,\ -\omega_2)} \right| \tag{3.87}$$

$$= \left| \frac{H_1^1(\omega_1)}{0.75 I_m^2 H_3^1(\omega_1,\ \omega_1,\ -\omega_2)} \right| \cdot \left| \frac{e^{j\omega_1\tau_1} + \dfrac{H_1^2(\omega_1)}{H_1^1(\omega_1)} e^{j\omega_1\tau_2}}{e^{j(2\omega_1-\omega_2)\tau_1} - e^{j(2\omega_1-\omega_2)\tau_2}} \right|$$

$H_i^2 \approx -H_i^1$ (I = 2, 3) as a consequence of the compensation effect. Equations (3.86) and (3.87) can further be simplified to:

$$\text{CIM}_{2,\text{dB}} = 20 \log \left\{ k_2 \frac{1 + k_1^2 + 2k_1 \cos[\omega_1(\tau_1 - \tau_2)]}{4 \sin^2\left[(\omega_1 + \omega_2)\dfrac{\tau_1 - \tau_2}{2} \right]} \right\} \tag{3.88}$$

$$\text{CIM}_{3,\text{dB}} = 20 \log \left\{ k_3 \frac{1 + k_1^2 + 2k_1 \cos[\omega_1(\tau_1 - \tau_2)]}{4 \sin^2\left[(2\omega_1 - \omega_2)\dfrac{\tau_1 - \tau_2}{2} \right]} \right\} \tag{3.89}$$

where k_2 and k_3 are the first factors on the right-hand side of (3.86) and (3.87), respectively, and k_1 is the ratio of the linear modal transfer functions $H_1^2(\omega_1)/H_1^1(\omega_1)$, which in general are complex. The transfer function $H_1^i(\omega_1)$ of mode i is the RF power at ω_1 of mode i divided by the RF current injected into the laser diode. Thus k_1 is the ratio of RF (detector current) amplitudes of modes 1 and 2 including the phase between them. If modes 1 and 2 carry the same RF signal amplitude, then k_1 = 1.

Figure 3.26 shows a plot of (3.88) and (3.89) for a laser-diode model with a high relaxation resonance (real k_i). The period of (3.88) is shorter than the period of (3.89), which means that the influence of the fiber dispersion on IM_2 can be remarkable even if the effect on IM_3 is negligible, particularly if the fiber length L is not more than some kilometers.

The difference $\tau_1 - \tau_2 = \sigma L \Delta\lambda$ can be calculated [75] using the dispersion parameter σ = 3 ps/(nm km) and the wavelength spacing of the modes $\Delta\lambda$ = 1.3 nm.

For two modes with equal RF amplitudes, there are certain lengths for which the two signals at the fiber end are exactly opposite and therefore cancel each other. This means that for k_1 = 1 the numerators of (3.88) and

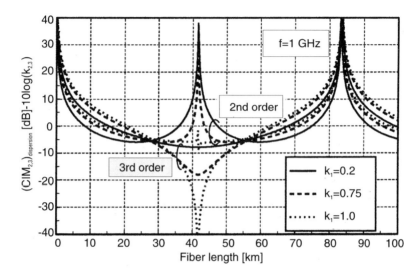

Figure 3.26 Suppression of the total second- and third-order intermodulation (IM) due to correlation between the modes. The vertical axis gives the enhancement with respect to the modal CIM ratios $k_{2,3}$. A high positive value means a low total intermodulation. (*Source:* [74], © 1997, IEEE. Reprinted with permission.)

(3.89) have a zero that leads to a zero CIM ratio. In practical situations it is more likely that k_1 is not exactly 1. This means that the signal component at the fiber end can only be decreased by some decibels, whereas the intermodulation is strongly increased because the modal intermodulation parts still have the same amplitudes due to the compensation in the laser diode. In a realistic fiber optic link the total intermodulation is a combination of the dispersion-induced intermodulation and the intrinsic laser-diode intermodulation of the total light power (without fiber). The latter has not been included in this simple two-mode model.

For a laser-diode lasing in more than two modes, an analytical treatment would be much more complicated. If the laser diode is emitting in more than two modes, the intermodulation curves are a combination of various two-mode intermodulation curves (3.88) and (3.89). A simulation of a five-mode laser diode together with a fiber including a realistic intrinsic laser-diode nonlinearity is shown in Figure 3.27 for σ = 3 ps/(nm km) and $\Delta\lambda$ = 1 nm (from mode to mode).

Qualitatively the CIM characteristics are as expected from the simple two-mode model. They are decreasing from $L = 0$ and increasing again but no longer periodic over the given range of fiber length. This is because there are more than two modes and in general the differences of dispersion between

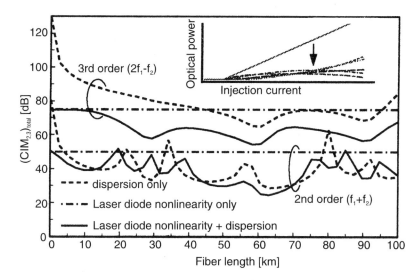

Figure 3.27 CIM calculation for a five-mode laser diode (LD) at 1,310-nm wavelength together with a standard single-mode fiber using a harmonic balance simulator. Shown are LI curves for the individual modes and the total emission (linear scale). The arrow marks the operating point. Optical modulation index OMI = 0.11, I/I_{th} = 3.3, f_1 = 0.999 GHz, f_2 = 1.001 GHz. Laser-diode parameters: $B = 2.75 \times 10^{-12}$ m^{3}s^{-1}, $n_t = 2 \times 10^{24}$ m^{-3}, τ_p = 1.5 ps, τ_e = 2.3 ns, Γ = 0.3, $V = 120 \times 10^{-18}$ m^3, $\epsilon = 2 \times 10^{-23}$ m^3, $\beta = 2 \times 10^{-4}$. (*Source:* [74], © 1997, IEEE. Reprinted with permission.)

pairs of modes are such that their CIM ratios have different periods with noninteger ratios. Therefore, the peaks in Figures 3.26 and 3.27 are not at the same positions. The intrinsic laser-diode nonlinearity breaks the peaks of the CIM characteristics, because it is constant with respect to the fiber length. The intrinsic intermodulation can even weakly compensate the dispersion-induced intermodulation if their phase difference is 180° as seen in Figure 3.27 for CIM$_2$ at $L \approx 19$ km.

3.7.2.3 Measurements

The influence of fiber dispersion on second- and third-order intermodulation of two (optically isolated) multiple-quantum-well Fabry-Perot laser diodes and one (optically isolated) single-longitudinal-mode laser diode has been measured at 1 GHz. Standard single-mode fiber sections of 2-, 5-, and 25-km length are combined to achieve lengths from 2 to 75 km. Intermodulation behavior and optical spectra are shown in Figure 3.28. The intermodulation of the multiple-quantum-well Fabry-Perot laser diodes shows a dependence

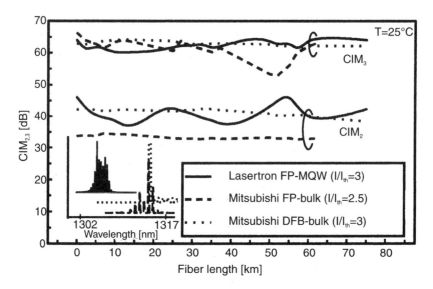

Figure 3.28 Measured dependence of carrier-to-second- and third-order intermodulation (IM) on the fiber length. Two multilongitudinal-mode Fabry-Perot (FP) laser diodes and one single-longitudinal-mode distributed feedback (DFB) laser diode have been measured. f_1 = 0.999 GHz, f_2 = 1.001 GHz. MQW: multiple quantum well. Optical modulation index OMI = 0.27 (Lasertron Fabry-Perot), 0.17 (Mitsubishi Fabry-Perot), and 0.27 (Mitsubishi distributed feedback). Threshold current I_{th} = 24.0 mA (Lasertron Fabry-Perot), 18.5 mA (Mitsubishi Fabry-Perot), and 27.7 mA (Mitsubishi distributed feedback). (*Source:* [74], © 1997, IEEE. Reprinted with permission.)

on the fiber length as expected. The Lasertron Fabry-Perot laser diode's IM_2 dependence on L is considerable, whereas the Mitsubishi Fabry-Perot laser diode has an almost constant CIM_2 but a strong dependence of CIM_3. This can be explained by the Mitsubishi Fabry-Perot laser diode's intrinsic second-order nonlinearity.

It is rather high in this laser diode and therefore masks the effect of fiber dispersion. At $L \approx$ 63–70 km and 54 km the Lasertron laser diode shows a weak compensation between intrinsic and dispersion-induced IM_3 and IM_2, respectively. For comparison, the chirp-induced intermodulation according to [76] has been calculated as well. For a fiber dispersion parameter of 3 ps/(nm km), the CIM_2 and CIM_3 due to laser-diode chirp are greater than 65/95 dB for L = 70 km, which is even much higher than the CIM due to the laser diode's intrinsic nonlinearities. If multilongitudinal mode instead of single-longitudinal mode laser diodes are used in 1,300-nm RF fiber optic links with single-mode standard fibers, an increase of intermodulation can occur if the fiber is longer than about 5–10 km.

3.7.3 Fiber Dispersion and Noise

Not only intermodulation, as discussed in Section 3.7.2, but also the noise contribution of a single mode in a multilongitudinal-mode laser diode is severe. Fortunately, this is not the case for the total light output. This implies that the noises of the modes must be correlated [10, 77] and that the noise contributions of the modes mutually compensate. This is only fully true at the output of the laser diode. At the output of a dispersive fiber, the modes are slightly delayed with respect to each other and therefore the efficiency of the compensation is decreased.

Wentworth et al. have analyzed this effect [75]. They found a relation for the minimum RIN that can be achieved for dispersion parameter σ, number of longitudinal modes and frequency:

$$\text{RIN}_{\Delta\tau}(\omega) = \frac{1}{m+1}\left[1 - \frac{\sin(m\Delta\tau\omega/2)^2}{m^2\sin(\Delta\tau\omega/2)^2}\right]\left[\frac{4\tau_0}{1+(\tau_0\omega)^2}\right]$$

$$\tau_0 = \frac{s_{\text{tot}}}{mR_{sp}} \tag{3.90}$$

$$\Delta\tau = \sigma L\Delta\lambda$$

with the total photon density at the laser-diode output s_{tot}, the number of laser-diode longitudinal modes m with spacing $\Delta\lambda$, the spontaneous emission rate per mode R_{sp}, and the fiber length L.

Curves for various fiber dispersion parameters in a realistic configuration for fiber length $L = 1$ km are plotted in Figure 3.29. A standard fiber in combination with a multiple-quantum-well Fabry-Perot laser diode has dispersion parameter σ of the order of 1 ps/(nm km).

In a practical link for cellular remote antenna feeding with standard fiber and a multiple-quantum-well Fabry-Perot laser diode the fiber dispersion has to be considered if the intrinsic laser-diode RIN is < -140 to -145 dB/Hz, which is usually the case.

3.7.4 Multicarrier Intermodulation

In an SCM system the CIM_3 generally decreases with increasing number of carriers. Two classes of third-order products are generated under multicarrier excitation: three-tone third-order intermodulation at $f_i + f_k - f_l$ and two-tone third-order intermodulation at $2f_i - f_k$.

The numbers of intermodulation products at frequency $f_1 + (r-1)\Delta f$ of each of the two classes generated from an equally distributed set of n

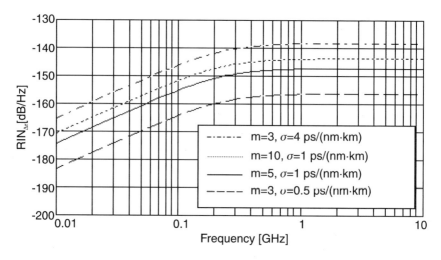

Figure 3.29 Estimation of the fiber dispersion-induced minimum RIN in a fiber optic link with a 1,310-nm Fabry-Perot laser diode with m longitudinal modes and standard fiber according to [75]. Fiber length $L = 1$ km.

equidistant carriers separated by Δf from a frequency f_1 up to a frequency $f_1 + n\Delta f$ are [78]:

$$Z_{2f_i - f_k} = \frac{1}{2}\left\{ n - 2 - \frac{1}{2}[1 - (-1)^n](-1)^r \right\} \tag{3.91}$$

$$Z_{f_i + f_k - f_l} = \frac{r}{2}\left(n - r + 1 + \frac{1}{4}[(n - 3)^2 - 5] - \frac{1}{8}[1 - (-1)^n](-1)^{r+n} \right) \tag{3.92}$$

The total intermodulation power of products of the two classes at frequency $f_1 + (r - 1)\Delta f$ is found by incoherent addition.

$$\left(\frac{3}{2}A^3 a_3\right)^2 Z_{f_i + f_k - f_l} \quad \text{for} \quad \text{the} \quad f_i + f_k - f_l \quad \text{products} \quad \text{and}$$

$\left(\frac{3}{4}A^3 a_3\right)^2 Z_{2f_i - f_k}$ for the $2f_i - f_k$ products with the coefficient of the third-order nonlinearity a_3 and the amplitude per carrier A. For six or more carriers, the $2f_i - f_k$ products can be neglected and only the $f_i + f_k - f_l$ products are significant.

For a high number of transmitted carriers, the maximum number of intermodulation products is generated in the center of the band ($r = n/2$):

$$Z_{f_i+f_k-f_l}\Big|_{\substack{n\to\infty\\ r=\frac{n}{2}}} = \frac{3}{4}n^2 \tag{3.93}$$

whereas on the band edges ($r = 1$ and n)

$$Z_{f_i+f_k-f_l}\Big|_{\substack{n\to\infty\\ r=1,n}} = \frac{1}{4}n^2 \tag{3.94}$$

If the two-tone intermodulation is known, the multicarrier intermodulation is found by relating the two-tone formulas from Section 3.2 with (3.91) and (3.92):

$$IM_{f_i+f_k-f_l} = IM_{2-\text{tone}} - 10\log(Z_{f_i+f_k-f_l}) - 6\text{ dB} \tag{3.95}$$

$$IM_{2f_i-f_k} = IM_{2-\text{tone}} - 10\log(Z_{2f_i-f_k}) \tag{3.96}$$

From (3.95) CIM_3 in the center of the band for a high number of carriers is found:

$$CIM_{3,\,n\text{carriers}}\Big|_{n\to\infty} = IM_{2-\text{tone}} - 10\log\left(\frac{3}{2}n^2\right) \tag{3.97}$$

The same calculation has been done numerically for fifth-order products. Figure 3.30 shows the results for this case as well as for the third-order intermodulation increase according to (3.97).

3.7.5 Optimization of Carrier Allocation

In a multicarrier transmission system, intermodulation can be minimized by distributing the carriers in a nonequidistant pattern or in such a way that the intermodulation products do not fall back on carrier frequencies. In the latter case, the resulting carrier distribution is called a "Golomb ruler" for N frequencies [79, 80]. Golomb rulers have been found for up to $N = 16$. The Golomb ruler leads to high bandwidth but it guarantees that no intermodulation is produced at carrier positions at all.

A tradeoff between intermodulation and bandwidth can be made by accepting a certain intermodulation level at carrier positions but finding a

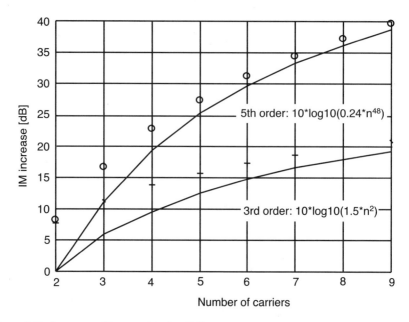

Figure 3.30 Increase of intermodulation (IM) power at the center of a transmission band due to multicarrier excitation with respect to the two-tone case. Key: —, exact values; *, approximation for third-order products; and o, approximation for fifth-order products. The formulas for a good fit for high n are given with the curves.

distribution that minimizes it. This can be done either by testing every possible carrier arrangement, which is time-consuming, or by applying genetic algorithms [81].

3.8 Low-Cost Laser Diode Driver and Photoreceiver Design for Fiber Optic Remote Antenna Feeding Links

3.8.1 Low-Cost Laser Diode Drivers

Laser diodes require biasing with a dc current. The bias network consists of a series inductor and a bypass capacitor. Coupling capacitors must be very low series R types, such as 1812 surface mount device capacitors with a series resistance below 1Ω. The laser-diode input impedance is of the order of some ohms. Standard 0805 or 0603 surface mount device capacitors with series resistances of the same order are used as coupling capacitors in RF designs. Some ohm series resistances are normally negligible in 50Ω environments but coupling a laser diode with such capacitors would increase the insertion loss significantly. Figure 3.31 shows laser-diode matching networks for bandwidth/center frequency ratio $\approx 1/3$.

Best matching can be achieved with a mixed lumped/distributed matching network [Figure 3.31(c)] [15, 82, 83]. In that case the OIP_3 requirement for the laser-diode driver amplifier is about 10 dB lower than for purely resistive matching [Figure 3.31(a)]. This means that with optimum matching a driver amplifier with much lower output power than with resistive matching can be used. This again decreases the costs. Considering cellular remote antenna feeding, in most cases an ordinary monolithic microwave integrated circuit amplifier is sufficient.

3.8.2 Example of a Low-Cost Laser Diode Driver

A realized laser-diode driver for a GSM remote antenna feeding link and the measured performance are shown in Figures 3.32 and 3.33. The driver amplifier is a low-cost monolithic microwave integrated circuit. The matching network is a mixed lumped/distributed type discussed above. The taper is realized on a *FR*-4 laser diode/driver printed circuit board.

3.8.3 Low-Cost Photoreceivers

3.8.3.1 Receiver Noise

To estimate the most important receiver property, namely, the C/N, a simple receiver model is considered. The noise sources are the same as those defined in Section 3.2.4 but here they are modeled not as equivalent input noise power levels but as spectral noise current densities. The receiver consists of a front end and one or more post amplifier(s). For the carrier-to-noise estimation only the front end has to be analyzed.

Several front-end topologies are known as high/low impedance, transimpedance, noise matched/resonant front ends [70, 84, 85]. For cellular remote antenna feeding links, the receiver is not very critical because due to the short fiber and relatively high laser-diode power (usually >1 mW), the receiver's own noise contribution is relatively low compared to shot noise and laser-diode RIN even if it is not a low noise design (see Section 3.2.4.2). A transimpedance amplifier—optionally with matching network at its input—seems appropriate because it combines simplicity with low cost and good noise properties even for lower laser diode powers. For received optical powers above 0.1 mW, p-i-n photodiodes with transimpedance amplifiers are superior to avalanche photodiodes.

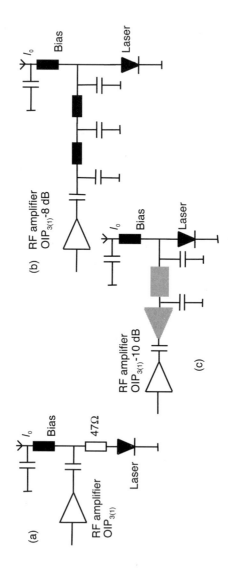

Figure 3.31 Laser-diode matching networks for bandwidth/center frequency ratio ≈ 1/3. (a) Resistive matching, (b) reactive matching with lumped elements, and (c) reactive matching with mixed lumped/distributed elements.

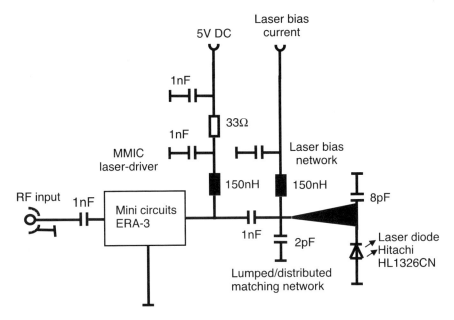

Figure 3.32 Example of a low-cost 900-MHz laser-diode driver for an uncooled, unisolated low-cost multiple-quantum-well laser diode with matching network [11, 14]. All passive elements including inductors are standard surface mount device types. (*Source:* [11]. © 1997 IOS Press.)

The model (Figure 3.34) of the receiver front end takes account of the following noise sources: shot noise current spectral density $I_{n,sh}$

$$I_{n,sh} = \sqrt{2eI_{PD}} = \sqrt{\frac{2eP_{opt}\, r_{PD}}{L_{opt}}} \tag{3.98}$$

with photodiode responsivity r_{PD}, electron charge e, optical loss L_{opt}, and laser-diode optical power P_{opt}.

Laser RIN current spectral density is given by

$$I_{n,RIN} = \sqrt{RIN} \cdot I_{PD} = \sqrt{RIN} \cdot \frac{P_{opt}\, r_{PD}}{L_{opt}} \tag{3.99}$$

with optical input power P_{opt} and laser-diode relative intensity noise $RIN_{laser,dB}$ [dB/Hz].

Receiver thermal noise current spectral density is given by

$$I_{n,th} \approx 4 - 20 \text{pA}/\sqrt{Hz} \tag{3.100}$$

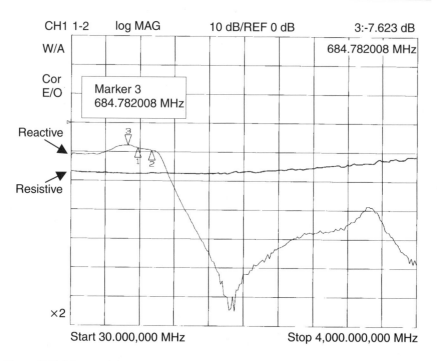

Figure 3.33 Measured improvement of link gain due to reactive matching using mixed lumped/distributed element matching network [11, 14]. (*Source:* [11]. © 1997 IOS Press.)

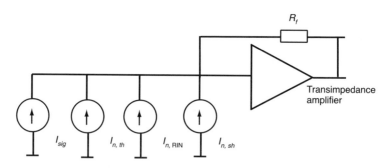

Figure 3.34 Simplified receiver front-end model for C/N estimation.

GaAs field effect transistor amplifiers are the preferred devices for narrowband low-noise RF noise applications. A GaAs field effect transistor transimpedance amplifier with matching network (narrowband case) or RF transformer (broadband case) can have $I_{n,th}$ as low as about 2–6 pA/$\sqrt{\text{Hz}}$ in the lower gigahertz range.

The C/N for bandwidth (BW) follows by dividing the squared signal current by the incoherently added noise spectral current densities:

$$\text{C/N} = 10 \log\left(\frac{I_{\text{sig}}^2}{I_{n,sh}^2 + I_{n,th}^2 + I_{n,\text{RIN}}^2} \frac{1}{\text{BW}}\right) \tag{3.101}$$

For link design and analysis, it is important to have an idea of the various receiver noise contributions. Figure 3.35 shows plots of the noise current spectral densities $I_{n,sh}$ and I_{n,R_f} for varying average photocurrent I_{PD} and transimpedance amplifier feedback resistor R_f.

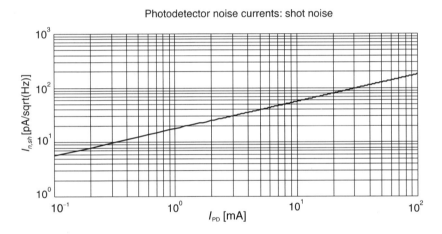

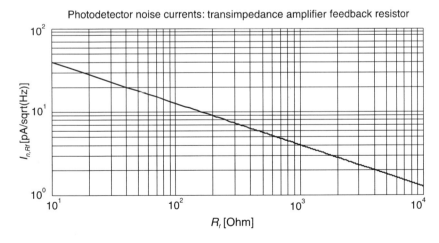

Figure 3.35 Dependence of noise current spectral densities $I_{n,sh}$ and I_{n,R_f} on average photocurrent I_{PD} and transimpedance amplifier feedback resistor R_f.

Comparing Figure 3.35 with Figure 3.7 in Section 3.2.4.2, we can also easily estimate the RIN_{laser} contribution to the total noise. RIN_Q in Figure 3.7 is replaced by RIN_{laser}. $I_{n,RIN}$ is then found by reading the upper curve of Figure 3.35 for the same abscissa current value.

3.8.3.2 Example of a Low-Cost Receiver for a Cellular Remote Antenna Link

The receiver shown in Figure 3.36 consists of an input matching network, a field effect transistor (FET) input stage, and a monolithic microwave integrated circuit (MMIC) post amplifier. The matching network together with the field effect transistor input acts as a parallel resonant circuit for the photodiode signal and shot noise current. Near resonance the photocurrent causes a high voltage level at the field effect transistor input. Shot noise and RIN components of the photocurrent then dominate the intrinsic amplifier noise [84–87]. All passive elements including inductors are standard surface mount device types. The tolerance is 5%. The bandwidth is chosen about 30% higher than required to absorb tolerances of the matching network components and parasitics.

A low-noise current below 4 $pA/\sqrt{Hz}$ (at 900 MHz) has been achieved with this receiver, which could still be reduced using low-loss RF components for the matching network. The receiver's transfer characteristic is shown in Figure 3.37.

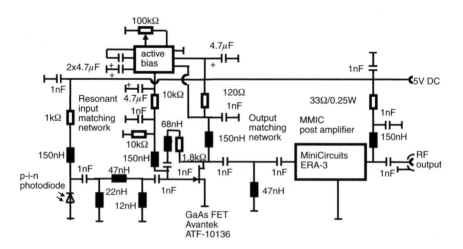

Figure 3.36 Example of a low-cost 900-MHz high-sensitivity optical receiver with input matching network. This receiver allows high dynamic range also for fiber optic links with low optical power [11, 14]. (*Source:* [11]. © 1997 IOS Press.)

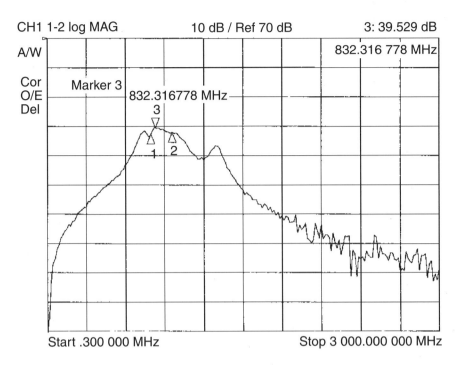

Figure 3.37 Measured response of matched low-cost, low-noise photoreceiver [11, 14]. (*Source:* [11]. © 1997 IOS Press.)

3.9 Alternative Low-Cost Fiber Optic Link Concepts: New Technologies

3.9.1 Lower Dynamic Range Requirements and Multimode Fibers

Apart from the presented high dynamic range low-cost links, links with lower dynamic range requirements can be realized at even lower cost. Laser diodes can be replaced by luminescent diodes and graded-index multimode fibers.[8] An SFDR of the order of 40–50 dB (BW = 200 kHz) can be achieved with a luminescent diode. Luminescent diode bandwidth is limited to some 100 MHz, which requires downconversion from the RF band (see below).

One strongly growing low-cost technology is vertical cavity surface emitting lasers [89–92]. Due to the high spontaneous emission rate into the

---

8. The bandwidth of multimode fibers as well as the C/N (lower connector mismatch-induced modal noise) can be increased significantly by selective fiber launch (e.g., coupling with a single-mode pigtail [88]). Selective launch means that only a small number of modes are excited in the multimode fiber.

lasing mode the vertical cavity surface emitting laser characteristic is—at least from the dynamic range point of view—between that of a laser diode and a luminescent diode. Therefore, the SFDR is some decibels higher than that of a luminescent diode. SFDR of 52 dB (BW = 200 kHz) has been measured with an 850-nm device and graded index multimode fibers [14]. Lee et al. [93] have measured SFDR values for vertical cavity surface emitting lasers up to 113 dB $Hz^{2/3}$ for a free-space link and >95 dB $Hz^{2/3}$ above 1 GHz for a short multimode link.

For lifetime reasons reliable broadband vertical cavity surface emitting lasers have large apertures and therefore emit in multiple spatial modes, which requires multimode fibers as transmission media for vertical cavity surface emitting lasers. The use of multimode fibers can significantly reduce packaging costs but limits the link span to below 1 km. Due to technological difficulties (Bragg mirrors) only 850-nm vertical cavity surface emitting lasers have been available commercially up to now. Recently vertical cavity surface emitting lasers for other wavelengths, such as 960 nm, 1,310 nm, or 1,550 nm, have been developed [90–92]. Therefore, 1,310- and 1,550-nm devices are expected to be commercially available soon. Results on long-wavelength vertical cavity surface emitting laser linearity have been reported [94].

3.9.2 Downconversion

In narrowband applications (total link BW < 100 MHz) downconversion of the RF signal from 900 MHz/1.8 GHz to less than 100 MHz could be applied. Coaxial cable loss is then lower than if the original RF signal were to be transmitted and slightly increases the suitability of coaxial cable for this application.

In fiber optic links downconversion is only appropriate if signals above 1–2 GHz have to be transmitted, because at frequencies below some 100 MHz, the SFDR decreases due to low frequency noise and intermodulation. Additional costs for mixers and oscillators have to be taken into account in any case and a synchronization signal for receiver local oscillator synchronization has to be transmitted. This is accomplished by dividing the transmitter oscillator signal down to a low frequency.

3.9.3 Hybrid Fiber Coaxial Architecture

To achieve best performance for lowest cost in distributed antenna systems, it is, from a system engineering point of view, best to combine optical fiber and coaxial cable technologies. Distant (>100m) hubs or single distant antennas can be fed by fiber links, perhaps with optical power splitters if multiple antennas at a great distance from each other have to be supplied. In cases where the antennas are spaced less than 100m from each other but more than 100m away from the base transceiver station, fiber links should feed a hub, but low-cost coaxial cables can be used for the fine distribution to the antennas.

3.9.4 Plastic Optical Fibers

In the 1990s some progress was made in the field of plastic fibers. New materials opened the range of applications from the visible to the infrared wavelength range. Fiber loss has been reduced and bandwidth has been dramatically improved [95, 96] and continuing progress is expected. A loss of 31 dB/km (with a potential of below 10 dB at 1,310 nm) and bit rates of 11 Gbps over 100m or 2.5 Gbps over 550m have been demonstrated with perfluorinated graded index polymer optical fibers. Even 3×2.5 Gbps WDM over 200m has been demonstrated [96].

Materials used in plastic fibers are very stable. Therefore the long-term stability of the loss and graded index profile is assumed to be good. Although up to now high-speed low-loss plastic fibers have been presented only in the laboratory, future commercial applications are expected for these reasons:

- Plastic fibers cost less than glass fibers.
- Graded index profiles are easier to realize with plastic than with glass fibers. Therefore, the bandwidth of high core diameter graded index fibers can be made very high.
- Packaging costs for multimode pigtails are much lower than for single-mode fibers. Bending radii for plastic fibers are lower than for glass fibers. This makes high-bandwidth multimode fibers attractive for low-cost applications in the RF range.

References

[1]   Meyers, R. A., *The Encyclopedia of Telecommunications,* San Diego, CA: Academic Press, 1989.

[2]   *RF/Microwave Fiber Optic Link Design Guide,* Alhambra, CA: Ortel Corp., 1988.

[3]   *A System Designer's Guide to RF and Microwave Fiber Optics,* Alhambra, CA: Ortel Corp., 1994.

[4]   Hunziker, S., and W. Baechtold, "Simple Model for Fundamental Intermodulation Analysis of RF Amplifiers and Links," *Electron. Lett.,* Vol. 32, No. 19, 1996, pp. 1826–1827.

[5]   Abramowitz, M., and I. Stegun, *Handbook of Mathematical Functions,* Applied Mathematics Series 55, Washington, D.C.: National Bureau of Standards, 1972.

[6]   *Noise Figure Primer,* Hewlett-Packard Application Note 57, Palo Alto, CA: Hewlett-Packard, 1965.

[7]   van der Ziel, A., *Noise in Solid State Devices and Circuits,* New York: John Wiley & Sons, 1986.

[8]   Browne, J. "Amps Raise Gain and Lower Costs in 4-GHz Designs," *Microwave RF,* Vol. 35, No. 4, 1996, pp. 117–120.

[9]   Ebeling, K. J., *Integrated Optoelectronics,* Berlin: Springer-Verlag, 1993.

[10]   Petermann, K., *Laser Diode Modulation and Noise,* Dordrecht, The Netherlands: Kluwer Academic Publishers, 1988.

[11]   Hunziker, S., et al., "Low Cost Fiber Optic Links for Cellular Remote Antenna Repeaters," *Proc. European Conf. on Networks and Optical Communication—NOC'97,* Antwerp, Belgium; Amsterdam, The Netherlands: IOS Press, Vol. 1, 1997, pp. 130–136.

[12]   Hunziker, S., and W. Baechtold, "Cellular Remote Antenna Feeding: Optical Fiber or Coax Cable?" *Electron. Lett.,* Vol. 34, 1997, pp. 1038–1040.

[13]   *Infineon Fiber Optic Data Book,* Berlin, Germany: Infineon Technologies AG, 1999 (BIDI Transceivers).

[14]   Hunziker, S., "Analysis and Optimization of Fiber Optic SCM Links," Thesis, ETH, Zurich, Switzerland, 1998 (in German).

[15]   Steiner, G., W. Baechtold, and S. Hunziker, "Bidirectional Single Fiber Links for Base Station Remote Antenna Feeding," *Proc. European Conf. on Networks and Optical Communication—NOC'00,* Stuttgart, Germany, June 6–9, 2000.

[16]   Sadler, A. A., "Blolite Blown Fiber System Performance," *Proc. European Conf. on Fiber Optic Communications and Local Area Networking—EFOC/LAN'91,* London, 1991, pp. 92–96.

[17]   Hunziker, S., "Volterra Analysis of Second- and Third-Order Intermodulation of InGaAsP/InP Laser Diodes: Theory and Experiment," *Opt. Eng.,* Vol. 34, No. 7, 1995, pp. 2037–2043.

[18]   Bose, A. G., *A Theory of Nonlinear Systems,* Technical Report No. 309, Research Lab of Electronics, MIT, 1956.

[19]   Frechet, M., "Sur les Fonctionelles Continues," *Annales Scientifiques de L'Ecole Normale Superieure,* Vol. 27, No. 5, 1910, pp. 193–216 (in French).

[20]   Rugh, W., *Nonlinear System Theory: The Volterra/Wiener Approach,* Baltimore, MD: The Johns Hopkins University Press, 1981.

[21]   Schetzen, M., *The Volterra and Wiener Theories of Nonlinear Systems,* New York: John Wiley & Sons, 1980.

[22]   Volterra, V., *Lecons sur les Fonctions De Lignes,* Paris, France: Gauthier-Villars, 1913 (in French).

[23]   Bussgang, A. J., L. Ehrman, and J. W. Graham, "Analysis of Nonlinear Systems with Multiple Inputs," *Proc. IEEE,* Vol. 62, No. 8, 1974, pp. 1088–1119.

[24]   Tucker, R. S., and I. P. Kaminow, "High Frequency Characteristics of Directly Modulated InGaAsP Ridge Waveguide and Buried Heterostructure Lasers," *J. Lightwave Technology,* Vol. LT-2, No. 8, 1984, pp. 385–393.

[25]   Tucker, R. S., "Large-Signal Circuit Model for Simulation of Injection-Laser Modulation Dynamics," *IEE Proc.,* Part I, Vol. 128, No. 5, 1981, pp. 180–184.

[26]   Tucker, R. S., "High-Speed Modulation of Semiconductor Lasers," *IEEE Trans. Electron. Devices,* Vol. 12, No. 12, 1985, pp. 2572–2584.

[27]   Helms, J., "Intermodulation and Harmonic Distortions of Laser Diodes with Optical Feedback," *J. Lightwave Technology,* Vol. LT-10, No. 11, 1991, pp. 1567–1575.

[28]   Helms, J., "Modulation Properties of Semiconductor Lasers with Optical Feedback," Thesis, Technical University Berlin, 1992 (in German).

[29]   Kikushima, K., and Y. Suematsu, "Nonlinear Distortion Properties of Laser Diode Influenced by Coherent Reflected Waves," *Trans. IECE (Japan),* Vol. E67, No. 1, 1984, pp. 19–25.

[30]   Achor, H. M., "An Invertible Volterra Model of Second-Order Distortion in DFB Lasers," *IEEE Phot. Technol. Lett.* Vol. 5, No. 3, 1993, pp. 294–297.

[31]   Biswas, T. K., and W. F. McGee, "Volterra Series Analysis of Semiconductor Laser Diode," *IEEE Phot. Technol. Lett.* Vol. 3, No. 8, 1991, pp. 706–708.

[32]   Czylwik, A., "Nonlinear System Modeling of Semiconductor Lasers Based on Volterra Series," *J. Opt. Commun.,* Vol. 7, No. 3, 1986, pp. 104–114.

[33]   Salgado, H. M., and J. J. O'Reilly, "Volterra Series Analysis of Distortion in Semiconductor Laser Diodes," *IEE Proc. J,* Vol. 138, No. 6, 1991, pp. 379–382.

[34]   Wang, J., M. K. Haldar, and F. V. C. Mendis, "Formula for Two Carrier Third-Order Distortion in Semiconductor Laser Diodes," *Electron. Lett.,* Vol. 29, No. 15, 1993, pp. 1341–1343.

[35]   Kuo, C. Y., "Frequency Modulation Responses of Two-Electrode Distributed Feedback Lasers," *J. Lightwave Technology,* Vol. 10, No. 2, 1992, pp. 235–243.

[36]   Kuo, C. Y., et al., "Frequency Modulation Responses of Two-Electrode Distributed Feedback Lasers," *Appl. Phys. Lett.,* Vol. 55, No. 13, 1989, pp. 1279–1281.

[37]   Vankwikelberge, P., et al., "Analysis of the Carrier-Induced FM Response of DFB Lasers: Theoretical and Experimental Case Studies," *IEEE J. Quantum Electronics,* Vol. QE-25, No. 11, 1989, pp. 2239–2253.

[38]   Aureli, G., et al., "Spectral Analysis of Clipping Impulsive Noise Affecting Hybrid Optical Fiber/Coaxial Network for Transmission of SCM (Subcarrier Multiplexing) Analog and Digital Channels," *Proc. European Conf. on Networks and Optical Communication—NOC'96,* Heidelberg, Germany, Vol. 1, 1996, pp. 22–29.

[39]   Frigo, N., "A Model of Intermodulation Distortion in Non-Linear Multicarrier Systems," *IEEE Trans. on Communications,* Vol. 42, No. 2–4, 1994, pp. 1217–1222.

[40]   Haldar, M. K., K. B. Chia, and F. V. C. Mendis, "Dynamic Considerations in Overmodulation of Semiconductor Laser Diodes," *Electron. Lett.,* Vol. 32, No. 7, 1996, pp. 659–661.

[41]   Le Bihan, J., "Approximate Dynamic Model for Evaluating Distortion in a Semiconductor Laser Under Overmodulation," *IEEE Phot. Technol. Lett.,* Vol. 9, No. 3, 1997, pp. 303–305.

[42]   Saleh, A. A. M., "Fundamental Limit on Number of Channels in Subcarrier-Multiplexed Lightwave Communications Systems," *Electron. Lett.,* Vol. 25, No. 1, 1989, pp. 79–80.

[43]   Hunziker, S., and W. Baechtold, "Classical and Nonclassical Intermodulation of Fabry-Perot Diode Lasers and Its Influence on Pre-Distortion Linearisation," *Proc. European Conf. on Networks and Optical Communication—NOC'96,* Heidelberg, Germany, Vol. 3, 1996, pp. 256–261.

[44]   Ogasawara, N., and R. Ito, "Longitudinal Mode Competition and Asymmetric Gain Saturation in Semiconductor Lasers. II. Theory," *Japan. J. Appl. Phys.,* Vol. 27, No. 4, 1988, p. 615.

[45]   Frankenberger, R., "Measurement of the Spectrum of Nonlinear Gain Saturation in InGaAsP-Laser Diodes," Ph.D. Thesis, Technical University, Munich, 1992 (in German).

[46]   Suematsu, Y., and A. R. Adams, *Semiconductor Lasers and Photonic Integrated Circuits,* London: Chapman & Hall, 1994.

[47]   Yamada, M., and Y. Suematsu, "Analysis of Gain Suppression in Undoped Injection Lasers," *J. Appl. Phys.,* Vol. 52, No. 4, 1981, pp. 2653–2664.

[48]   Lu, X. L., C. B. Su, and R. B. Lauer, "Increase in Laser RIN Due to Asymmetric Nonlinear Gain, Fiber Dispersion, and Modulation," *IEEE Phot. Technol. Lett.,* Vol. 4, No. 7, 1992, pp. 774–777.

[49]   Morton, P. A., et al., "Large-Signal Harmonic and Intermodulation Distortions in Wide-Bandwidth GaInAsP Semiconductor Lasers," *IEEE J. Quantum Electronics,* Vol. QE-25, No. 6, 1989, pp. 1559–1567.

[50]   Alalusi, M. R., and R. B. Darling, "Effects of Nonlinear Gain on Mode Hopping in Semiconductor Laser Diodes," *IEEE J. Quantum Electronics,* Vol. 37, No. 7, 1995, pp. 1181–1192.

[51]   Peters, F. H., and D. T. Cassidy, "Effect of Scattering on the Longitudinal Mode Spectrum of 1.3-mm Semiconductor Diode Lasers," *Appl. Phys. Lett.,* Vol. 57, No. 4, 1990, pp. 330–332.

[52]   Peters, F. H., and D. T. Cassidy, "Model of the Spectral Output of Gain-Guided and Index-Guided Semiconductor Diode Lasers," *J. Opt. Soc. Am. B,* Vol. 8, No. 1, 1991, pp. 99–105.

[53]   Peters, F. H., and D. T. Cassidy, "Spectral Output of 1.3-mm InGa-AsP Semiconductor Diode Lasers," *IEE Proc. J.,* Vol. 138, No. 3, 1991, pp. 195–198.

[54]   Peters, F. H., and D. T. Cassidy, "Strain and Scattering Related Spectral Output of 1.3-mm InGaAsP Semiconductor Diode Lasers," *Appl. Opt.,* Vol. 30, No. 9, 1991, pp. 1036–1041.

[55]   Peters, F. H., and D. T. Cassidy, "Scattering, Absorption, and Anomalous Spectral Tuning of 1.3-mm Semiconductor Diode Lasers," *J. Appl. Phys.,* Vol. 71, No. 9, 1992, pp. 4140–4144.

Low-Cost Fiber Optic Links for Cellular Remote Antenna Feeding   179

[56]   Cassidy, D. T., and F. H. Peters, "Spontaneous Emission, Scattering, and the Spectral Properties of Semiconductor Diode Lasers," *IEEE J. Quantum Electronics,* Vol. QE-28, No. 4, 1992, pp. 785–791.

[57]   Peters, F. H., and D. T. Cassidy, "Spatially and Polarization Resolved Electroluminescence of 1.3-mm InGaAsP Semiconductor Diode Lasers," *Appl. Opt.,* Vol. 28, No. 17, 1989, pp. 3744–3750.

[58]   Dentan, M., and B. de Cremoux, "Numerical Simulation of the Nonlinear Response of a P-I-N Photodiode Under High Illumination," *J. Lightwave Technology,* Vol. LT-8, No. 8, 1990, pp. 1137–1144.

[59]   Hayes, R. R., and D. L. Persechini, "Nonlinearity of p-i-n Photodetectors," *IEEE Phot. Technol. Lett.,* Vol. 5, No. 1, 1993, pp. 70–72.

[60]   Williams, K. J., and R. D. Esman, "Observation of Photodetector Nonlinearities," *Electron. Lett.,* Vol. 28, No. 8, 1992, pp. 731–732.

[61]   Williams, K. J., R. D. Esman, and M. Dagenais, "Nonlinearities in P-I-N Microwave Photodetectors," *J. Lightwave Technology,* Vol. 14, No. l, 1996, pp. 84–96.

[62]   Haken, H., *Synergetics,* Berlin: Springer-Verlag, 1983.

[63]   Henry, C. H., "Line Broadening of Semiconductor Lasers," *Coherence, Amplification, and Quantum Effects in Semiconductor Lasers,* Y. Yamamoto (Ed.), New York: John Wiley & Sons, 1991.

[64]   Schimpe, R., "Intensity Noise Associated with the Lasing Mode of a (GaAl)As Diode Laser," *IEEE J. Quantum Electronics,* Vol. 19, No. 6, 1983, pp. 895–897.

[65]   Lu, X., et al., "Analysis of Relative Intensity Noise in Semiconductor Lasers and Its Effect on Subcarrier Multiplexed Lightwave Systems," *J. Lightwave Technology,* Vol. LT-12, No. 7, 1994, pp. 1159–1165.

[66]   Su, C. B., J. Schlafer, and R. B. Lauer, "Explanation of Low-Frequency Relative Intensity Noise in Semiconductor Lasers," *Appl. Phys. Lett.,* Vol. 57, No. 9, 1990, pp. 849–851.

[67]   Ohtsu, M., Y. Otsuka, and Y. Teramachi, "Precise Measurements and Computer Simulations of Mode-Hopping Phenomena in Semiconductor Lasers," *Appl. Phys. Lett.,* Vol. 46, No. 2, 1985, pp. 108–110.

[68]   Chinone, N., et al. "Mode-Hopping Noise in Index-Guided Semiconductor Lasers," *IEEE J. Quantum Electronics,* Vol. QE-21, No. 8, 1985, pp. 1264–1270.

[69]   Lau, K. Y., and H. Blauvelt, "Effect of Low-Frequency Intensity Noise on High-Frequency Direct Modulation of Semiconductor Injection Lasers," *Appl. Phys. Lett.,* Vol. 52, No. 9, 1988, pp. 694–696.

[70]   Unger, H. G., *Optical Communications, Part II,* Heidelberg, Germany: Dr. Alfred Huthig Verlag, 1992 (in German).

[71]   Saleh, B. E. A., and M. C. Teich, *Fundamentals of Photonics,* New York: John Wiley & Sons, 1991.

[72]   Bergmann, E. E., C. Y. Kuo, and S. Y. Huang, "Dispersion-Induced Composite Second-Order Distortion at 1.5 μm," *IEEE Phot. Technol. Lett.,* Vol. 3, No. 1, 1991, pp. 59–61.

[73] Ih, C. S., and W. Gu, "Fiber Induced Distortion in a Subcarrier Multiplexed Lightwave System," *IEEE J. Selected Areas of Communication,* Vol. 8, No. 9, 1990, pp. 1296–1303.

[74] Hunziker, S., and W. Baechtold, "Fiber Dispersion Induced Nonlinearity in Fiber Optic Links with Multimode Laser Diodes," *IEEE Phot. Technol. Lett.,* Vol. 9, No. 3, 1997, pp. 371–373.

[75] Wentworth, R. H., G. E. Bodeep, and T. E. Darcie, "Laser Mode Partition Noise in Lightwave Systems Using Dispersive Optical Fiber," *J. Lightwave Technology,* Vol. LT-10, 1992, pp. 84–89.

[76] Gao, Y., and E. Reuss Keller, "Nonlinear Distortion by Laser Frequency Chirping and Fiber Chromatic Dispersion in Multichannel Subcarrier Intensity Modulation Systems," *Proc. European Conf. on Fiber Optic Communication and Networks—EFOC/ LAN'92,* 1992, pp. 432–435.

[77] Jäckel, H., "Light Emission Noise and Dynamics of GaAlAs Heterostructure Diode Lasers in the Frequency Range 10 MHz to 8 GHz," Thesis, ETH, Zurich, Switzerland, 1980 (in German).

[78] Westcott, R. J., "Investigation of Multiple FM/FDM Carriers Through a Satellite TWT Operating Near to Saturation," *Proc. IEE,* Vol. 114, No. 6, 1967, pp. 726–740.

[79] Gardner, M., "The Graceful Graphs of Solomon Golomb, or How to Number a Graph Parsimoniously," *Scientific American,* Vol. 226, No. 6, 1972, pp. 108–112.

[80] Shearer, J. B., "Some New Optimum Golomb Rulers," *IEEE Trans. on Information Theory,* Vol. IT-36, No. 1, 1990, pp. 183–184.

[81] Erni, D., et al., "Application of Evolutionary Optimization Algorithms in Computational Optics," *ACES J. Special Issue on Genetic Algorithms,* Vol. 15, No. 2, 2000, pp. 43–60.

[82] Ghiasi, A., and A. Gopinath, "Novel Wide-Bandwidth Matching Technique for Laser-Diodes," *IEEE Trans. on Microwave Theory and Techniques,* Vol. MTT-38, 1990, pp. 673–675.

[83] Goldsmith, C. L., and B. Kanack, "Broad-Band Reactive Matching of High-Speed Directly Modulated Laser Diodes," *IEEE Microwave Guided Wave Lett.,* Vol. 3, 1993, pp. 336–338.

[84] Alexander, S. B., *Optical Communication Receiver Design,* Bellingham, WA: SPIE Press, 1997.

[85] Wilson, B., Z. Ghassemlooy, and I. Darwazeh, *Analogue Optical Fibre Communications,* London: IEE Press, 1995.

[86] Darcie, T. E., et al., "Resonant P-I-N-FET Receivers for Lightwave Subcarrier Systems," *J. Lightwave Technology,* Vol. LT-6, 1988, pp. 582–589.

[87] Gimlett, J. L., "Low Noise 8GHz Optical Receiver," *Electron. Lett.,* Vol. 23, 1987, pp. 281–283.

[88] Wood, T. H., and L. A. Ewell, "Increased Received Power and Decreased Modal Noise by Preferential Excitation of Low-Order Modes in Multimode Optical-Fiber Transmission Systems," *J. Lightwave Technology,* Vol. LT-4, No. 4, 1986, pp. 391–395.

[89] Sale, T. E., *Vertical Cavity Surface Emitting Lasers,* Taunton, U.K.: Research Studies Press Ltd., 1995.

[90] Wilmsen, C. W., H. Temkin, and L. A. Coldren, *Vertical-Cavity Surface-Emitting Lasers: Design, Fabrication, Characterization, and Applications,* Cambridge, U.K.: Cambridge University Press, 2000.

[91] Streubel, K., et al., "Novel Technologies for 1.55-mm Vertical Cavity Lasers," *Opt. Eng.,* Vol. 39, No. 2, pp. 488–497, 2000.

[92] Zhang, S. Z., et al., "1.54-mm Vertical-Cavity Surface-Emitting Laser Transmission at 2.5Gb/s," *IEEE Phot. Technol. Lett.,* Vol. 9, No. 3, 1997, pp. 374–376.

[93] Lee, H. L. T., et al., "Dynamic Range of Vertical-Cavity Surface-Emitting Lasers in Multimode Links," *IEEE Phot. Technol. Lett.,* Vol. 11, No. 11, 1999, pp. 1473–1475.

[94] Piprek, J., et al., "Harmonic Distortion in 1.55-mm Vertical-Cavity Lasers," *IEEE Phot. Technol. Lett.,* Vol. 12, No. 12, 2000, pp. 374–376.

[95] Giaretta, G., et al., "Demonstration of 500-nm-Wide Transmission Window at Multi-Gbit/s Data Rates in Low-Loss Plastic Optical Fiber," *Proc. European Conf. on Optical Communications—ECOC'99,* Nice, France, Vol. 2, 1999, pp. 240–241.

[96] Khoe, G. D., et al., "Status of GIPOF and Related Technologies," *Proc. European Conf. on Optical Communications—ECOC'99,* Nice, France, Vol. 2, 1999, pp. 274–277.

4

Radio over Fiber Technology for the Next Generation

Hamed Al-Raweshidy

4.1 Introduction

Chapters 1 and 2 have presented the main elements of the optical devices and the parameters related to radio over fiber: laser diode performance, intermodulation, RIN, and clipping noise. This chapter discusses in more detail the system performance of radio over fiber on Universal Mobile Telecommunication System/wideband code-division multiple access (UMTS/WCDMA). We start with an introduction to UMTS and WCDMA with an overview of radio over fiber technology for the UMTS system. The advantages and disadvantages of radio over fiber technology for third-generation (3G) and fourth-generation (4G) systems are also discussed. In addition, a simulation for both radio and optical systems is presented. Furthermore, an analytical model for evaluating the performance of WCDMA-based radio over fiber systems with numerical results showed an improvement in performance when the effect of voice activity monitoring on intermodulation distortion and clipping noise was taken into account.

4.2 Radio over Fiber Systems

Recently, optical fiber microcellular systems, in which microcells in a wide area are connected by optical fibers and radio signals are transmitted over an optical fiber link among base stations and control stations, has attracted much attention [1]. This is because of the low loss and enormous bandwidth of optical fiber, the increasing demand for capacity/coverage, and the benefits it offers in terms of low-cost base station deployment in microcellular systems, all of which make it an ideal candidate for realizing microcellular networks [1, 2]. In such a system, each microcell radio port would consist of a simple and compact optoelectronic repeater connected by an RF fiber optic link to centralized radio and control equipment, possibly located at a preexisting macrocell site.

Use of RF antenna remoting allows changes to the system frequency plan or modulation format to be done at a central location, without the need to modify any radio port equipment. Antenna remoting should also simplify the provision of system features such as rapid handover, dynamic channel assignment, and diversity combining.

This system will make extensive use of microcells and picocells in order to deliver high bandwidth. Such microcell systems can solve the frequency limitation problem because a number of base stations can be installed, the zone radius can be reduced, and the radio frequencies can be reused effectively in many radio zones [3]. The much lower power level eliminates the need for the expensive frequency multiplexes or high-power amplifiers currently employed at base stations. The limited coverage due to low antenna height greatly reduces the cochannel interference from other cells [4]. Radio over fiber (RoF) systems are now being used extensively for enhanced cellular coverage inside buildings such as offices, shopping malls, and airport terminals [1, 2, 5].

The WCDMA air interface can now be regarded as a mature technology that is ready to provide the basis for the third-generation wireless personal communication system known as the UMTS [3, 6]. These systems will make extensive use of microcells and picocells in order to deliver high-bandwidth services to customers. The benefit of using RoF for WCDMA distributed antenna systems is expected to be even more important, partly because of their higher frequency and bandwidth requirements [1, 2].

Two key features are expected to be employed in the UMTS system to minimize multiple-user interference: adaptive antenna arrays and fast closed-loop forward and reverse power control techniques [7; see also http://www.3gpp.org]. Other important techniques that are used to reduce multiple-user interference are cell sectorization and voice activity monitoring, particularly in speech-oriented cellular systems [8, 9].

The UMTS is designed to support simultaneous transmission of multiple services and data rates including video. One of the major drawbacks in RoF systems is laser diode nonlinearity, which gives rise to intermodulation distortion and clipping noise. It is well known that intermodulation distortion and clipping noise are signal level dependent [10]. We envisage that in WCDMA RoF systems, voice activity monitoring will have an impact not only on multiple-user interference, but also on intermodulation distortion and clipping noise power. A full analysis for intermodulation distortion was given in Chapters 1 and 2, and Chapter 3 provides details of the dynamic range of microcellular systems.

4.3 Cellular Architecture

Increases in demand and the poor quality of existing service led mobile service providers to research ways to improve the quality of service and to support more users in their systems. In modern cellular telephony, rural and urban regions are divided into areas according to specific provisioning guidelines. Deployment parameters, such as amount of cell splitting and cell sizes, are determined by engineers experienced in cellular system architecture.

4.3.1 Cell

A cell is the basic geographic unit of a cellular system. The term *cellular* comes from the honeycomb shape of the areas into which a coverage region is divided. Cells are base stations transmitting over small geographic areas that are represented as hexagons. Each cell size varies depending on the landscape. Because of constraints imposed by natural terrain and man-made structures, the true shape of cells is not a perfect hexagon. Based on the radius of the cells, there are three types of cellular networks: macrocells, microcells, and picocells.

4.3.2 Macrocellular

A macrocellular network is deployed using relatively large cells with a diameter of 16 to 48 km. This creates a substantial footprint with significantly fewer sectors. A regional switching center controls all of the traffic within a market and interconnection with the Public Switched Telephone Network (PSTN). Capacity can be modularly upgraded by adding sectors to existing sites to facilitate subscriber growth.

4.3.3   Microcellular

Microcellular radio networks are used in areas with high traffic density, like suburban areas. The cells have radii between 200m and 1 km. For such small cells, it is hard to predict traffic densities and area coverage.

4.3.4   Picocells

Picocells or indoor cells have radii between 10 and 200m. Today, picocellular radio system are used for wireless office communications.

4.4   UMTS Architecture

4.4.1   General Architecture

Figure 4.1 shows the assumed UMTS architecture as outlined in European Telecommunication Standards Institute (ETSI)/SMG. The figure shows the architecture in terms of its user equipment (UE), UMTS Terrestrial Radio Access Network (UTRAN), and core network (CN). The respective reference points are Uu (radio interface) and Iu (CN-UTRAN interface) [6].

4.4.2   WCDMA for 3G Cellular Systems

Third-generation mobile radio networks have been under intense research and discussion recently and were scheduled to emerge in 2001. Emerging requirements for high-rate data services and better spectrum efficiency are the main drivers identified for these 3G mobile systems [6]. In the International Telecommunication Union (ITU), 3G networks are called International Mobile Telecommunications 2000 (IMT 2000), and in Europe, Universal Mobile Telecommunications Systems. The main objective of the IMT 2000 air interface can be summarized as follows:

- Full coverage and mobility for 144 Kbps, preferably 384 Kbps;
- Limited coverage and mobility for 2 Mbps;
- High spectrum efficiency compared to existing systems;
- High flexibility to introduce new services.

Wideband direct sequence code-division multiple access (DS-CDMA) has emerged as the mainstream air interface solution for the 3G networks

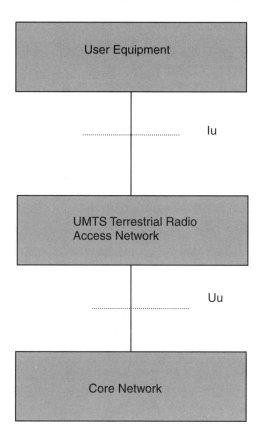

Figure 4.1 UMTS architecture.

due to its numerous advantages over TDMA and frequency-division multiple access (FDMA). These advantages include the use of soft handover for macrodiversity, the exploitation of multipath fading through RAKE combining, and a direct capacity increase as a result of the use of cell sectorization. However, the air interface standardization seems to focus on two main types of WCDMA: network asynchronous and network synchronous.

In network asynchronous schemes, the base stations are not synchronized, whereas in network synchronous schemes, the base stations are synchronized to each other within a few microseconds. ETSI and the Association of Radio Industries and Businesses in Japan have chosen to develop a common WCDMA air interface standard, based on an asynchronous network for frequency-division duplex (FDD) bands.

The important concept of WCDMA is the introduction of intercell asynchronous operation and a user-dedicated pilot channel. Intercell asynchronous operation facilitates continuous system deployment from outdoors to indoors. Other technical features of WCDMA are summarized next [6, 7; see also http://www.3gpp.org]:

- Fast cell search under intercell asynchronous operation;
- Fast transmit power control on both reverse (mobile-to-base station) and forward (base station-to-mobile) links;
- Coherent spreading code tracking;
- Coherent RAKE reception on both links;
- Orthogonal variable spreading factors in the forward link;
- Variable-rate transmission with blind rate detection;
- Data-rate-dependent transmit power.

The introduction of the data channel-associated pilot channel allows WCDMA to support interference cancellation and adaptive antenna array techniques that can significantly increase the link capacity and coverage. Table 4.1 lists the parameters of the WCDMA air interface standard [7].

Another major contender for the 3G WCDMA air interface standard, known as CDMA2000, is being developed by the Telecommunication Industry Association (TIA) in the United States. This chapter describes mainly the ETSI's WCDMA air interface. However, a comparison of the WCDMA, CDMA2000, and the second-generation (2G) IS-95 CDMA systems is given in Table 4.2. The main differences between WCDMA and CDMA2000 systems are the chip rate, downlink channel structure, and network synchronization. Because WCDMA has an asynchronous network, different long codes rather than phase shifts of the same code are used for the cell and user separation. The code structure further impacts how synchronization, cell acquisition, and handover synchronization are performed.

4.5 Radio over Fiber Concept

A microcellular network can be implemented by using a fiber-fed distributed antenna network as shown in Figure 4.2. The received RF signals at each remote antenna are transmitted over an analog optical fiber link to a central base station where all the demultiplexing and signal processing are done. In this way, each remote antenna site simply consists of a linear analog optical transmitter, an amplifier, and the antenna. The cost of the microcellular

Table 4.1
Parameters of WCDMA

Channel bandwidth	(1,25), 5, 10, 20 MHz
Downlink R_f channel structure	Direct spread
Chip rate	(1.024)/4.096/8.192/16.384 Mcps
Roll-off factor for chip shaping	0.22
Frame length	10 mc/20 ms (optional)
Spreading modulation	Balanced QPSK (downlink)
	Dual channel (uplink)
	Complex spreading circuit
Data modulation	QPSK (downlink)
	BPSK (uplink)
Coherent detection	User-dedicated time-multiplexed pilot (downlink and uplink), common pilot in downlink
Channel multiplexing in uplink	Control and pilot channel time-multiplexed I and Q multiplexing for data and control channel
Multirate	Variable spreading and multicode
Spreading factors	4–256
Power control	Open and fast closed loop (1.6 kHz)
Spreading (downlink)	Variable-length orthogonal sequence for channel separation, Gold sequences 2^{18} for cell and user separation (truncated cycle 10 ms)
Spreading (uplink)	Variable-length orthogonal sequences for channel separation, Gold sequences 2^{41} for user separation (different time shift in I and Q channel, truncated cycle 10 ms)
Handover	Soft handover, interfrequency handover

antenna sites must be greatly reduced before the deployment of these networks is practical [11]. (More details are provided in Chapters 3 and 5.)

4.5.1 Advantages of Using RoF in Mobile Communications Networks

The radio network is a distributed antenna system, with the potential for adaptive antenna selection, as well as adaptive channel allocation to increase the spectrum efficiency. The distributed antenna system provides an infrastructure that brings the radio interface very close to the users and has the following benefits:

- Low RF power remote antenna units (RAUs);
- Line-of-sight (LOS) operation (multipath effects are minimized);

Table 4.2
Comparison of 2G and 3G CDMA Systems

	IS-95 CDMA	WCDMA	CDMA2000
Bandwidth	1.25 MHz	5 MHz	5 MHz
Data rate	9.6 Kbps	384 Kbps/2 Mbps	384 Kbps/2 Mbps
Chip rate	1.228 Mcps	4.096 Mcps	3.6864 Mcps
Frame length	20 ms	10 ms	20 ms
Packet and circuit switch data	No	Yes	Yes
Simultaneous data and voice	No	Yes	Yes
Intercell synchronization	Synchronous	Asynchronous	Synchronous
Spreading codes	Walsh + Long m-sequence	Variable-length orthogonal sequence + Gold sequence (10 ms)	Variable-length orthogonal sequence (Walsh) + M-sequence
Multirate capability	No	Yes	Yes
Coherent detection	—	User-dedicated time-multiplexed pilot (downlink and uplink), common pilot in downlink	Pilot symbols time-multiplexed with PC bits (uplink), common continuous pilot channel and auxiliary pilot
Handover	Soft handover	Soft handover, interfrequency handover	Soft handover, interfrequency handover
Power control	Reverse open and fast closed PC, forward slow control loop	Forward and reverse, open and fast closed loop (1.6 kHz)	Forward and reverse, open and fast closed loop

- Enabling of mobile broadband radio access close to the user in an economically acceptable way;
- Reduced environmental impact (small RAUs);
- Good coverage;
- Capacity enhancement by means of improved trunking efficiency;
- Dynamic radio resource configuration and capacity allocation;
- Alleviation of the cell planning problem;

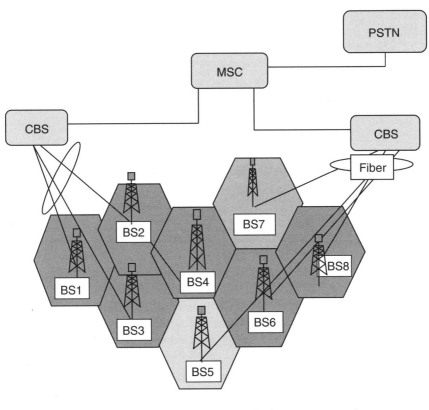

CBS = central base
station

MSC = mobile switching
center

BS = base station

Figure 4.2 Optically fed remote antenna network for microcellular RoF systems.

- Reduction in the number of handovers;
- Centralized upgrading or adaptation;
- The potential to deploy precision tracking of user equipment for safety/first aid and other purposes;
- Higher reliability and lower maintenance costs;
- Support for future broadband multimedia applications;
- Better coverage and increased capacity;
- High-quality signals;
- Support for macrodiversity transmissions;
- Low fiber attenuation (up to 0.2 dB/km);

- Reduced engineering and system design costs;
- Multiple services on a single fiber;
- Lightweight fiber cables;
- No electromagnetic interference;
- Reliability.

The use of low RF power RAUs has these benefits:

- Low generated interference;
- Increased spectrum efficiency;
- Easier frequency/network planning;
- Increased battery lifetime of mobile terminals;
- Relaxed human health issues;
- The potential to use RF complementary metal oxide semiconductor technology in mobile terminals.

A multioperator shared infrastructure allows for equal performance and power levels (the problem with uncoordinated mobiles would be alleviated).

4.5.2 Radio over Fiber Application in Microcellular Systems

Recently, the use of optical fiber feeders for alleviating the problems inherent in microcellular systems has attracted much attention. The goals of the microcellular mobile communication systems include service availability over an extremely high percentage of user environments and provision of a combination of services such as voice, data, and multimedia. Provision of such enhanced services at a high-quality level of service requires an expensive remote antenna (base station) density of tens, hundreds, or even thousands of antennas per square kilometer. This antenna density enables more subscribers to be accommodated per unit service area and allows for the use of smaller and lower power handsets.

To make the base stations compact and cost-effective, it has been proposed that the RoF technique be used to transfer the complicated RF modem and signal processing functions from the base stations to a centralized control station. The proposed fiber-feeder microcellular system is shown in Figure 4.3. The microbase stations will consist mainly of a photodiode circuitry (O/E), RF power amplifier (PA) in the forward link, and a low-noise amplifier (LNA) laser diode circuitry (E/O) in the reverse link [12, 13].

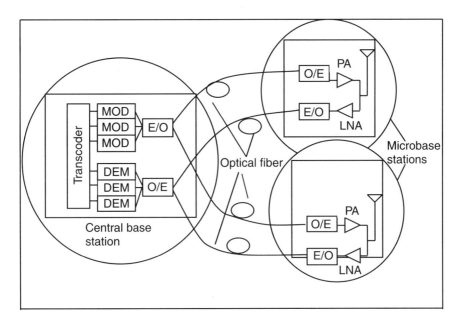

Figure 4.3 Fiber-feeder microcellular network for mobile communication systems.

4.5.3 Simulation Approach and Model Components

In this section, a simulation approach is used for the handover performance evaluation. Figure 4.4 shows the components of the proposed simulation model for the analysis. A microcellular cell plan in the Manhattan-type environment [14], as illustrated in Figure 4.5 (i.e., full-square cell plan), is used for the WCDMA RoF systems. The proposed cell plan will offer the following benefits:

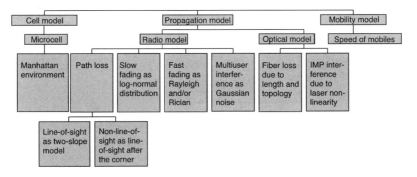

Figure 4.4 Proposed simulation model components for analysis of soft handover algorithms in WCDMA RoF networks.

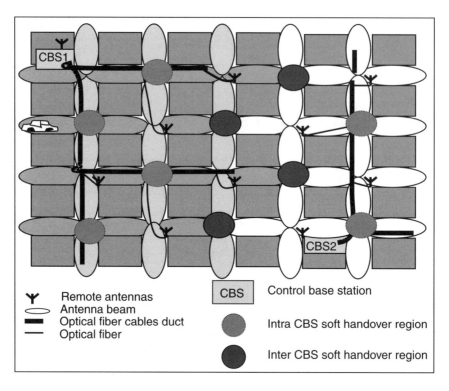

Figure 4.5 Manhattan-type environment.

- Reduced rate of handover;
- Reduced number of base stations and signaling load for RoF systems employing network control handover with soft handover;
- Enhanced handover execution delay, based on the fact that handover will occur only at street junctions where high-speed mobiles are assumed to have slowed down.

4.6 Simple Simulation Model of WCDMA RoF System

The simulation model of the downlink WCDMA RoF system using SystemView is as shown in Figure 4.6. It consists of one dedicated physical data channel and the synchronization channels, of which the primary SCH level is used to select the strongest base station using a match filter in the mobile unit. The transmitting filter is a raised cosine filter with a roll-off factor of 0.22. The laser diode is modeled as a third-order polynomial. The

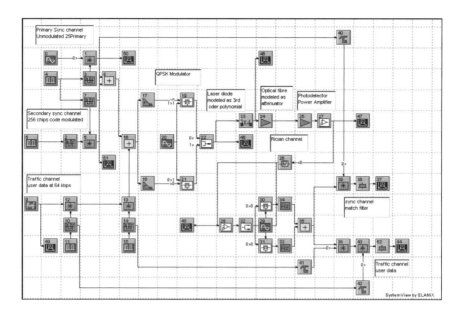

Figure 4.6 Simulation model of WCDMA RoF system using SystemView.

optical fiber is modeled as an attenuator at 0.2 dB/km. The maximum fiber length is assumed to be 1 km (i.e., the furthest remote antenna location from the serving central base station). A p-i-n photodiode responsivity of 0.8 is assumed. The RF power amplifier gain is set at 10 dB. The radio channel is modeled as Rician for the desired signal and the interfering signals.

4.6.1 The Optical Fiber Link Model

A model of the laser that predicts laser performance in an analog transmission environment has been developed as shown in Figure 4.7.

The combination of WCDMA and RoF represents a first demonstration of new broadband applications using the full radio over fiber environment. A key component in the system is the laser diode. The nonlinear nature of the light source is obtained from modeling of both memoryless and laser rate equations. The simulation of a single microcell WCDMA system with a power control and soft handover subsystem is used to determine the effects of the optical link on the performance of the power control and soft handover subsystem. The simulation model has provided accurate results and is simple in order to be efficient in simulating such a complex environment.

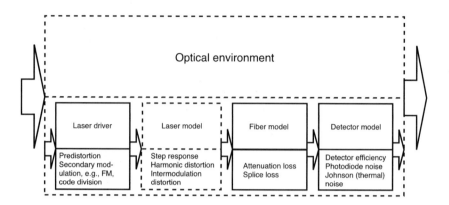

Figure 4.7 The optical fiber link model.

The simulation package SystemView has been used to construct a hierarchical simulation model. The simulation model consists of mainly four parts:

1. The RF WCDMA network consists of one microcell base station and a number of uniformly distributed mobile stations where the air channel will be modeled as a Rician fading channel with log-normal shadowing and fourth-power path loss.

2. The calls generator randomly generates requests for mobile connections and can be specified by a number of random parameters: the speed and direction of the mobile station(s) (Gaussian), the location of the mobile stations relative to the microcell (uniform), and the call interval (Gaussian).

3. The optical link model uses parameters that have already been identified from experimental work.

4. The power control subsystem model is based on the Third-Generation Partnership Project (3GPP) standards.

The aim of the simulation model is to have an integrated model for both wireless and optical systems that would be transparent to a different system such as GSM, UMTS, and High-Performance Radio Local Area Network type 2 (HiperLAN2) as shown in Figure 4.8. The interface between the core network and the optical premises network (the multiaccess optical network) would be an office or a residential gateway with a common access platform, which would include the radio base station (RBS).

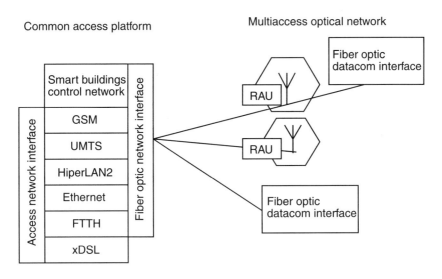

Figure 4.8 The transparent optical network.

4.7 Macrodiversity Versus Microdiversity

In microdiversity, multiple antennas are employed at the receiver to form the branches for the diversity combiner. Microdiversity is used to combat small-scale fading induced by the channel at the receiver. Microdiversity has also been used to realize space-division multiple access (SDMA) communications systems. In these systems, all users transmit simultaneously on a single channel, and a user's physical location is used to separate the desired signal from the interference. The diversity antennas are used to effectively create a steerable beam that selects the desired user's signal while suppressing interference from the other users.

SDMA systems involving microdiversity techniques have been successfully developed and demonstrate the potential of using microdiversity to improve capacity on a single channel. The distance between the antennas in the microdiversity system is of the order of only a few wavelengths [15]. Macrodiversity is another realization of space diversity in cellular systems. Like microdiversity, macrodiversity is used to improve the S/N of the received signal. However, the diversity branch signals in a macrodiversity system originate from the base station antennas, instead of an array of antennas at the receiver as in a microdiversity system. Hence, the main difference between branch signals received in a microdiversity system versus those in a macrodiversity system is the amount of attenuation and time delay on the received signals.

Macrodiversity is useful for combating shadowing of a user's signal caused by obstructions in the environment. To this end, macrodiversity has been employed in CDMA systems to improve signal quality near cell boundaries during soft handover. However, macrodiversity has not been widely targeted as a method for reducing interference in communications systems. Furthermore, systems employing macrodiversity have not considered diversity techniques involving combining of signals on the reverse link [16, 17].

4.7.1 Macrodiversity Combining for the Uplink

Although macrodiversity techniques have been employed in modern cellular systems, techniques that employ combining have not been explored as a means of reducing interference in these systems. In WCDMA, the use of more sophisticated techniques could significantly reduce multiple-access interference and extend the capacity of current and next-generation systems [17].

To combat the interference problem, power control of the mobiles is employed so that users close to the base station do not overwhelm those who are farther away [9]. Soft handover is employed whereby a mobile communicates with two or more base stations when making a transition from one cell to another. Soft handover improves the quality of the link by using signal diversity to improve the S/N of the received signal. On the forward link, multiple base stations transmit the same signal to a mobile terminal, which aligns, cophases, and combines the signals to improve the S/N. On the reverse link, the quality of the link is improved by selecting the best signal received from a set of base stations in the vicinity of the mobile terminal [18].

The use of selection diversity on the reverse link suggests that there could be a substantial capacity improvement if a combining technique were used, because selection diversity offers a limited increase in signal quality compared to these techniques.

4.7.2 Macrodiversity Architecture in Radio over Fiber

The design of a radio over fiber microcellular system with macrodiversity schemes should consider the effect of the optical fiber link. The noise of the optical fiber link is relatively small compared to that of a wireless link. Because the processing will be carried out at the central base station, the macrodiversity can be realized very easily, as shown in Figure 4.9. From that

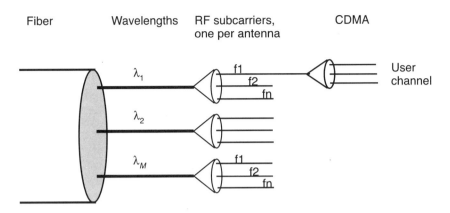

Figure 4.9 CDMA signal in an optical fiber link.

the degradation caused by fading is expected to be improved. The signal received by the mobile station is the sum of the faded transmitted signal and the optical fiber link noise. The parameters that need to be considered in implementing an analog fiber optic system are C/N, bandwidth, and signal distortion resulting from nonlinearity in the transmission system.

4.7.3 Simulation Model

The main components, which are essential for the investigation of the macrodiversity architecture in radio over fiber, are shown in Figure 4.10.

The propagation model for the radio link must consider the path loss, signal fading, and multiuser interference. For the fiber link, it is considered the loss in the fiber and laser nonlinearity.

4.8 Estimation of Number of Downlink Traffic Channels per RF Channel

From spread-spectrum communications theory and in an ideal case, the number of CDMA channels that can be multiplexed in a given bandwidth

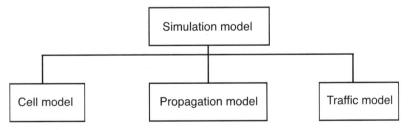

Figure 4.10 Simulation model components.

is equal to the ratio of the spreading code chip rate to the data rate. For a chip rate of 3.84 Mcps and data rate of 384 Kbps, for example, this would result in 10 possible CDMA channels per carrier—theoretically. Beyond this limit the multiplexing capability of the system collapses catastrophically unless special techniques are used such as multiuser detection. In practice, especially for a multicell microcellular mobile communications system, the situation is far from ideal.

It may well be close to impossible to actually determine the number of possible traffic channels per RF channel. However, simulation techniques can be utilized to find a rough estimate based on orthogonality factor α, which is given by [6]:

$$\alpha = 1 - \frac{E_b}{I_0}\left(\frac{E_b}{N_0}\right)^{-1} \tag{4.1}$$

where I_0 is intracell interference and N_0 is intercell interference. The block diagram of such a downlink simulation is shown in Figure 4.11.

From such simulations that can be found in the literature, the orthogonality factor is found to be equal to 0.7 in the microcellular environment. From other spectrum efficiency simulations for the downlink utilizing the CODIT microcell channel model, the corresponding spectrum efficiency is found to be 300 Kbps/MHz/carrier. For a 5-MHz bandwidth, this is 1.5 Mbps/carrier, which is equal to 3.9 users with 384-Kbps service. In our calculation five simultaneous users are assumed in order to count for the pilot, synchronization, and paging channels in the downlink. These simultaneous users are distributed arbitrarily throughout a given cell. For the purpose of calculating the required transmission (Tx) power per channel, the worst situation is assumed. All users are situated at the edge of the cell, hence requiring maximum transmission power. Therefore:

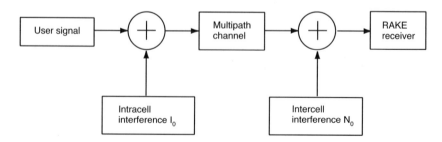

Figure 4.11 Block diagram of a downlink simulation.

Tx power per channel = 5 (average transmitter power per traffic channel)

or, in decibels,

$$\text{Tx power per channel} = P_{\text{TX,avg}} + 7$$

The resulting figures are 19.34 and 38.05 dBm for the urban and suburban scenarios, respectively.

4.9 Spectral Efficiency, Power Level, and Projected Number of Users

In microcellular systems the number of users N is given by [6]:

$$N = F\left(\frac{G_p\left(\frac{E_b}{N_0}\right)^{-1} - (\beta - 1)}{1 - F\beta}\right) \tag{4.2}$$

where F is the fraction of the intracell interference to the total interference, G_p is the processing gain of the link (= 10 dB in the 384-Kbps case), β is the multiuser detection efficiency (= 0% if no multiuser detection), and E_b/N_0 is the energy per bit to power spectral density ratio.

In the case of RoF the noise power spectral density will increase by an amount that is proportional to the optical link NF, forcing a higher required E_b (power level). In our calculations the difference in E_b between the wireless and RoF systems is less than 1 dB.

4.9.1 Network Capacity

The capacity is strongly dependent on the radio environment, which is defined by the path loss attenuation factor, shadowing, and wideband channel. The capacity gain depends on the intracell to total interference ratio (F), as in (4.2).

4.9.2 Dynamic Range

In our opinion we anticipate that the dynamic range required for the WCDMA system is much lower than the required dynamic range for the

GSM system. This is due to the high-performance power control system in the WCDMA. This system is much faster (1,500 updates per second) and more comprehensive than any other power control system. In addition, the required power level in WCDMA is much lower than in the GSM system.

4.10 WCDMA-Based RoF System Performance with Multiple-User Interference

This section is intended to investigate the effect of voice activity monitoring on multiple-user interface, intermodulation distortion, and clipping noise power and the overall performance of WCDMA RoF systems in multiple-service transmission. In addition, this section provides a simple analytical model for estimating intercell interference power density (I_{ter}), using minimum path attenuation criteria. The model proposed in this section is based on a valid assumption that poses much less computational complexity compared with other models used elsewhere in the literature.

4.10.1 WCDMA RoF System Configuration

The uplink of the WCDMA RoF system configuration is shown in Figure 4.12. It consists of the radio link, which is based on the WCDMA air interface, and the optical link. A detailed description of the uplink WCDMA

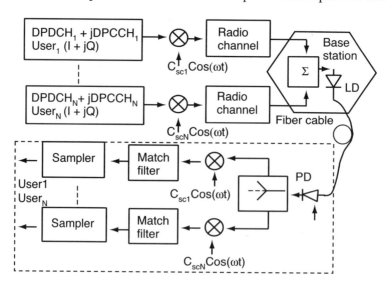

Figure 4.12 WCDMA RoF system. (LD: laser diode; PD: photodiode.)

system is not given in this chapter, but can be found in [6, 7]. Basically, the WCDMA is a multiple-service and multiple-rate system. Thus, each user can transmit a number of different services with variable data rates on dedicated physical data channels (DPDCH), plus a common dedicated physical control channel (DPCCH). The data symbols on DPDCH and DPCCH are independently multiplied (spread) with an orthogonal variable spreading factor on the so-called I and Q branches. The resultant signals on the I and Q branches are further multiplied by a user-unique but complex-valued long/short scrambling code C_{sc}, where I and Q denote real and imaginary parts, respectively. The complex-valued chip sequence (i.e., 3.84 Mcps) generated by the spreading process is QPSK modulated and transmitted on carrier frequency ω_c, in the radio channel.

By neglecting multipath propagation effects on the transmitted signals, we can express the total received signal at the base station from N users as

$$I(t) = \sum_{i=1}^{N} \sum_{s}^{M} \varphi_s^i d_s^i(t - \tau_{i,s}) C_{sc,i}(t - \tau_i)\sqrt{2P_i}\, \cos[\omega_c(t + \tau_i) + \theta_i]$$

$$(4.3)$$

where M is the number of ith user's multiple services, φ_s^i is is the weighted gain factor on each multiple service channel, P_i and θ_i are the carrier power and phase, respectively, of the ith user. Also, θ_i is assumed to be uniformly distributed in $[0, 2\pi]$. τ_i, $\tau_{I,s}$ are random variables that model the asynchronous transmission of users and their respective service type signals to the base station, respectively, which are also assumed to be uniformly distributed in $[0, T]$. $d_s^i(t) = \sum_{j=-\infty}^{\infty} d_{s,j}^i P_T(t - jT)$ is the data stream of the ith user's sth service type. The term $d_{s,j}^i$ takes the binary values ± 1 with equal probability in the interval $[jT, (j + 1)T]$, where T is the symbol period and P_T is a rectangular pulse, which equals unity for $0 \leq t \leq T$ and zero otherwise. $C_{sc,i}(t)$ and $C_{ch,s}^i(t)$, the orthogonal variable spreading factors, are given here, respectively, as

$$C_{sc,i}(t) = \sum_{l=-\infty}^{\infty} c_{sc,i,l} P_{T_c}(t - lT_c) \qquad (4.4)$$

$$C_{ch,s}^i(t) = \sum_{l=-\infty}^{\infty} c_{ch,s,l}^i P_{T_c}(t - lT_c) \qquad (4.5)$$

where $C^i_{ch,s,l}$ and $C_{sc,i,l}$ are real and complex-valued chip sequences that take values ±1 in the interval $[lT_c, (l+1)T_c]$, and $P_{T_c}(t)$ is a rectangular pulse, which equals unity for $0 \leq t \leq T_c$, and zero otherwise. T_c is the chip rate. With a spreading factor (i.e., processing gain) of G, $T = GT_c$. At the base station the signal in (4.3) is used to direct modulate the optical intensity of the laser diode (i.e., electrical-to-optical conversion) and transmitted through the fiber cable to the central base station for processing. The laser diode nonlinearity can be modeled by a third-order polynomial without memory to give the output optical signal as [2, 19, 20]:

$$P(t) = P_t[1 + I'(t) + a_2 I'^2(t) + a_3 I'^3(t)] + \eta(t) \qquad (4.6)$$

Here P_t is the average transmit optical power, $I'(t) = I(t)/(I_b - I_{th})$, where I_b and I_{th} are the laser-diode bias and threshold current, respectively; a_2 and a_3 are the second- and third-order coefficients, respectively; and $\eta(t)$ is background noise.

4.10.2 Intermodulation Distortions and Clipping Noise

The CDMA system's capacity is known to be interference limited; therefore, any reduction in interference will lead to direct increase in capacity. In WCDMA RoF systems, apart from intracell interference power (I_{tra}) and intercell interference power (I_{ter}), another source of interference is due to laser diode nonlinearity. The laser diode nonlinearity causes mixing of users' signals, resulting in generation of harmonics and intermodulation distortions. Because the second-order intermodulation terms generate zero frequency and the double-frequency components, and the third-order term generates the common frequency and the triple-frequency components, only one harmonic of the third-order term influences the system performance (i.e., $2f_1 - f_2 = f$, since $f_1 = f_2 = f$) [8]. Therefore, the intermodulation distortion (IMD) noise can be assumed to be part of the multiple-user interface. As a result the total intracell interference power can be given as $I_{traT} = I_{tra} + \sigma^2_{imd}$. An expression for third-order intermodulation distortion power of random sequence CDMA in optical transmission systems has been derived in [10] as:

$$\sigma^2_{imd} = \frac{(\rho P_r m)^2}{2}\left\{\frac{a_3 m^2(2N-1)(N-1)}{2G} + \frac{a_3^2 m^4(N-1)}{64}\right.$$
$$\left.\left(9 + \frac{126N^2 + 30N - 264}{5G} - \frac{15(N-2)}{G^2}\right)\right\} \qquad (4.7)$$

where ρ is photodiode responsivity, m is the optical modulation index, and P_r is the average received optical power. When considering a perfect power control, I_{tra} can be given as [10]:

$$I_{tra} = (\rho m_0 P_r)^2 \frac{N-1}{12G} \qquad (4.8)$$

In [5] the expression for the variance of clipping impulsive noise was assumed to be

$$\sigma_{clip}^2 = \frac{I_p^2 m^6 N^3}{27.2} \exp\left(-\frac{1}{2m^2 N}\right) \qquad (4.9)$$

4.10.3 Intercell Interference Model

In cellular systems even cell ownership for users is difficult to determine due to shadowing effects. As a result two types of intercell interference scenarios may be possible, as shown in Figure 4.13: interference from mobiles located in their home cells to adjacent cells, and interference from mobiles located in one cell but communicating with a base station in another cell. This could happen, for instance, during handover initiation. Assuming perfect power control, the power control algorithm is such that the transmitting power of mobile unit m2 in Figure 4.13 to the base station B2 will be

$$S_t = S_r r^\beta e^{-\xi \ln(10)/10} \qquad (4.10)$$

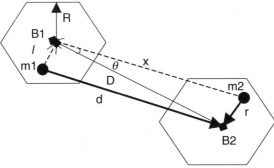

Figure 4.13 Intercell interference scenarios. The bold lines and broken lines indicate the desired links and interference links, respectively.

where S_r is the desired receive power at B2, r is the distance from the m2 to B2, β is the path-loss exponent, and ξ is a Gaussian random variable with zero mean and standard deviation σ, representing the log-normal shadowing. Supposed that ζ is the shadowing effect on the path m2 to B1, then the normalized interference power generated at B1 due to m2 will be given as

$$\frac{I_{\text{ter}}}{S_r} = y^{-\beta} e^{-\zeta \ln(10)/10} < 1 \tag{4.11}$$

Otherwise, m2 should switch its transmission to B1. We also assumed a circular cell structure with uniform distribution of users in the cells and user density/cell given as

$$\eta = N / \pi R^2 \tag{4.12}$$

where R is the cell radius. The total normalized intercell interference power is obtained by integrating over the total area of the interfering cells:

$$I_{\text{ter}}/S_r = \eta \alpha q \int\int (y)^{-\beta+1} e^{-\zeta \ln(10)/10} dA \tag{4.13}$$

$$= 2\eta \alpha q \int_0^{(D+R)} \int_0^{\pi} (y)^{-\beta+1} e^{-\zeta \ln(10)/10} dx d\theta$$

where q is the number of interfering cells.

The expected intercell interference power can be expressed similarly to [9] as

$$E\left(\frac{I_{\text{tra}}}{S_r}\right) = 2\eta \alpha q \int_0^{(D+R)} \int_0^{\pi} (x)^{-\beta+1} E[e^{-\zeta \ln(10)/10} f(x, \zeta)] dx d\theta \tag{4.14}$$

From the condition in (4.11) we can assume $\zeta = 40 \log(y)$ as the worst case interference. Therefore,

$$f(\zeta, y) = \begin{cases} 1 & \text{if } \zeta \le 40 \log(y) \\ 0 & \text{otherwise} \end{cases} \tag{4.15}$$

$$E[e^{-\zeta \ln(10)/10} f(\zeta, y)] = 1/\sqrt{2\pi}\sigma \int_{-\infty}^{40 \log(y)} e^{-\zeta \ln(10)/10} e^{-\zeta^2/2\sigma^2} d\zeta$$

where

$$= \frac{e^{[\sigma \ln(10)/10]^2}}{\sqrt{2\pi}\sigma} \int_{-\infty}^{40 \log(y)} Exp\left[-1/2(y/\sigma - \sigma \ln(10)/10)^2\right] dy$$

$$\quad (4.16)$$

$$= e^{(\sigma \ln(10)/10)^2} \left\{ 1 - \frac{1}{2}\left[1 - \text{erf}\left(\frac{40 \log(y)}{\sigma\sqrt{2}} - \frac{\sigma \ln(10)}{10\sqrt{2}}\right)\right]\right\}$$

where erf$(\cdot)$ is the error function. Substituting (4.16) in (4.14), the resulting equation can be integrated numerically.

4.10.4 System Performance

In WCDMA RoF systems the transmission quality specification can be expressed in terms of energy per bit over the total interference and noise power density (ENR) of the dedicated traffic channel as

$$\frac{E_b}{N_t} = \frac{G(m\rho P_r)^2/2}{I_{tra}(1 + I_r)\alpha + \eta} \quad (4.17)$$

where $\alpha = 3/8$ is the voice activity factor; I_r is the intercell interference factor, defined as the ratio of I_{ter} and I_{tra} are background noise, η in RoF systems include RIN I_p^2, laser relative intensity noise (taken to be -155 dB/Hz); $2eI_p$, shot noise due to photodetector; σ_{clip}^2, clipping noise variance; σ_{imd}^2, intermodulation noise variance; and $<I_t^2>$, thermal noise equivalent power referred to the optical receiver front end ($I_t = 10$ pA/$\sqrt{Hz}$) [2]. Thus, $\eta = [(\text{RIN} \cdot I_p^2 + 2eI_p + <I_t^2>)B + \sigma_{imd}^2 + \sigma_{clip}^2]$, with $B = 5$ MHz as the bandwidth and $I_p = \rho P_t 10 - \gamma L/10$, where $P_t = 5$ mW is the average transmitting optical power, $\gamma = 0.2$ dB/km is fiber loss [21], and $L = 10$ km is the fiber length used in this analysis. When using voice activity monitoring, the probability that k out of n interfering users are active can be described by a binomial distribution [8, 9]:

$$P(n, k) = \binom{n}{k} \alpha^k (1 - \alpha)^{n-k} \qquad (4.18)$$

Note that here n represents the number of voice channels out of the total number of multiple service channels, N (voice + data). Now, considering the effect of voice activity monitoring on intermodulation distortion and clipping noise power, (4.17) can be rewritten as

$$\frac{E_b}{N_t} = \frac{G(m_o \rho P_r)^2/2}{\{[(I_{tra}T + \sigma^2_{imd}) + \sigma^2_{clip}](I + I_r)\}\delta + [\text{RIN} \cdot I_p^2 + 2eI_p + <I_{th}^2>]B} \qquad (4.19)$$

Here $\delta = [\alpha + (1 - \alpha)x]$ is a factor that accounts for the effective voice activity factor due to multiple-service transmission, where $0 \leq x \leq 1$ is the percentage ratio of data channels out of the total number of service channels, N.

4.10.5 Results and Discussions

The results presented in this analysis are based on a three-tier ($q = 36$ cells) microcellular system with $R = 100$m, $\sigma = 10$ dB, and $\beta = 4$, giving $I_r = 0.1$. The ENR dependence on the number of data channels x and modulation index m, with and without the voice activity effect on intermodulation distortion and clipping noise, is shown in Figures 4.14 and 4.15, respectively. We also assumed that the data channels, unlike the voice channel, are transmitted continuously. We can see from Figure 4.14 that up to a 4.5-dB improvement in ENR performance can be achieved when taking into account the effect of voice activity on intermodulation distortion and clipping noise. It is worth noting, however, that the performance decreases with decreasing m and increasing x, whereas in Figure 4.15 the ENR remains constant, irrespective of x, at high values of m. The effect of voice activity becomes perceptive only at low m. The reason for this behavior is that, at $x = 0$, although only voice channels are being transmitted, the effect of voice activity on the multiple-user interface is being compensated for by high intermodulation distortion and clipping noise due to a high modulation index. Also, at $x = 1$ there is no voice activity monitoring or only data channels are being transmitted, but a high modulation index with high intermodulation distortion and clipping noise generally results in a low ENR. Clearly, it can be seen that this noise compensation behavior changes with decreasing modulation index.

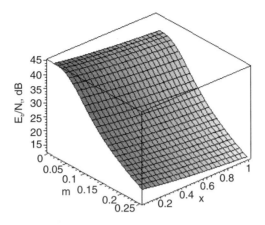

Figure 4.14 ENR dependence on *m* and *x* for *N* = 30 with voice activity monitoring effect on intermodulation distortion and clipping noise at *G* = 256.

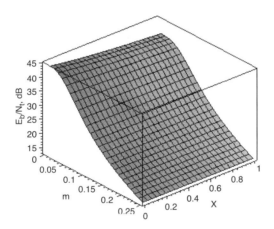

Figure 4.15 ENR dependence on *m* and *x* for *N* = 30 without voice activity monitoring effect on intermodulation distortion and clipping noise at *G* = 256.

It is obvious from Figure 4.15 that, in WCDMA RoF systems voice activity monitoring becomes ineffective when operating at a high modulation index if we ignore the effect of voice activity monitoring on intermodulation distortion and clipping noise. These results can be misleading because they can lead to overestimation of ENR requirements for voice and data services, which could consequently impact the link budget.

Results of ENR dependence on processing gain G and x, with and without voice activity effect on intermodulation distortion and clipping noise, are shown in Figures 4.16 and Figure 4.17, respectively. We can see from Figure 4.16 that ENR decreases with decreasing G and increasing x, whereas in Figure 4.17, again the ENR increases with G, but remains constant with respect to x at $m = 0.25$ for the same reasons given above.

4.11 Radio over Fiber for HiperLAN2 Microcellular Communication Networks

Today's typical local area network (LAN) environment requires costly planning and investments to build and maintain. Data exchange between computers occurs at present over wireline LANs. Wireless technology will make it less demanding to move computer equipment to new locations that lack easy access to LAN cables. Wireless technology has enjoyed increased demand from the general public as well as from business and other professional users. The demands for the wireless technology range from use in cellular phones

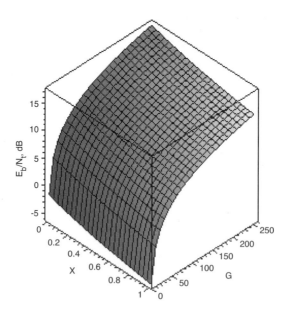

Figure 4.16 ENR dependence on G and X for $N = 30$ with voice activity monitoring effect on intermodulation distortion and clipping noise at $m = 0.25$.

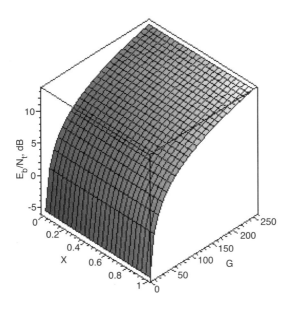

Figure 4.17 ENR dependence on *G* and *X* for *N* = 30 without voice activity monitoring effect on intermodulation distortion and clipping noise at *m* = 0.25.

to high-speed digital networks supporting high-speed computer communications. At present, standards for broadband wireless multimedia communications in the 5-GHz band are being developed in Europe as well as in the United States and Japan. HiperLAN2 is an upcoming standard that is being specified by the ETSI/BRAN project. HiperLAN2 is one of the wireless broadband access networks and will provide high-speed communications with a bit rate of at least 20 Mbps between mobile terminals and various broadband infrastructure networks.

Currently a strong alignment exists among the three standardization bodies, IEEE 802.11a (United States), ETSI/BRAN (Europe), and MMAC (Japan) on the physical layer. This agreement provides a worldwide platform for broadband wireless multimedia communications, whereas the upper layer protocols are different. The three systems in Europe, the United States, and Japan will operate at the 5-GHz band. In addition, all have adopted orthogonal FDMA with 64 subcarriers, where 48 subcarriers are used for data and 4 are used for pilot signals (which eases phase tracking for coherent demodulation); the remaining 12 subcarriers are set to zero [22–27].

HiperLAN2 is intended to provide local wireless access to, for example, IP, ATM, and UMTS infrastructure networks by using both moving and stationary terminals that interact with access points, which, in turn, are usually connected to an IP, ATM, or UMTS backbone network, as shown in Figure 4.18. A typical HiperLAN2 system consists of a number of access points connected to a backbone network, for example, an Ethernet LAN. An access point can use an Omni antenna, a multibeam antenna, or a number of distributed antenna elements. The system supports mobility between access points on the same backbone network; that is, handover is made between access points. Figure 4.18 shows the HiperLAN2 system architecture. Mobile terminals are associated with access points, where each access point is connected to a backbone network; for this example it is an Ethernet backbone [26–28; see also http://ieeexplore.ieee.org/lpdocs/epic03/, http://www.hiperlan2.com/web/, and http://www.networkcomputing.com]. Such wireless access networks will be able to provide the quality of service that users expect from a wired IP or ATM network.

The introduction of the fiber into HiperLAN2 will mean that the antenna of the HiperLAN2 access point could be distributed, which leads to an increased number of coverage areas and, hence, increased capacity and improved performance from a single access point. The coverage areas of both systems (with and without the fiber) will be different; in fact, the system with the fiber link will have less coverage area per antenna, but the area per access point will be larger, as shown in Figure 4.19.

As the mobile terminal moves out from the serving access point toward the neighboring cell, it will establish a second parallel link, at a different

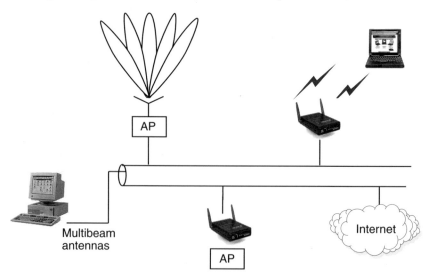

Figure 4.18 HiperLAN2 system overview.

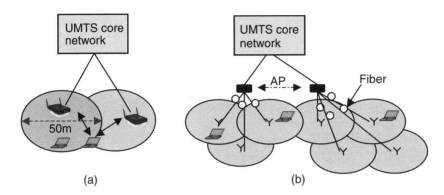

Figure 4.19 HiperLAN2 system model: (a) system without the fiber and (b) system with the insertion of the fiber.

time slot allocation, from the ongoing access point before releasing the old link with the previously serving access point [http://www.hiperlan2.com/web, http://www.networkcomputing.com, http://www.telecomnames.com/, and http://www.nokia.com/corporate/wlan/index.html]. At the moment when new parallel link is received, the mobile terminal can then communicate with both access points, receiving data from the two access points, by allocating time slots by each access point. The phenomenon of receiving the data from two neighboring access points is called *macrodiversity*. The advantage that can be obtained from macrodiversity is the increase of the S/N at the single access point, resulting in a corresponding capacity increase.

The structure of the HiperLAN2 system without the fiber is shown in Figure 4.19(a). The problem with this structure is that, as the mobile terminal communicates with two access points, the signals cannot be combined due to the separation of the access point's transceivers. If the fiber is included, the antennas can be distributed and controlled by a single access point. This is a great advantage of the mobile terminal communicating with two access points; therefore, a maximum ratio combining could be obtained in the uplink to improve system capacity. More investigations are under way to forward our understanding of the requirement for this 4G technology.

4.12 Conclusions

The general concepts of CDMA/WCDMA radio over fiber systems have been presented. A simple simulation model of the downlink segment of a

WCDMA radio over fiber system was presented using SystemView. A definition for the macrodiversity architecture and radio over fiber was also presented. The concepts of the cellular architecture and the UMTS were explained along with the role of the macrodiversity function in the UMTS architecture. A model for the macrodiversity function and algorithms has been developed.

A simulation model of the multipath WCDMA using SystemView to investigate the performance of the macrodiversity function was developed. An analytical model for evaluating the performance of WCDMA-based radio over fiber systems has been presented. We have shown that up to a 4.5-dB improvement in ENR can be achieved if we take into account the effect of voice activity monitoring on intermodulation distortion and clipping noise due to laser diode nonlinearity. The performance variation with respect to modulation index and processing in multiple-service transmission gain has also been investigated. The results are useful for determining the optimum modulation index and for conducting capacity analyses and link budget estimations. Finally, the 4G applications of radio over fiber with HiperLAN2 were presented.

References

[1]   Kajiya, S., K. Ksukamoto, and S. Komaki, "Proposal of Fiber-Optic Radio Highway Networks Using CDMA Method," *IEICE Trans. on Electronics,* Vol. E79-C, No. 1, Jan. 1996, pp. 496–500.

[2]   Koshy, B. J., and P. M. Shankar, "Spread-Spectrum Techniques for Fiber-Fed Microcellular Networks," *IEEE Trans. on Vehicular Technology,* Vol. VT-48, No. 3, May 1999, pp. 847–857.

[3]   Dahlman, E., et al., "UMTS/IMT 2000 Based on Wideband CDMA," *IEEE Communications Magazine,* Sept. 1998, pp. 70–80.

[4]   Adachi, F., M. Sawahashi, and H. Suda, "Wideband DS-CDMA for Next-Generation Mobile Communications Systems," *IEEE Communications Magazine,* Sept. 1998, pp. 40–45.

[5]   Betti, S., et al. "Effect of Semiconductor Laser Intrinsic Dynamic Distortions in Hybrid Fiber-CDMA Cellular Radio Networks," *Microwave and Optical Technology Lett.,* Vol. 20, No. 3, Feb. 1999, pp. 207–211.

[6]   Holma, H., and A. Toskala (Eds.), *WCDMA for UMTS,* New York: Wiley, 2000.

[7]   Ojanpera, T., and R. Prasad, *Wideband CDMA for Third Generation Mobile Communications,* Norwood, MA: Artech House, 1998.

[8]   Jansen, M. G., and R. Prasad, "Capacity, Throughput, and Delay Analysis of a Cellular DS CDMA System with Imperfect Power Control and Imperfect Sectorization," *IEEE Trans. on Vehicular Technology,* Vol. VT-44, No. 1, Feb. 1995, pp. 67–75.

[9]   Gilhousen, K. S., et al., "On the Capacity of a Cellular CDMA System," *IEEE Trans. on Vehicular Technology,* Vol. VT-40, No. 2, May 1991, pp. 303–312.

[10] Huang, W., and M. Nakagawa, "Nonlinear Effect of Direct-Sequence CDMA in Optical Transmission," *IEICE Trans. on Communications,* Vol. E78-B, May 1995, pp. 1185–1189.

[11] Cutrer, D. M., et al., "Dynamic Range Requirements for Optical Transmitters in Fiber-Fed Microcellular Networks," *IEEE Photonics Technology Lett.,* Vol. 7, No. 5, 1995, pp. 564–566.

[12] Way, W. I., "Optical Fiber-Based Microcellular Systems: An Overview," *IEICE Trans. on Communications,* Vol. E76-B, No. 9, 1993, pp. 1091–1102.

[13] Chu, T. S., and M. J. Gans, "Fiber-Optic Microcellular Radio," *IEEE Trans. on Vehicle Technology,* Vol. VT-40, No. 3, 1991, pp. 599–607.

[14] Tripathi, N. D., J. H. Reed, and H. F. Vanlandingham, "Handoff in Cellular Systems," *IEEE Personal Communications,* Dec. 1998, pp. 26–37.

[15] Chambert, G., "Uplink Macrodiversity Method and Apparatus in a Digital Mobile Radio Communication System," U.S. Patent 5,867,791, 1999.

[16] Kim, J. Y., G. L. Stuber, and I. F. Akyildiz, "Macrodiversity Power Control in Hierarchical CDMA Cellular Systems," *49th Vehicular Transport Conf. VTC,* Holland, May 1999, pp. 279–285.

[17] Weiss, U., "Improving Coverage and Link Quality by Macroscopic Diversity," *Proc. EPMCC'99—3rd European Personal Mobile Communications Conference,* Paris, France, March 9–11, 1999, pp. 268–273.

[18] Third Generation Partnership Project, Technical Specification Group Radio Access Network, "Physical Channels and Mapping of Transport Channels onto Physical Channels (FDD)," 3G TS 25.211 V3.2.0 (2000–03), released 1999.

[19] Daher, A. M., and H. S. Al-Raweshidy, "Design Considerations of Radio on Fiber for Microcellular CDMA Systems," *Proc. Third Communication Networks Symp.,* Manchester, England, July 8–9, 1996, pp. 94–97.

[20] Al-Raweshidy, H. S., and R. Prasad, "Spread Spectrum Technique to Improve the Performance of Radio over Fiber for Microcellular GSM Networks," *Wireless Personal Communications,* Special Issue on Radio over Fiber Systems Technologies and Applications, Vol. 14, Sept. 2000, pp. 133 145.

[21] Schuh, R. E., and D. Wake, "Distortion of W-CDMA Signals over Optical Fiber Links," *IEEE MWP99 Digest-Post-Deadline Session,* 1999, pp. 9–12.

[22] ETSI, Broadband Radio Access Networks (BRAN), "HiperLAN Type 2; Data Link Control (DLC) Layer; Part 2: Radio Link Control (RLC) Sublayer," TS 101 761-2, April 2000.

[23] Edfors, O., and M. Sandell, "OFDM Channel Estimation by Singular Value Decomposition," *IEEE Trans. on Communications,* Vol. 46, No. 7, July 1998, pp. 931–936.

[24] ETSI, Broadband Radio Access Networks (BRAN), "HiperLAN Type 2: Physical (PHY) Layer," TS 101 475, April 2000.

[25] Universal Mobile Telecommunications System, "RRC Protocol Specification," GPP TS 25.331, version 3.4.1, released 1999; ETSI TS 125 331.

[26]   Torsner, J., and G. Malmgren, "Radio Network Solutions for Hiperlan/2," *IEEE Communications Magazine,* Feb. 1999, pp. 1217–1221.

[27]   ETSI, Broadband Radio Access Networks (BRAN), "HiperLAN Type 2; Data link Control (DLC) Layer," ETSI TS 101 761–1, v1.2.1, Technical Specification.

[28]   Mohr, W., "Broadband Radio Access for IP-Based Networks in the IST BRAIN," presented at IEEE ICT 2000 Conf., Acapulco, Mexico, May 22–25, 2000.

5

Radio over Fiber Systems for Mobile Applications
David Wake

5.1 Introduction

The use of RoF technology for cordless or mobile communications systems was first proposed and demonstrated in 1990 by Cooper [1]. Since then the international research community has spent much time investigating the limitations of RoF and trying to develop new, higher performance RoF technologies. Many laboratory demonstrations and field trials have been performed, but currently RoF still remains a niche application within the broad remit of optical fiber technology. RoF is fundamentally an analog transmission system because it distributes the radio *waveform,* directly at the radio carrier frequency, from a central unit to an RAU. Note that although this transmission system is analog, the radio system itself may be digital (such as GSM).

Further details on the RoF system, together with its advantages, are given in later sections, but it is appropriate at this stage to compare it with mainstream optical fiber technology. Mainstream optical fiber technology is digital. Telecommunication networks use synchronous digital hierarchy transmission technology in their cores. Fiber-based data networks (such as fiber distributed data interface and gigabit Ethernet) all use digital transmission. Fiber transmission links to base stations in mobile communications systems are digital. Digital optical fiber transmission links are therefore ubiquitous in telecommunications and data communications, constituting a high-volume market worth billions of dollars worldwide. This high volume drives down cost, which means that digital transmission technology is relatively cheap. By contrast, analog optical fiber links are low-volume, niche products, used for applications such as satellite Earth station remoting and the cellular radio antenna feeding that is the subject of this chapter. Furthermore, analog transmission suffers from impairments such as noise and distortion that degrade the signal and cannot be regenerated. Special low-noise, low-distortion components (particularly the optical source) must be designed and used for acceptable performance in many cases.

With their relatively high cost and the need for careful engineering, why are analog links considered at all? The reason is that they add flexibility, antenna site simplicity, and tailored coverage to radio networks (again, this is discussed more fully later). The balance between these advantages and the high system cost has been shifting recently with the advent of low-cost products aimed at the cellular communications market. These products typically use cheap components developed for digital applications but are engineered in such a way that they meet the more stringent performance specifications for analog applications. Several companies are now competing in this market and the principal application is in-building distributed antenna systems for large corporate office buildings, shopping malls, airports, and so forth. Volumes are increasing and this market is expected to be a major growth area for cellular operators over the coming years as they compete ever harder for customers. More details on this are given in Chapters 3 and 4.

This chapter discusses the use of RoF for cellular in-building distributed antenna systems. It looks at the problems associated with providing cellular coverage within buildings and assesses the competing technologies and architectures that can be used. Both current and emerging RoF technologies are included and the suitability of these systems as we evolve from 2G to 3G cellular networks is discussed.

5.2 Cellular Communications

Before discussing RoF systems, technologies, and architectures for cellular communications, it is worthwhile to briefly describe some of the principles and systems of cellular radio. The purpose of this section is to introduce the cellular concept and some of the systems in use today and the near future.

In these systems, the coverage area is divided into cells and access to the network is achieved via an RBS located in each of these cells. Cellular systems allocate radio communication channels by assigning a unique combination of carrier frequency and time slot or code. Frequency spectrum is a very precious and scarce resource, and cellular systems are given very small amounts because of competition with many other users. To use this small allocation of spectrum efficiently means that the same spectrum must be used over and over again, hence the cellular architecture. The other reason for dividing a large area into smaller cells is the reduced base station and mobile terminal power requirements that result from the shorter transmission distances.

The most widespread cellular system in use today is Global System for Mobilecommunications (GSM). GSM was used by more than 350 operators in 143 countries with a user base of 275 million subscribers as of February 2000. Each frequency carrier is divided into eight time slots, and each time slot constitutes a communication channel in a scheme known as TDMA. The link directions are differentiated in terms of frequency; that is, the downlink channels are in one frequency band and the uplink channels are in another frequency band, referred to as FDD.

GSM uses a carrier spacing of 200 kHz, so an operator with 25 MHz of spectrum will have 125 frequency carriers. Without frequency reuse, there would only be 1,000 simultaneous channels in total (125 frequencies × 8 time slots). If the coverage area is split into a large number of relatively small cells, then the same carriers can be reused in many cells as long as the same frequencies are not used too close together (because they would cause mutual interference). Frequencies are allocated in a repeat pattern to reduce this interference to acceptable levels. This concept is illustrated in Figure 5.1. The cells are represented as hexagons in a cluster with seven different frequencies, so in this example the reuse factor is 7. The shaded cells, with frequency 2, are separated by a certain minimum distance that is required to give acceptably low interference levels. In practical networks, a reuse factor of around 15 or even higher is often required to achieve acceptable interference.

When a cellular system is first deployed, the main goal of the network is to achieve a large coverage area. This is done by installing high-power base stations on tall masts inside relatively large cells (macrocells). This is possible because there are not many users of the system and therefore capacity is not a major issue. As the network matures and more users join, capacity starts to become the dominant factor. To increase capacity, each large cell is split into several smaller cells so that the frequency carriers can be reused. In densely populated areas such as city centers, the cells must be very small

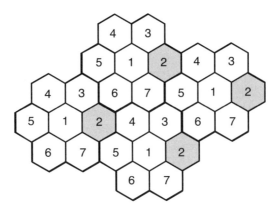

Figure 5.1 Schematic diagram illustrating the principle of cellular communications. Frequencies are reused in this case in a repeating pattern of 7.

(a few hundred meters) to cope with the demand. These cells are known as microcells and are typified by small base stations mounted below roof height on buildings. During the last few years, coverage inside buildings has become more of an issue and many base stations have been installed within the building itself to achieve the required coverage and capacity. These cells are known as picocells.

GSM is an example of a 2G cellular communication system and operates at a frequency of around 900 or 1,800 MHz (1,900 MHz in the United States). The 2G systems are characterized by digital transmission, which gives many advantages such as higher quality and security and the option of including data transmission as well as voice. By contrast, older first-generation (1G) systems, such as Total Access Communication System (TACS) in the United Kingdom, used analog transmission for speech. Data rates for 2G systems are usually less than 10 Kbps, which is not adequate for many of the multimedia applications in use today.

To address this problem, standards bodies such as the ITU and ETSI are defining new 3G systems that will be able to provide data rates of a few hundred kilobits per second to users and will operate at frequencies of around 2 GHz. These systems will use more advanced radio interfaces involving WCDMA as the means of addressing each user. In these systems, such as the UMTS, the frequency carriers have a bandwidth of around 4 MHz within a 5-MHz spacing and can therefore provide total bit rates of more than 1 Mbps. (More details are given in Chapter 4.) In WCDMA, each cell can use the same carrier frequency since the users are differentiated by code, which means that frequency planning becomes much less important than with GSM, for example.

For WCDMA systems to work efficiently, each mobile terminal must control its transmitter power so that all of the power levels received at the base station are equal. This has important implications for RoF since the optical link will not need to have a large dynamic range. RoF systems may be better suited to 3G systems, compared to 2G. The 3G networks will be deployed within the next couple of years around the world so it is important to understand now the performance and architecture issues associated with using RoF systems for distributing the WCDMA signals within buildings. This is discussed in more detail in Section 5.6.

5.3 What Is RoF?

RoF uses highly linear optical fiber links to distribute RF signals from a central location to RAUs. This allows the RAUs to be extremely simple since they only need to contain optoelectronic conversion devices and amplifiers. Functions such as coding, modulation, multiplexing, and upconversion can be performed at a central location. A simple RAU means small and light enclosures (easier and more flexible installation) and low cost (in terms of equipment cost and maintenance costs). Centralization results in equipment sharing, dynamic resource allocation, and more effective management. All of this adds up to an access technology that makes life easier and cheaper for operators.

The reason why RoF is able to shift system complexity away from the antenna is that optical fiber is an excellent low-loss (0.2 dB/km optical loss at 1,550 nm), high-bandwidth (50-THz) transmission medium. Transmission takes place at the radio carrier frequency rather than the more conventional digital baseband systems. The optical links in RoF are therefore analog in nature, in that they reproduce the carrier waveform. The radio carrier can, of course, be modulated with a digital modulation scheme such as GMSK (as in GSM) or QPSK (as in UMTS).

Figure 5.2 shows the layout of a simple bidirectional directly modulated RoF link. In each direction the input RF signal is applied to a laser diode where it modulates the intensity of the output light. In most cases this light will have a wavelength of either 1,300 or 1,550 nm for low transmission loss in silica fiber. The fiber may be multimode or single mode, although the latter is preferred for link spans of more than a few tens of meters for its low dispersion properties. The optical receiver usually consists of a p-i-n photodiode, which provides an RF power output proportional to the square of the input optical power. This type of optical link is known as intensity

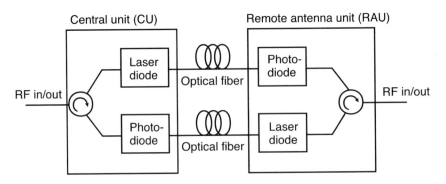

Figure 5.2 Typical layout of a bidirectional analog optical link using direct modulation of laser diodes.

modulated-direct detection (IM-DD). Other types of link are possible involving frequency or phase modulation, but for cellular applications the IM-DD links are used for reasons of simplicity and cost.

As a result of the analog nature of the optical links, they suffer from impairments such as noise and distortion unless great care is taken in their design. These impairments tend to limit the dynamic range of the links. The dominant noise source in the laser is RIN. The photodiode produces shot noise and the receiver amplifier produces thermal noise. The dominant distortion in the system comes from the laser in most cases. The most important type of distortion in narrowband systems is third-order intermodulation distortion (IM3) since the distortion products occur inside the band. The fiber impairments include chromatic dispersion and polarization mode dispersion, although the latter is not a problem for relatively short fiber spans. For systems with high optical power, the fiber and photodiode will also exhibit nonlinear behavior but this is not the usual case. A good way of characterizing an analog optical link is the two-tone intermodulation measurement. Two continuous-wave tones at frequencies f_1 and f_2 are fed into the link. The output consists of these tones in addition to the IM3 products at $2f_1 - f_2$ and $2f_2 - f_1$.

Figure 5.3 shows how the fundamental signal and IM3 signal vary with input power. The fundamental signal varies linearly with input power (slope of 1), whereas the IM3 signal varies with the cube of the input power (slope of 3). The point at which these lines meet is the third-order intercept point and is an important figure of merit for the optical link. The other important figure of merit is the SFDR, which can be defined as the range of input power bounded by the points at which the fundamental signal and the IM3 are intersected by the noise floor. Dynamic range is a very important parameter for cellular systems (especially 2G systems) due to the wide range

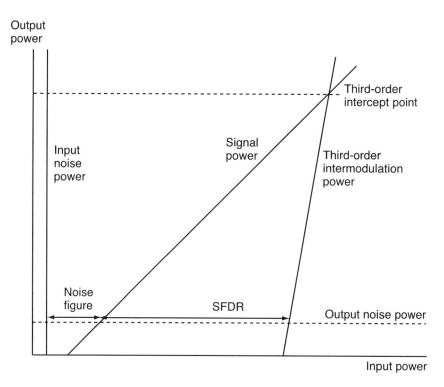

Figure 5.3 Output power as a function of input power for analog optical links, illustrating system impairments caused by noise and distortion.

of input powers that can be received at the base station from mobile terminals. For example, a mobile terminal could be very close to the base station or at the edge of the cell and the difference in received power can be greater than 80 dB.

In spite of these problems, RoF has emerged during the last few years as a commercial success for certain applications. In particular, RoF links are used increasingly for in-building coverage systems for current 2G networks in large office blocks, shopping malls, and airports, for example. These picocell and microcell applications are ideal for RoF because they do not have demanding dynamic range requirements. Figure 5.4 shows how RoF can be used for this type of application. In this case, a roof-top antenna picks up a signal over the air from a macrocell base station, which is amplified and fed into the RoF system. The RoF system converts this RF signal to light, which is distributed by the optical fibers to the RAUs. The RAUs convert the light back to the original RF signal, which is amplified and radiated into the picocells by low profile antennas.

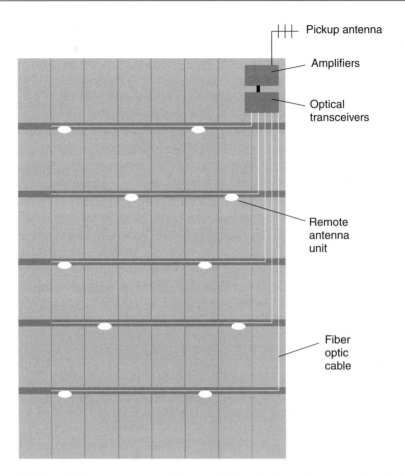

Figure 5.4 How RoF systems are used in a typical application today. (Reproduced with permission of Tekmar Sistemi S.R.L.)

A specific example of the deployment of RoF systems is in the Bluewater shopping center near London, which is the largest in Europe with a retail floor space of 1.6 million square feet. A RoF system was installed that provides service for all four U.K. cellular operators (two at 900 MHz and two at 1,800 MHz). The RoF system uses more than 10 km of single-mode fiber and has 41 antennas distributed throughout the center.

RoF systems may also be used for outdoor applications, for example, for feeding microcells, although getting access to fiber cables becomes the critical issue. Inside buildings, the fiber cables may be easily installed, for example, by the owners of the building. Installation of fiber along streets is not so easy, however, and leasing "dark fiber" (i.e., unterminated fiber) from a network operator may not be possible. These problems have led to the dominance of the indoor market for RoF systems, but this may change in the future if dark fiber becomes easier to obtain.

5.4 RoF Technologies

Products available today use simple IM-DD links for reasons of cost. During the last few years several companies have been created to serve the market for RoF cellular products. These products use either Fabry-Perot lasers or distributed feedback lasers depending on the link length and dynamic range requirements. Fabry-Perot lasers are the cheaper option and have acceptable performance for most cellular applications, especially if constructed using multiple quantum well technology. The optical receivers used for these products are standard p-i-n photodiodes.

Although lasers have been designed specifically for analog applications with low noise and low distortion [2], most current cellular products use mass-market lasers that were originally designed for digital applications in data communications and telecommunications. These mass-market devices are relatively cheap due to the high volumes produced and provide acceptable performance for all but the most demanding applications. Typical products have spurious-free dynamic range values of $100–110$ dB $\cdot$ Hz$^{2/3}$, which translates to between 65 and 75 dB for GSM. Typically, output noise is below -120 dBm/Hz for these links.

As discussed earlier, the remote antenna units in these conventional RoF systems contain a laser, photodiode, circulator, amplifiers, control circuits, and power supplies. Although these can be housed in quite small enclosures, they are reasonably complex. A recently proposed new approach replaces the laser, photodiode, and circulator with a single optoelectronic device, an electroabsorption modulator that acts as a transceiver [3]. This device consists of a semiconductor optical waveguide inside a pseudonoise (PN) junction, where the waveguide core is electroabsorptive, that is, absorption of light in the waveguide can be controlled by a dc bias voltage.

Because photocarriers are generated in this process, the device can be used as a photodiode as well as its conventional use as an optical modulator in a configuration as shown in Figure 5.5. Here the electroabsorption modulator (EAM) acts as a photodiode for the downlink and as a modulator (of the downstream light) for the uplink. The advantage of this approach is that no light source is needed at the RAU (and therefore no control circuitry), which makes it much simpler and therefore cheaper.

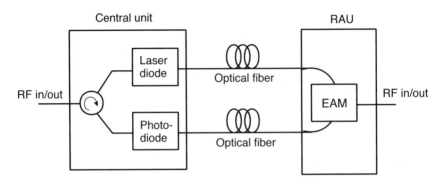

Figure 5.5 Layout of bidirectional analog optical link using an electroabsorption transceiver at the RAU.

The RAU can be made even simpler if the radio coverage is restricted to small open spaces. In this case no amplifiers are required and the electroabsorption modulator can be operated without a bias voltage so that the RAU consists only of the electroabsorption modulator and an antenna [4]. Amplifiers, control circuits, and power supplies can be removed from the RAU, which takes simplification to the limit. Because we are relying solely on the RF signal power generated by the electroabsorption modulator from the downstream light, the range of the radio link is confined to around 10–100m depending on the available optical power, propagation environment, antenna type, and radio system.

A demonstration system has been set up at BT Adastral Park using radio LAN (2.4-GHz) and digital enhanced cordless telephone (DECT) (1.9-GHz) radio systems in a layout shown in Figure 5.6. The inset photograph shows the RAU. At the central hub, DECT and radio LAN signals are multiplexed together and the composite signal modulates the intensity of the optical source. In the picocell, the electroabsorption modulator demodulates the optical carrier and the DECT and radio LAN signals are radiated to their respective terminals (cordless telephone for DECT and wireless laptop computer for radio LAN). The return signals are picked up by the RAU antenna, where they remodulate the remaining light from the optical source, and are transported back to the central hub by the return fiber. Here they are demultiplexed and fed to their respective base units.

As mentioned above, the radio range depends on a large number of factors associated with the optical link parameters, the radio system, and the radio propagation characteristics. Radio range has been calculated for a single UMTS carrier as a function of optical power and bit rate assuming realistic values of the optical link parameters, as listed in Table 5.1.

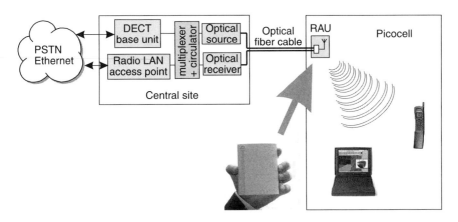

Figure 5.6 Demonstration system for passive RAU, using DECT and radio LAN standards simultaneously.

Table 5.1
Parameters Used for Radio Range Calculations

Parameter	Value
Electroabsorption modulator responsivity	0.15 A/W at 2 GHz
Electroabsorption modulator insertion loss	7 dB
Electroabsorption modulator impedance	50Ω
Electroabsorption modulator efficiency	0.5/V
Photodiode responsivity	1 A/W
Photodiode impedance	50Ω
One-way fiber insertion loss	3 dB
Propagation path-loss slope	20 and 43 dB/decade
Antenna gain	6 dBi
Antenna heights	5m, 1.5m
Downstream modulation depth	0.5
Fade margin	10 dB
Carrier frequency	2 GHz
Mobile transmit power	+20 dBm
Receiver sensitivity (for both	−115 dBm (144 Kbps)
base and mobile stations)	−110 dBm (384 Kbps)
	−103 dBm (2,048 Kbps)

A dual-slope radio propagation model was used that is valid for relatively open environments. The path loss slopes have been included in Table 5.1. Figure 5.7 shows the results of this calculation. For a bit rate of 384 Kbps, which is likely to be the maximum data rate offered in 3G systems, an optical power of 10 mW will provide a radio range of over 100m.

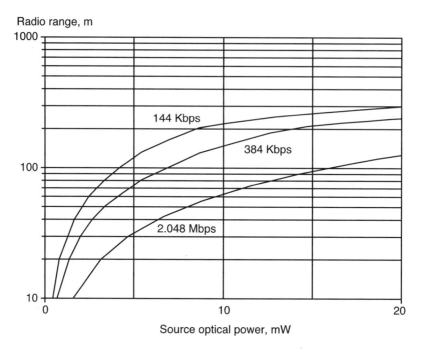

Figure 5.7 UMTS radio range as a function of optical power and data rate for passive RAU.

Apart from limited radio range, another potential problem with the passive optical link is intermodulation distortion, especially since the electroabsorption modulator is operated without bias. Figure 5.8 shows the carrier to third-order intermodulation power ratio as a function of bias voltage and input RF power for a typical electroabsorption modulator (transverse magnetic polarization). This plot was obtained using a two-tone measurement setup. The intermodulation power is highest at zero bias and goes through several nulls as the reverse bias is increased. The intermodulation power also increases significantly at higher input RF power levels. The input RF power in a real situation is likely to be −10 dBm or lower since there is no amplifier at the RAU. Intermodulation power is therefore at least 50 dB lower than the carrier power even at zero bias, which will not cause significant problems for the radio systems of interest.

5.5 In-Building Coverage

5.5.1 Coverage Problems

Buildings are generally challenging propagation environments for cellular radio systems. Most buildings are currently served from signals penetrating

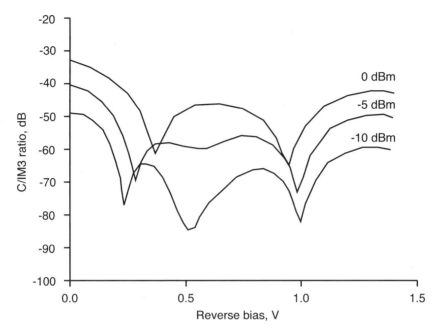

Figure 5.8 CIM power ratio as a function of bias voltage for electroabsorption transceiver uplink.

the walls from the outside since this is the lowest cost solution. However, the building penetration loss is usually very high and has great variability. Chapters 7 and 8 discuss several applications in Japan/Asia Pacific, United States, and Europe.

The situation is even worse for areas deeper inside the building, where signal levels will in general be very low and with large fluctuations in time and precise location. Coverage in this case is unreliable, depending on building materials, wall positions, and furniture. There is also competition in capacity with the outside cell since the same spectrum resource is being shared.

In-building coverage can be improved by using a pickup antenna mounted on the roof connected to a repeater amplifier and an indoor antenna system. This solution will not overcome the capacity problems involved in sharing with the outdoor cell, however. Dedicated in-building systems can overcome these problems. These differing solutions are illustrated in Figure 5.9. The situation where the radio signals penetrate from the outside is illustrated in Figure 5.9(a). A better solution is shown in Figure 5.9(b), where a pickup antenna and a distributed antenna system provide dependable coverage. For situations where both coverage and capacity are important, Figures 5.9(c) and 5.9(d) give better performance. In both of these cases,

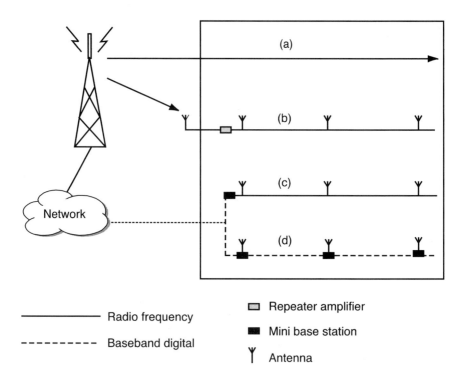

— Radio frequency

- - - - - - - Baseband digital

▢ Repeater amplifier

▬ Mini base station

Y Antenna

Figure 5.9 In-building coverage options: (a) penetration of outdoor signals into building, (b) pickup antenna linked to distributed antenna system, (c) internal RBS linked to distributed antenna system, and (d) distributed RBSs linked by baseband digital transmission systems.

dedicated mini base stations are set up inside the building so that capacity does not need to be shared with the outside cell. The distributed antenna system in Figure 5.9(c) is similar to that in Figure 5.9(b) because the radio frequency signals are being distributed directly from a centralized RBS. In Figure 5.9(d), however, the signals are transmitted at baseband and several RBSs are distributed around the building (distributed radio). A comparison between these last two approaches is considered in more detail in Section 5.5.2.

5.5.2   Coverage Solutions

5.5.2.1   Distributed Radio or Distributed Antenna?

Distributed radio is the architecture used for wide-area cellular radio networks. The RBSs contain all of the RF transceiver hardware and are linked to the core network using digital baseband transmission systems. The advantage of this approach for in-building applications is that existing data network cables can be used to connect the base stations to the network controller. The disadvantage with distributed radio is that each base station in the building must be equipped with sufficient capacity to meet peak traffic demands with a given quality of service, which leads to inefficiency and high cost.

An alternative solution is to employ a centralized base station with distributed antennas connected by analog RF links. In this case the total base station capacity can be reduced compared to the distributed radio architecture. To illustrate this, consider a large building that requires four radio cells to achieve acceptable coverage. Figure 5.10 shows how this could be accomplished using a distributed radio or distributed antenna system.

If we assume that this building has 150 network users with an average traffic of 0.1 Erlangs per user, then the total traffic is 15 Erlangs. The number of channels required to meet this capacity with a given blocking probability is given by the well-known Erlang B formula. If the users are equally distributed among the radio cells, then each cell in a distributed radio system must have nine channels for a 1% blocking probability, making a total of 36 channels for the four-cell building.

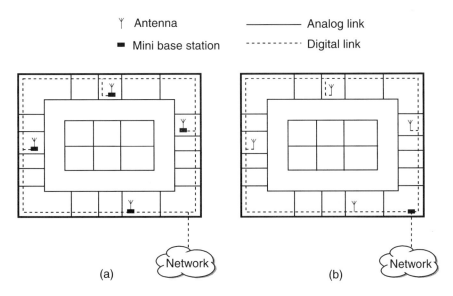

Figure 5.10 Building floor plans illustrating the difference between (a) distributed radio and (b) distributed antenna systems.

In the case of the distributed antenna system, the building can be regarded as one large cell served by a central base station. In this case 24 channels must be provided to meet the same quality of service. The saving in terms of channels is more than 30% for the distributed antenna system. If the network is GSM, then a single RF transceiver will accommodate seven user channels. The distributed radio system will require two transceivers per cell, making a total of eight. The distributed antenna system will require four transceivers. The saving in transceivers is therefore 50%. This is illustrated in Figure 5.11, which also shows the situation for 50 and 250 users. If the network is UMTS, then the savings with distributed antennas are even greater due to the larger channel granularity and the need for only a single RF head for the whole building.

5.5.2.2 Technology Options for Distributed Antenna Systems

Distributed antenna systems can be constructed using either coaxial cable or optical fiber (RoF). Coaxial cable has traditionally been preferred, mainly on grounds of cost, but suffers from high attenuation. The coaxial cable can be increased in diameter to reduce the attenuation or else repeater amplifiers

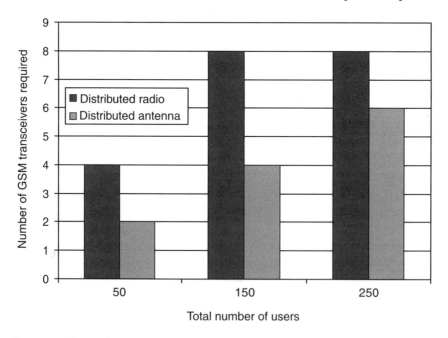

Figure 5.11 Comparison between the number of GSM transceivers required when using distributed radio and distributed antenna systems for various numbers of users.

must be used to achieve the same effect. Thick coaxial cable is difficult to install due to its size, weight, and large bend radius. Furthermore, coaxial systems must be planned carefully due to the strong dependence of the link performance on length. This also makes them difficult to change or extend later.

RoF systems by contrast use optical fiber as the transmission medium and therefore have virtually no dependence on link length if single-mode fiber is used. The fiber cables are thin, light, and flexible. All of these factors mean that RoF systems are easy to install and extend. RoF systems are usually considered to be more expensive than coaxial cable, however. A recent analysis by Hunziker and Baechtold showed that this is only true for very short link lengths [5]. They analyzed the cost as a function of link length for single-mode fiber and three types of coaxial cable: thin (flexible with low cost but high attenuation), medium (medium flexibility, cost, and attenuation), and thick (low attenuation but inflexible with high cost).

An example of thin coaxial cable is RG223U, which has an attenuation of 500 dB/km. By contrast, single-mode fiber cable has an attenuation of 0.4 dB/km and is approximately half the cost of this coaxial cable. RoF systems, however, have a fixed cost for the optical transceivers, which may result in higher total cost for short lengths. For an optical transceiver cost of a few hundred dollars, the break-even length is around 100m [5]. As the cost of optical transceivers falls lower still, the break-even length will be even shorter. RoF systems are therefore gaining popularity, especially for larger installations.

5.6 Evolution to 3G Systems

The 3G mobile systems are due to be deployed during the early to mid-2000s using spectrum allocated around 2 GHz on a worldwide basis. These systems will support voice and high-speed data applications such as Internet browsing and video streaming. A number of radio standards have been selected for these systems that fit into a family known as IMT 2000. The most important of these standards use WCDMA as the multiple-access mechanism, where the data are spread over a bandwidth of a few megahertz using spreading codes.

An example of this is the UMTS Terrestrial Radio Access (UTRA) standard used for UMTS, where the data are spread to a bandwidth of just under 4 MHz within a 5-MHz channel spacing. RoF systems must be developed for these new radio standards that have the required frequency response and dynamic range performance. Fortunately, the dynamic range requirements of the RoF system are relaxed due to the nature of CDMA (all transmitted power from the mobile terminals must be received by the base station at the same level). Distortion in the RoF link can be very serious, however, so care must be taken to ensure that the input power does not exceed a predetermined value. Distortion has been shown to increase with fiber length, caused by the interaction between frequency modulation in the laser (chirp) and chromatic dispersion in the fiber [6], so there are also limits to fiber length that have to be considered.

The limits have been assessed by measurement and simulation [7]. IM3 on a WCDMA signal causes spectral regrowth into adjacent channels. This is shown in Figure 5.12. The signal in Figure 5.12(a) shows the effect of low distortion (input power = 9 dBm and fiber length = 2 km), whereas the signal in Figure 5.12(b) shows the effect of high distortion (input power = 15 dBm and fiber length = 25 km). The ratio of the main channel power to the power leaked into adjacent channels is known as the adjacent channel leakage power ratio (ACLR). This is specified within the UTRA standard, for example, as having to be less than −45 dB for the forward link and less than −33 dB for the reverse link.

Figure 5.13 shows how ACLR varies with fiber length for a range of input powers. Plots are given for a single channel and 32 channels (each at

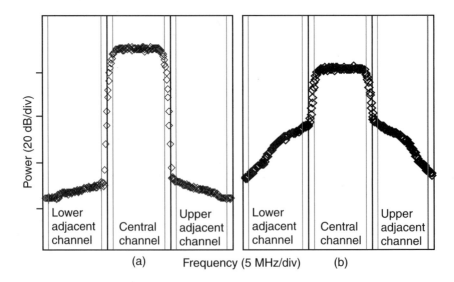

Figure 5.12 Signal spectrum traces: (a) input power = 9 dBm, fiber length = 2 km; and (b) input power = 17 dBm, fiber length = 25 km.

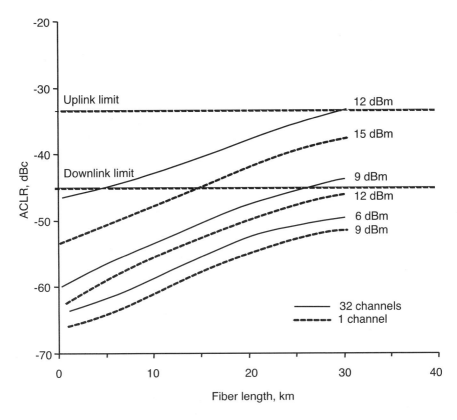

Figure 5.13 ACLR as a function of fiber length and input power for 1 and 32 channels (32 Kbps per channel).

32 Kbps). These results were obtained by simulation using the following parameters: a line-width enhancement factor of 5, a 1-dB compression point of 21 dBm, and a third-order intercept point of 33 dBm. Good agreement was obtained with measurements on a link consisting of an Ortel model 10370A transmitter and model 10450B receiver. The limits to ACLR for UTRA are shown in this figure. If we take the 32-channel case, then the maximum forward link fiber span is 7 km for an input power of 12 dBm and 27 km for an input power of 9 dBm. The reverse link limit for an input power of 12 dBm is more than 30 km. For most of the applications for distributed antenna systems, the fiber span will be below 10 km, which means that an input power of 10 dBm or below will not cause unacceptable distortion for either link direction.

Distortion also affects the baseband signal. The IQ constellation diagram for the demodulated UTRA signal after fiber transmission is shown in Figure 5.14. UTRA uses QPSK modulation, so the decision points occur in four quadrants. As was the case for the previous figure, Figure 5.14(a) shows the effect of low distortion (input power = 9 dBm and fiber length = 2 km) and Figure 5.14(b) shows the effect of high distortion (input power = 15 dBm and fiber length = 25 km). The most widely used modulation quality metric in digital RF communication systems is the error vector magnitude (EVM). EVM provides a way to quantify the errors in digital demodulation and is sensitive to any signal impairment that affects the magnitude and phase trajectory of a demodulated signal.

Figure 5.15 shows measured EVM results as a function of input power and fiber length for a single 32-Kbps channel. As for ACLR, we can observe that EVM is degraded by a combination of high input power and long fiber length. For QPSK modulation, an EVM of more than 6% can be tolerated to achieve a BER of less than 10^{-9}, so we would not expect the degradation in EVM to cause problems. The measurements discussed here show that the limits to input power and fiber length come mainly from ACLR considerations rather than from EVM. As long as these limits are observed, then the use of RoF transmission links for UTRA distributed antenna systems should not pose any significant technical problems.

5.7 Conclusions

This chapter has looked at the use of RoF technologies for cellular radio communications systems. It has described the generic advantages and disadvantages of RoF and has discussed the characteristics and requirements of cellular radio. We have seen that RoF technology can result in substantial cost savings for in-building coverage compared to a conventional distributed radio architecture or to distributed antenna systems based on coaxial cable. The performance of RoF systems for next-generation cellular networks has been analyzed and the limits evaluated. The main conclusions from this chapter are summarized:

- Simplification of the RAU leads to reduced installation and maintenance costs compared to conventional distributed radio.
- Centralization of the system complexity leads to significant efficiency savings through resource sharing. For example, savings of 50% have been calculated for GSM based on a four-cell installation and a 1% call blocking probability.
- Fiber cables are easier and cheaper to install than coaxial cable.

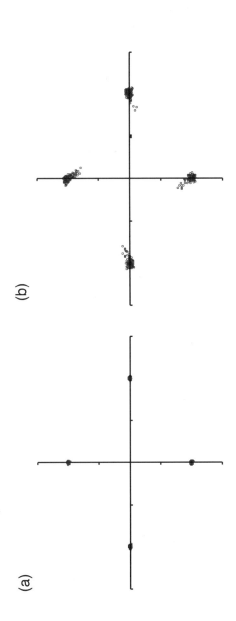

Figure 5.14 IQ constellation diagrams: (a) input power = 9 dBm, fiber length = 2 km; and (b) input power = 17 dBm, fiber length = 25 km.

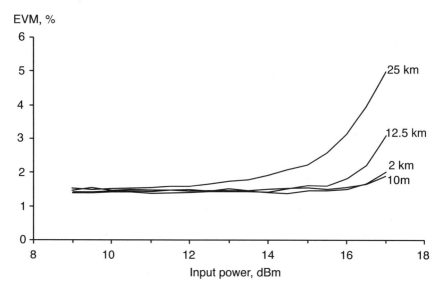

Figure 5.15 EVM as a function of input power and fiber length (for a single 32-Kbps channel).

- Analog system impairments (noise, distortion, and dispersion) will limit the dynamic range, which will be important for TDMA systems and multiple-operator installations. These impairments can be managed successfully.

- The cost of RoF is perceived to be high. Current RoF products tend to use cheap mass-market optical components to keep costs competitive.

- Current RoF products are active, which may incur higher maintenance costs. Future products based on passive EAM technology may alleviate this problem if the cost of the underlying technology can be reduced sufficiently.

- RoF will be well suited to 3G mobile networks due to the cost savings associated with providing many small cells compared to conventional approaches. The input power and fiber length limits have been established for WCDMA transmission and will not impose undue restrictions on the use of RoF.

Overall, we can see that RoF has a very important role in cellular communications today and is set to become even more prevalent in the future as more in-building systems are required.

Acknowledgments

David Wake would like to acknowledge Ralf Schuh, of Telia Research, for a major contribution to the work on WCDMA transmission over optical fiber. He would also like to thank Andrea Casini, of Tekmar Sistemi S.R.L., for his assistance in the area of current RoF systems. Some of the work presented here was performed as part of EURESCOM Projects P816 and P921.

References

[1] Cooper, A. J., "Fiber-Radio for the Provision of Cordless/Mobile Telephony Services in the Access Network," *Electron. Lett.,* Vol. 26, 1990, pp. 2054–2056.

[2] Chen, T. R., et al., "Wide Temperature Range Linear DFB Lasers with Very Low Threshold Current," *Electron. Lett.,* Vol. 33, 1997, pp. 963–965.

[3] Westbrook, L. D., and D. G. Moodie, "Simultaneous Bi-Directional Analog Fiber-Optic Transmission Using an Electroabsorption Modulator," *Electron. Lett.,* Vol. 32, 1996, pp. 1806–1807.

[4] Wake, D., D. Johansson, and D. G. Moodie, "Passive Picocell—A New Concept in Wireless Network Infrastructure," *Electron. Lett.,* Vol. 33, 1997, pp. 404–406.

[5] Hunziker, S., and W. Baechtold, "Cellular Remote Antenna Feeding: Optical Fiber or Coaxial Cable," *Electron. Lett.,* Vol. 34, 1998, pp. 1038–1039.

[6] Charles, S., and G. Wanyi, "Fiber Induced Distortions in a Subcarrier Multiplexed Lightwave System," *J. on Selected Areas in Communications,* Vol. 8, 1990, pp. 1296–1303.

[7] Wake, D., and R. E. Schuh, "Measurement and Simulation of WCDMA Signal Transmission over Optical Fiber," *Electron. Lett.,* Vol. 36, 2000, pp. 901–902.

6

Fiber Optic Radio Networking: The Radio Highway

Shozo Komaki

6.1 Introduction to the Radio Highway

Fiber optic radio access networks are optical backbone networks for radio access systems, where fiber optic links have the function of transferring radio signals into remote stations without destroying their radio format, such as RF, modulation format, and so on. For that purpose, the transmission format in the fiber optic networks is typically based on analog optical modulation techniques [1–7].

The concept of fiber optic radio access networks is illustrated in Figure 6.1.

The radio zone architecture may follow that of conventional microcellular or picocellular radio systems. However, the interface receiving or radiating radio signals in each radio zone, called the RBS, equips only the converter between radio signals and optical signals. The RBS requires neither the modulation functions nor demodulation functions of radio. The radio signals converted into optical signals are transferred via a fiber optic link with the benefit of its low transmission loss. Therefore, the architecture of fiber optic radio access links can be independent of the radio signal format and can provide many universal radio access links that are available to any type of radio signal. This means that such radio access links are very flexible to the modification of radio signal formats or the opening of new radio services.

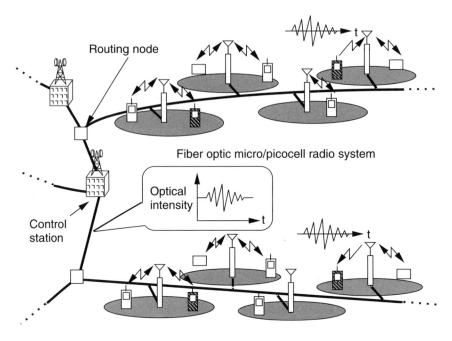

Figure 6.1 Concept of radio highway networks.

A remote central station, called the radio control station, executes the functions of modulation and demodulation of radio and other controls such as channel allocations. Such concentrated execution of these troublesome functions provides a much simplified and cost-effective radio access network and promises easy realization of recent advanced demodulation techniques, such as macrodiversity and handover controls.

Consequently, fiber optic radio access networks are considered hopeful candidates for various future cellular radio access networks, such as third generation mobile communication systems, fixed wireless access systems, wireless LAN, roadside-to-vehicle radio access links in intelligent transport systems (ITS), or distribution systems of CATV signals.

In terms of mobile radio terminal development, the concept of software radio technology has been researched for public mobile communications. The software radio is classified into the three types discussed next, as shown in Figure 6.2.

One is a software radio terminal, defined by a terminal that has a digital signal processor and some type of firmware in it, and the air interfaces are defined by the firmware selection. Software may be downloaded from RBS equipment if the downloading protocol is standardized and radio resources are

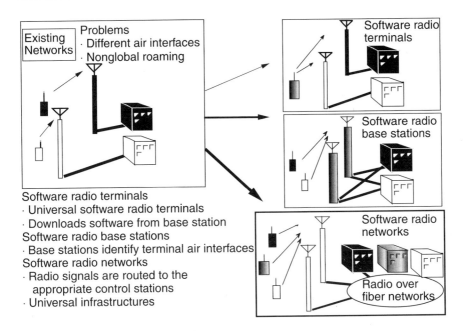

Figure 6.2 Various types of software radios.

sufficient for downloading. In this case, the mobile terminal may be rather complicated compared to traditional single-mode terminals.

Another type is software radio base stations, which are defined by the software-definable RBSs. This type has various types of software and radio equipment in the station. The unit selects appropriate RF equipment and software according to the target mobile terminals. This type allows for rather complicated and large power dissipation. In addition, various types of RF bands can be realized, because the limitations on size, power dissipation, and costs are rather lenient compared to the software radio terminals.

The last type is software radio networks, which are defined by transparent RBSs and some central stations. Transparent RBSs are constructed by the radio over fiber technology, and the fiber networks are constructed by the radio highway networks described in this chapter. Central stations, called radio control stations (RCSs), are constructed by the conventional RBS equipment, and some RCSs may be equipped with the same software radio equipment as that in the software RBS. In this case, the radio air interfaces are more flexible compared to the other two types of software radios, because the networks are transparent. So we can use a variety of frequency bands without replacing the base station equipment.

In the latter two types of software radio, mixed use of various types of air interfaces is necessary, so it is necessary to consider the interference and coexistence conditions, and the microcell technology and unlicensed band "listen before talk" rule may be one of the solutions.

Many subjects must be studied in order to realize fiber optic radio access networks that connect RBSs to RCSs. One basic subject is which photonic link configuration and multiplexing scheme are suitable for the construction of fiber optic radio access networks. This subject is treated in Section 6.2 in detail. The section introduces various types of link configurations and photonic multiplexing schemes, such as subcarrier multiple access, photonic frequency-division multiple access or WDM access, photonic time-division multiple access, photonic code-division multiple access, and chirp multiplexing transform access and routing. We then go on to compare these access methods.

6.2 Various Types of Radio Highway

6.2.1 Network Topology

Three candidates are under consideration for the link configuration topology in constructing networks: star configuration, ring configuration, and bus configuration. Figure 6.3 illustrates these three configurations.

The star configuration is the most popular one because of its easy maintenance, high reliability, and simple construction. The most common type of star configuration is the passive double star. However, it is difficult to construct or extend star networks cost effectively or quickly because the same fiber counts are required as the number of RBSs, and it is difficult to increase the number of RBSs when cell splitting is required.

However, the bus configuration or the ring configuration can reduce the fiber counts by quite a bit; thus, they offer cost-effective and quick construction of networks and also easy extension of RBSs. These capabilities are very important in constructing fiber optic radio access networks because the density of RBSs increases according to the number of users, and becomes very high in recent microcellular requirements. For indoor use, network flexibility and fiber count reduction are very important matters, because the reinstallation of fibers is wasteful.

Therefore, we should study the fiber optic links that use the bus or the ring configuration. These two configurations are quite similar except for the difference of whether the network is terminated at a certain RBS or at

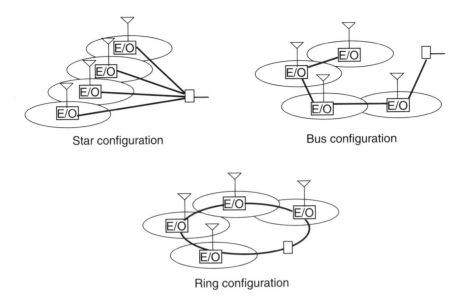

Figure 6.3 Various network configurations.

the RCS. This chapter concentrates mainly on the discussion of a realization of the bus-type configuration.

6.2.2 Various Types

Now we turn to the study of the photonic access scheme, another important subject in our discussion of the construction of the bus link system. The candidates for multiplexing schemes are SCM, TDMA, code-division multiple access (CDMA), and wavelength-division multiple access (WDMA). The features of each scheme are also summarized in Table 6.1.

In discussing which photonic multiplexing scheme is suitable for radio access networks, we should make much of the cost efficiency, the ease of use, and the simplicity of network construction because these are the most important benefits offered by fiber optic radio access networks. The optical communication systems are generally classified as direct detection systems and coherent detection systems. The former have many advantages in terms of the simplicity of the system, whereas the latter increases the complexity of the system in general. However, the ability to demultiplex signals at the optical stage provides an important benefit. The fiber optic radio access networks have the potential to be universally used among different radio

Table 6.1
Performance Comparison of Advantages and Disadvantages
of Various Multiple Access Schemes

Multiplexing Scheme	Advantages	Disadvantages
Time-division multiplexing (TDM)	Intermodulation/direct detection link configuration is allowed; Easy photonic routing by photonic time switching; No generation of optical beat noise.	Requirement of fast photonic switching; Requirement of time synchronization control among base stations.
Subcarrier multiplexing (SCM)	Intermodulation/direct detection link configuration is allowed.	Occurrence of optical beat noise; Radio signals must be frequency-division multiplexing format; Photonic routing is impossible.
Frequency-division multiplexing	Effective utilization of optical frequency; Robustness to fiber dispersion if single-sideband modulation is used; High receiver sensitivity due to coherent detection.	Requirement of coherent detection; Requirement of very narrow optical and frequency shifter for photonic routing.
Code-division multiplexing (CDM)	Easy realization of random access.	Requirement of fast code synchronization; Requirement of much fast operation of photonic device for coding; Feasibility of photonic routing is unknown.
Wavelength-division multiplexing (WDM)	Intermodulation/direct detection link configuration is allowed.	Requirement of many wavelengths; Requirement of many wavelength filters and wavelength converter for photonic routing.

services or those operated by different carriers, because the configuration of RBSs and fiber optic networks can be independent of the radio signal format.

Such advanced purpose requires further subjects to be studied. It is expected that the different types of radio services, or those operated by different carriers, will need to be delivered to different RCSs in different locations in most cases. This means that the networks must include the routing node that executes routing of radio signals, that is, distinguishing and switching of radio signals. Therefore, in selecting a particular photonic multiplexing scheme, it is also important to make much of the feasibility for the routing of radio signals, and it is desirable for the routing process to be executed at the optical stage. Such routing of radio signals at the optical stage is referred to as the *radio highway*. This is nearly the same word as photonic switching and photonic networks; however, the target of the radio highway is radio signal routing on the fiber optics and the different technologies are applied to the radio highway.

Attractive features of the TDMA scheme are that the intermodulation/direct detection link can be applied and a simple photonic switch can easily realize the routing function of radio signals at the optical stage. The SCM scheme can also apply an intermodulation/direct detection link; however, it is impossible to realize the routing function of radio signals at the optical stage because signals are multiplexed in frequency at the electrical stage. A more detailed comparison of the photonic TDMA bus link system and SCM bus link system is given in Section 6.3.1.

The FDMA scheme, in particular the scheme that uses the optical single-sideband modulation technique, has been investigated thoroughly because of its effective optical frequency utilization and its robustness to optical fiber dispersions. However, such an FDMA scheme requires a high stabilized coherent laser diode and very narrow optical filters to be demultiplexed at the optical stage. So it introduces many problems into fiber optic radio access networks from the viewpoint of cost-effective and easy construction of networks.

The several CDMA schemes such as frequency spread code-division multiplexing and time spread code-division multiplexing have also been investigated because of their easy applicability to random access networks. They require high-speed code synchronization between a transmitter and a receiver at the optical stage. In addition, the possibility of photonic routing of signals presents another difficulty.

The WDM scheme has appeared to be promising in recent very high bit rate digital transmission systems. However, it requires preparing as many wavelengths as the number of RBSs.

From the above perspective on cost efficiency and the feasibility of photonic routing of signals, the TDMA bus link systems are considered first in Section 6.3 as the best choice for constructing fiber optic radio access networks. In Section 6.3, the principle of natural sampling, self-synchronous

TDMA bus link systems, and asynchronous TDMA bus link systems are described.

Section 6.4 describes the photonic CDMA architecture, in particular direct optical switch CDMA. This system requires no optical phase switch and mask, so the code switch is very simple and easy to realize. In Section 6.5, the chirp multiple access method is described. This method is based on the chirp multiplexing transform and can transform an FDM signal into a TDM signal. So, the different radio services operated in different frequency bands can be routed to the different central stations by means of a conventional photonic time switch. In Section 6.6 photonic WDM technologies are reviewed and radio signal routing and multiple access on WDM switching are discussed.

6.3 Photonic TDMA Highway

6.3.1 Natural Sampling and Basic Configuration of Photonic TDMA

Section 6.3.1.1 introduces the principle of the photonic natural bandpass sampling technique, which realizes the conversion of a radio signal into the photonic TDMA signal format. Section 6.3.1.2 introduces the basic configuration of the photonic TDMA bus link system, and specifies the subjects that need study in order to realize the TDMA bus link systems.

6.3.1.1 Principle of Photonic Natural Bandpass Sampling

Pulse amplitude modulation systems are promising systems with which to realize TDM systems, where multiplexing of several signals is achieved by interleaving the samples of the individual signals. According to the Nyquist sampling theorem, the signal whose highest frequency spectrum component is f_M is determined at regular intervals separated by times $T_s \leq 1/2f_M$, that is, the signal has to be periodically sampled every T_s seconds. T_s and $f_s = 1/T_s$ are referred to as the sampling interval and sampling frequency, respectively. If this theory is applied to the sampling of radio signals whose highest frequency spectrum component is a few or dozens of gigahertz, ultra-high-speed sampling frequency is required. However, the radio signals are fortunately band-limited in general. Let the bandwidth of a radio signal be B_{RF} and you can find that the required condition about sampling interval can be reduced to

$$T_s < 1/2B_{RF} \tag{6.1}$$

Such a sampling technique to reduce sampling frequency is referred to as bandpass sampling [8].

Another important concept is photonic natural sampling. Photonic sampling means that the sampling process is executed at the optical stage. Figure 6.4 shows the fundamental architectures of two different ways of achieving a photonic natural sampling system. One way is for a photonic switch to sample the optical carrier whose intensity was directly modulated by radio signals [Figure 6.4(a)]. The second way is for the optical pulses generated by a mode-locked laser diode to be modulated in intensity by means of an external optical modulator [Figure 6.4(b)]. The optical pulses obtained by such operations are referred to as pulse amplitude modulated/intensity modulated optical pulses. From the viewpoint of cost efficiency, the former is superior to the latter because a mode-locked laser diode is still comparatively expensive. However, the latter scheme has the possibility of very high-speed sampling operation and broadband modulation due to the EOM utilization.

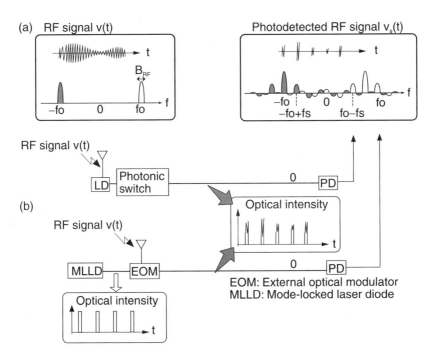

Figure 6.4 Principle of photonic natural bandpass sampling: (a) direct intensity modulation type, and (b) external intensity modulation type. (LD: laser diode; PD: photodiode.)

Let $v(t)$ be a radio signal that is band-limited to B_{RF} and has the center frequency of f_o. The pulse amplitude modulated/intensity modulated optical pulses obtained by the photonic sampling are written by

$$P_r(t) = P_s(1 + mv(t)) \sum_{n=-\infty}^{\infty} p(t - nT_s) \qquad (6.2)$$

where $P_r(t)$, P_s, m, and $p(t)$ are optical intensity, average transmitting optical power, modulation index, and sampling pulse waveform, respectively, and n is an integer. The output current after photodetection of this pulse amplitude modulated/intensity modulated optical pulses, $i_s(t)$, is given by

$$i_s(t) = \alpha_o P_s mv(t) \sum_{n=-\infty}^{\infty} p(t - nT_s) \qquad (6.3)$$

where α_o is the responsivity of the photodetector. The spectrum of $i_s(t)$ is given by Fourier transforming of $i_s(t)$.

$$I_s(f) = \frac{\alpha_o P_s m}{T_s} \sum_{n=-\infty}^{\infty} P(nf_s)V(f - nf_s) \qquad (6.4)$$

where $P(f)$ and $V(f)$ are Fourier transforms of $p(t)$ and $v(t)$, respectively. The waveform $i_s(t)$ and the spectrum $I_s(t)$ have the appearance as illustrated in Figure 6.4 for the case in which $p(t)$ is a rectangular pulse waveform and $T_s = 1/2 B_{RF}$. Figure 6.4 shows that photodetected pulse amplitude modulated signals contain the original radio spectrum, and its replicas at intervals of f. The overall spectrum is symmetrical about the frequency f_o. Therefore, the original radio signal can be obtained by means of a bandpass filter with a center frequency of f_o and a bandwidth of B_{RF}. In this example case, $P(f)$ has a sinc function characteristic and infinite bandwidth. So you can understand from the figure that the condition of (6.1) is required in order to avoid interference due to image frequency component.

Now we further extend our discussion to the special case in which a band-limited sampling pulse waveform is used instead of a rectangular pulse. Such a case seems to be encountered often in actual systems. If the highest frequency component of the pulse waveform is limited to f_u and the condition of $f_u < 2f_o - B_{RF}$ is satisfied, the image spectrum component does not appear in the RF band. This result means that the required condition about sampling frequency can be further reduced to $f_s > B_{RF}$. However, we must not forget that the band-limited pulse waveform has a wider pulse width than the rectangular one, which affects the capacity of TDMA systems, because the allowable number of multiplexing signals is determined by the sampling interval T_s divided by the pulse width.

6.3.1.2 Photonic TDMA Bus Link Systems

Figure 6.5 shows the basic architecture of the photonic TDMA bus link system using the photonic natural bandpass sampling technique. Several RBSs are connected to one fiber optic bus link by means of a photonic switch. The photonic switch performs the two important functions of photonic sampling and multiplexing of several signals at the same time. Such a multiplexing scheme using a photonic switch can seriously reduce the transmission loss of the signal in comparison with a multiplexing scheme that uses a directional coupler because the use of a photonic switch can avoid a 3-dB coupling loss, which is unavoidable when using a directional coupler.

The theoretical maximum number of multiplexed RBSs is given by

$$N = T_s / T \tag{6.5}$$

where T_s and T are sampling interval and pulse width, respectively. However, to attain this number of multiplexed RBSs implies the difficulty of time synchronization control among RBSs because each RBS must drive the photonic switch strictly in given time slots. The required accuracy of the time synchronization increases in proportion to the bandwidth of the radio signals or the number of connected RBSs, and the required order of accuracy becomes on the order of nanoseconds when the bandwidth of the radio signals is of the order of megahertz. Then schemes to solve the difficulty of time synchronization have to be investigated. The solutions to this problem are discussed in Sections 6.3.2 and 6.3.3 in detail.

In the subcarrier multiple access (SCMA) system described in Section 6.6, the optical beat noises are caused by photodetecting more than two optical carriers with the frequency difference being less than the detection bandwidth. Therefore, the TDMA bus link systems experience less performance degradation due to optical beat noise, and signal performance is much improved.

Another important advantage of the TDMA bus link system is the fact that the system allows all the connected RBSs to use the same RF because the TDMA system gives an independent optical time channel to each RBS. This means that the TDMA system can be applied to the single-frequency

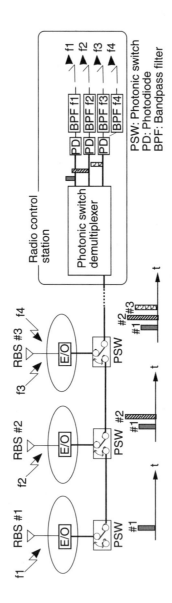

Figure 6.5 Configuration of photonic TDMA bus link system.

network such as CDMA radio systems to be introduced as the radio access format for IMT 2000 and other systems. In addition, it becomes easier to execute macrodiversity because the radio signals of all RBSs are concentrated at the RCS.

Finally in this section, we discuss the capabilities of the photonic routing of radio signals. Suppose that the jth RBS and the kth RBS receive different types of radio service signals, respectively, which have to be transferred to different RCSs in separated locations. Because the signals are multiplexed in the TDM format in the fiber optic bus link, the separation and routing of two radio services can be easily performed by means of a photonic switch. This capability of photonic routing is also another advantage of the photonic TDMA bus link system. However, the above supposition means to construct one RBS for each radio service and it reduces the possibility of universal use of RBSs among several radio services. In general situations, different radio services are operated under different frequency bands. Therefore, if an RBS is shared among several radio services, received radio services are signals in the FDMA format unless conversion is performed. Consequently, it is impossible to separate and route them at the optical stage once they are converted into optical intensity signals, because such multiple radio services are transferred using the SCMA format. This subject and the solution are treated in detail in Section 6.5.

6.3.1.3 Summary

This section introduced the concept of fiber optic radio networking, the radio highway, and several candidates for network configurations and photonic multiple access schemes applicable to the networks. The suitability of each network configuration and multiplexing scheme was discussed from the viewpoints of simplicity of the network configuration and easy realization of demultiplexing of signals at the optical stage.

As a result, the following advantages of a photonic TDMA bus link system were confirmed:

- A photonic TDMA bus link system allows for the cost-effective construction of radio access networks by means of an intermodulation/direct detection fiber optic link.

- A photonic TDMA bus link system offers higher C/N performance of received radio signals than the conventional SCM bus link system because performance deterioration due to optical beat noises is reduced.

- A photonic TDMA bus link system is available for single RF.

- The photonic routing of signals is easily executed by means of a photonic switch because of the TDM format.

Furthermore, we concluded that the following problems have to be studied and solved in order to realize photonic TDMA bus link system and advance the functions of it.

- The time synchronization control for TDMA of signals is difficult, and the required accuracy of the synchronization increases in proportion to the bandwidth of radio signals and the number of connected RBSs.
- The photonic routing of radio service is impossible when an RBS is universally used among different radio services.

6.3.2 Self-Synchronous TDMA Highway

A photonic TDMA bus link system requires strict time synchronization control among all RBSs in order to realize TDMA of signals. The difficulty surrounding such synchronization control increases as the bandwidth of the radio signal or the number of connected RBSs increases, because the required pulse width of TDMA signals is inversely proportional to both of them.

To solve the difficulty of time synchronization control in TDMA bus link systems, this section proposes a self-synchronous TDM bus link system. The proposed system needs no synchronization control to assign time slots to RBSs, and it allows all RBSs to use the identical optical source provided from the end of the bus link. However, the system severely degrades the signal power received at an RCS due to the splitting of the optical source power and the insertion of many optical devices. Therefore, this section investigates expected system performance by delineating a theoretical analysis of the C/N performance of radio signals detected at an RCS. Furthermore, this section proposes the application of an optical preamplifier in order to improve C/N performance, and the improvement is theoretically investigated.

The proposed system needs two independent optical channels. One is used to transfer unmodulated optical pulses and the other is for transferring modulated pulses. Such independent optical channels can be prepared by means of various optical multiplexing schemes. However, this section discusses two multiplexing schemes, fiber-space multiplexing utilization and polarization multiplexing utilization, from the viewpoint of feasibility [9, 10].

6.3.2.1 Self-Synchronous TDMA Bus Link System

Figure 6.6(a) illustrates the configuration of the uplink for the proposed self-synchronous TDM bus link system using a fiber-space multiplexing scheme.

N RBSs and one RCS are connected with two independent fiber optic bus links. One of the links is used to transfer unmodulated optical pulses (carrier link) and the other is used to transfer the pulses modulated at each RBS (signal link). Unmodulated optical pulse trains are provided from the

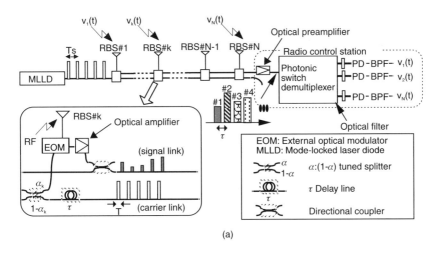

(a)

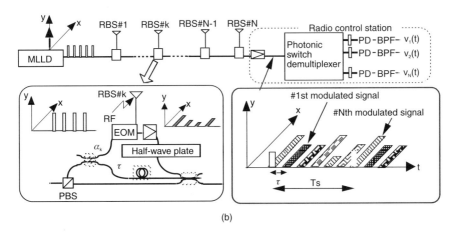

(b)

Figure 6.6 System configuration of self-synchronous TDM bus link: (a) fiber-space multiplexing utilization, and (b) polarization multiplexing utilization.

end of the bus link using a mode-locked pulsed laser diode or the combination of a laser diode and a photonic switch. The optical pulse trains are used to sample and transfer the radio signal, and their repetition rate has to be more than double the bandwidth of the radio signal in order to regenerate it at an RCS, as described in Section 6.3.1.

The kth RBS splits the power of received unmodulated optical pulses with the ratio of $\alpha_k:(1 - \alpha_k)$ and intensity modulates them by radio signals using an external optical modulator as shown in Figure 6.6(a). A fiber optic delay line equipped with a carrier link generates the appropriate time difference, τ, between the unmodulated pulses to be relayed to the next RBS and the pulses that were already modulated. The purpose of such delay processing is to prevent their overlaps on the time axis at the time when they reach the next RBS. Therefore, the generated time difference must be more than the pulse width at least. If such processing is executed at all of the RBSs in series, TDM of the signals of all RBSs can be automatically realized and any other synchronization control signaling is not required. The pulse width and the delay time must satisfy the following condition in order to obtain successful TDM signals from all RBSs:

$$T \leq \tau = \frac{T_s}{N} \tag{6.6}$$

where T, T_s, and N are the pulse width, the pulse repetition interval, and the number of connected RBSs, respectively.

Figure 6.7 shows the required pulse width and that translated to the fiber delay length versus sampling frequency, where we assume 10 RBSs are connected. For example, if we assume that the bandwidth of a radio signal is 200 MHz, which is the service bandwidth of the Japanese PHS, the required repetition interval of pulses becomes $T_s = 2.5$ ns (sampling frequency = $1/T_s = 400$ MHz) and the required delay time becomes 0.25 ns, which corresponds to 50 mm of fiber optic delay line. However, this condition is valid in the ideal system where there is no time difference error between the signal link and the carrier link. In the real system where there is a time difference error between the two links due to the insertion of external optical modulator, optical amplifier, and the fibers connecting such devices, we have to compensate for the time difference error by adjusting the delay time τ; otherwise, we have to use a narrower pulse width than the delay time in order to avoid the overlaps between pulses in the obtained TDMA signals.

At the optical receiver in the RCS, received optical signals are amplified by the use of an optical preamplifier. The optical preamplification can much

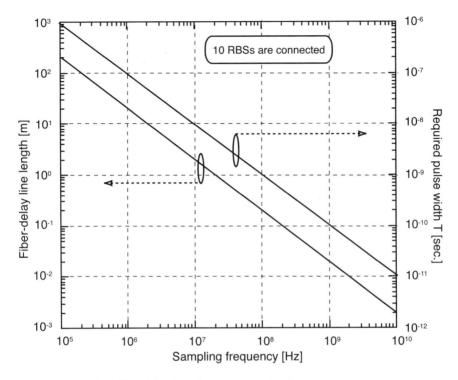

Figure 6.7 Required fiber delay length versus sampling frequency.

improve the signal performance as described in the following sections. Then the TDMA pulses are demultiplexed by means of a photonic switch after an optical bandpass filter, which removes the excess optical noise. Finally, photodetection and bandpass filtering centered at RF can regenerate the original radio signal of each RBS.

The self-synchronous TDMA bus link needs two independent optical links to transfer unmodulated pulses and modulated pulses as already introduced. It can be realized not only by fiber-space multiplexing, but also by other optical multiplexing schemes such as polarization multiplexing. Figure 6.6(b) illustrates the self-synchronous TDMA bus link system using polarization multiplexing. The difference from the system using fiber-space multiplexing is that modulated and unmodulated pulses are multiplexed into a single fiber optic bus link with orthogonal polarization states. Therefore, each RBS needs a polarization beam splitter to extract unmodulated pulses, a half-wave plate to rotate the polarization state by $\pi/2$, and a 3×1 directional coupler. Furthermore, it is necessary to use polarization-maintaining fiber such as a PANDA fiber in order to maintain the polarization state in the signal transmission. Thus, although the system using polarization multiplexing has an advantage in that it needs only one fiber optic link, it is obvious that the signal loss and the system complexity increases in comparison to a system that uses a fiber-space multiplexing scheme.

6.3.2.2 Theoretical Analysis of Received C/N

In this part, the received C/N is theoretically analyzed.

The optical peak power fed into the kth RBS's external optical modulator is given by

$$P_{\text{in}}(k) = P_s L_{\text{bef}}(k) \tag{6.7}$$

where P_s and $L_{\text{bef}}(k)$ are the transmitting optical peak power of the mode-locked laser diode and the total loss from which the optical carrier suffers until it reaches the kth external optical modulator. According to the multiplexing scheme used, it is differently given by

$$L_{\text{bef}}(k) = \begin{cases} \Pi_{i=1}^{k-1}(1 - \alpha_i)\alpha_k L_s^k \\ \Pi_{i=1}^{k-1}\left[\dfrac{1}{3}(1 - \alpha_i)\right]\alpha_k (L_d L_s)^k L_c^{k-1} \end{cases} \tag{6.8}$$

where α_i, L_d, L_s, and L_c are the power ratio of the tuned splitter in the ith RBS, and the fiber-to-fiber insertion losses of the polarization splitter, tuned power splitter, and directional coupler. The output signal power from the external optical modulator is given by

$$P_o(k, t) = L_m \frac{P_{\text{in}}(k)}{2}(1 + m\nu_k(t)) \tag{6.9}$$

where L_m, m, and $\nu_k(t)$ are the fiber-to-fiber insertion loss of the external optical modulator, modulation index, and modulating signal of the kth RBS, respectively. When the optical signal of the kth RBS reaches the RCS, the received optical power is reduced to

$$P_r(k, t) = L_m G_p \frac{P_{\text{in}}(k)}{2}(1 + m\nu_k(t))L_{\text{after}}(k) \tag{6.10}$$

where G_p is the optical preamplifier gain, and $L_{\text{after}}(k)$ is the total loss from which the modulated optical signal of the kth RBS suffers until it reaches the RCS. $L_{\text{after}}(k)$ is given by

$$L_{\text{after}}(k) = \begin{cases} \left(\dfrac{1}{2}\right)^{N-k+1} L_c^{N-k+1} L_f & \text{fiber multiplexing} \\[4mm] \left(\dfrac{1}{3}\right)^{N-k+1} L_t L_c^{N-k+1} L_d^{N-k} L_f & \text{polarization multiplexing} \end{cases}$$

$$(6.11)$$

where L_t and L_f are the fiber-to-fiber insertion losses of the half-wave plate and optical filter setup before the photodetector at the RCS.

The expression of actual received optical signal in the proposed system equals the above, which are sampled with the pulse duty of T/T_s (T = pulse width; T_s = sampling interval). Such sampled signal causes the power penalty of $(T/T_s)^2$ to be applied to the photodetected signal [10]. Consequently, the photodetected radio signal power is given by

$$\langle i_s^2(k) \rangle = \frac{1}{2}\left(\frac{e\eta}{h\nu}\right)^2 \left[mG_p L_m \frac{P_{\text{in}}(k)}{2} L_{\text{after}}(k) \right]^2 \left(\frac{T}{T_s}\right)^2 \qquad (6.12)$$

where e, η, and $h\nu$ are the electron charge, the quantum efficiency of photodetector, and the photon energy, respectively.

As for noise factors treated in the analysis of the C/N performance, the amplified spontaneous emission should be considered when an optical amplifier is used. Then, the following noise factors are considered: beat noise current among amplified spontaneous emissions and signal; beat noise current among amplified spontaneous emissions; shot noise current from amplified spontaneous emissions; shot noise current from signals; the RIN current; and the receiver thermal noise current.

The noise power due to the amplified spontaneous emission is studied precisely and it can be modeled as a white Gaussian noise process. Considering the splitting and coupling loss of the signal and amplified spontaneous emission power, the power spectrum density level of each noise current is given by

$$n_{s-sp}(k) = 4\frac{(\eta e)^2}{h\nu}(G_p - 1)\chi n_{sp} L_f \cdot \left[G_p L_m \frac{P_{\text{in}}(k)}{2} L_{\text{after}}(k) \right] \qquad (6.13)$$

$$n_{sp-sp} = 2[\eta e(G_p - 1)\chi n_{sp} L_f]^2 W \qquad (6.14)$$

$$n_{sp_{shot}} = 2e^2 \eta (G_p - 1) \chi n_{sp} L_f W \tag{6.15}$$

$$n_{s_{shot}}(k) = 2e^2 \left(\frac{\eta}{h\nu}\right) \left[G_p L_m \frac{P_{in}(k)}{2} L_{after}(k) \right] \tag{6.16}$$

$$n_{RIN}(k) = RIN \left(\frac{\eta e}{h\nu}\right)^2 \left[G_p L_m \frac{P_{in}(k)}{2} L_{after}(k) \right]^2 \tag{6.17}$$

$$n_{th} = \frac{4k_b T_{th}}{R} \tag{6.18}$$

where χn_{sp}, W, k_b, T_{th}, and R are the spontaneous mission factor, the bandwidth of optical filter, Boltzmann constant, noise temperature, and load resistance of the receiver, respectively. Note that each power spectrum density level for the case in which an optical preamplifier is not used is given by substituting 0 for G_p. Considering that the white Gaussian noise current that is sampled with pulse duty T/T_s degrades its power spectrum density level by T/T_s, the C/N of the detected radio signal is finally given by

$$C/N_k = \frac{\langle i_s^2(k) \rangle}{\left[(n_{s-sp} + n_{sp-sp} + n_{sp_{shot}} + n_{s_{shot}} + n_{RIN}) \frac{T}{T_s} + n_{th} \right] B_{RF}} \tag{6.19}$$

where B_{RF} is the bandwidth of the radio signal.

From (6.12), (6.13) through (6.18), and (6.19), you can see that the equalization of $P_{in}(k)L_{after}(k)$ for all value of k realizes the equal C/N performance for signals of all RBSs. There exists the set of the power ratio of the splitter, α_k, that satisfies this condition. Solving the simultaneous equations, $P_{in}(1)L_{after}(1) = P_{in}(2)L_{after}(2) = \ldots = P_{in}(N)L_{after}(N)$, and the use of the condition, $\alpha_N = 1$, each α_k in the case of fiber-space multiplexing is given as

$$\alpha_k = \begin{cases} \dfrac{\left(\dfrac{2L_s}{L_c}\right)^{N-k}\left(\dfrac{2L_s}{L_c} - 1\right)}{\left(\dfrac{2L_s}{L_c}\right)^{N-k+1} - 1} & L_c \neq 2L_s \\[2em] \dfrac{1}{N - k + 1} & L_c = 2L_s \end{cases} \qquad (6.20)$$

and in the case of polarization multiplexing as

$$\alpha_k = \begin{cases} \dfrac{(3L_s)^{N-k}(3L_s - 1)}{(3L_s)^{N-k+1} - 1} & L_s \neq \dfrac{1}{3} \\[2em] \dfrac{1}{N - k + 1} & L_s = \dfrac{1}{3} \end{cases} \qquad (6.21)$$

6.3.2.3 Numerical Results of C/N Performance

In the following calculations, we use the parameters shown in Table 6.2 unless a value is specified in each figure.

Figure 6.8 shows the received C/N performance versus the number of connected RBSs in cases where an optical preamplifier is not used. It includes the performance in fiber-space multiplexing utilization (fiber-space MUX) and that in polarization multiplexing utilization (polarization MUX) for different values of transmitting optical peak power P_s. You can see that the C/N in both schemes decreases as the number of RBSs increases. However, the degradation slope in polarization multiplexing utilization is more rapid than that in fiber-space multiplexing utilization because of the additional optical power loss due to the insertions of RBSs, half-wave plates, and couplers. In a small number of RBSs, the increase of P_s cannot much improve

Table 6.2
Default Parameters Used in Calculations

RIN	−152 dB/Hz	T_{th}	300K
α_0	0.8 A/W	R	50Ω
P_s	20 dBm	χn_{sp}	2.0
m	0.5	B	300 kHz
T/τ	1.0	W	1 THz
L_m	3 dB	L_s, L_c	0.5 dB
L_d, L_t	0.5 dB	L_f	0.5 dB

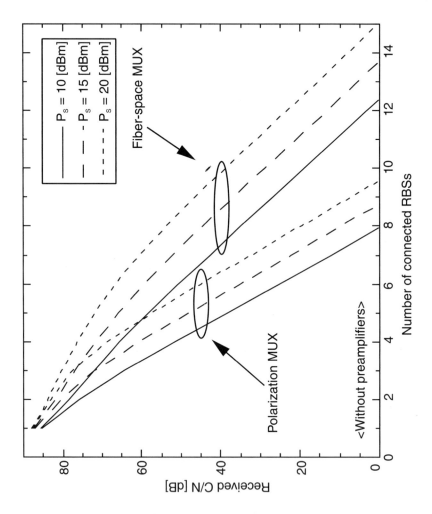

Figure 6.8 Received C/N versus the number of connected RBSs in case of no preamplification.

C/N performance because RIN dominates the performance. However, in a large number of RBSs, a 5-dBm addition in P_s can improve the C/N performance by 10 dB because the thermal noise dominates the performance. This figure shows that the system using fiber-space multiplexing utilization allows only seven RBSs to be connected at P_s = 10 dBm if 40 dB of C/N is required.

In regions where thermal noise dominates the performance, optical preamplification before photodetection is effective to improve C/N performance. Figure 6.9 shows the received C/N performance of a fiber-space multiplexing utilization system versus preamplifier gain.

You can see that the preamplification can much improve C/N performance, especially for the condition in which a large number of RBSs are connected. For example, in the case of N = 15 and P_s = 20 dBm, the preamplification improves the C/N by 28 dB. You can also find the saturation characteristic of the C/N performance at the large preamplifier gain because the optical beat noise between signal and amplified spontaneous emission dominates the C/N performance.

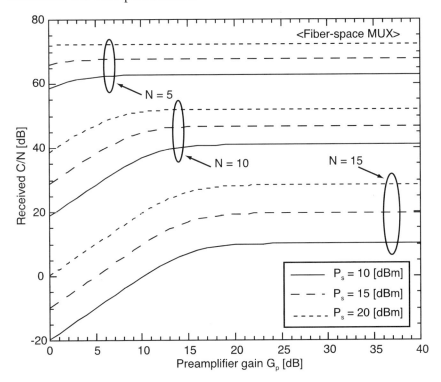

Figure 6.9 Received C/N versus preamplifier gain.

Figure 6.10 shows the relationship between the number of RBSs connected and preamplifier gain.

The number of RBSs connected is the maximum number of RBSs that can satisfy the threshold C/N, C/N_{th}. You can see that preamplification can increase the number of connectable RBSs. For example, the system using fiber-space multiplexing utilization allows 12 RBSs to be connected at $G_p = 22$ dB under the requirement of $C/N_{th} = 40$ dB.

As described in Section 6.3.2.1, we may have to reduce the pulse width required in the ideal system in order to avoid the stringent requirement of accuracy in delay length adjustment. Therefore, we should investigate how pulse duty reduction affects C/N performance. Figure 6.11 shows the relationship between received C/N versus pulse duty. The pulse duty is defined here as the ratio of the pulse width in the real system, T, and that in an ideal system, τ, that is, T/τ.

The figure shows that the C/N is degraded as the pulse duty decreases in spite of the value of G_p. You can see that for the case in which G_p is small, C/N degradation is proportional to the pulse duty to the second

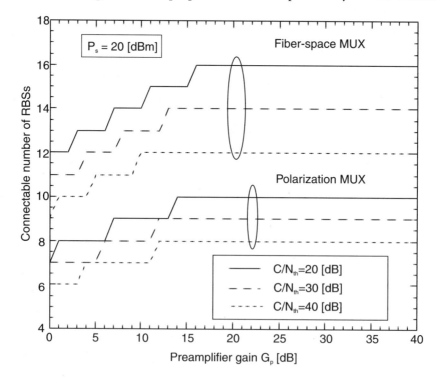

Figure 6.10 Number of RBSs connected versus preamplifier gain.

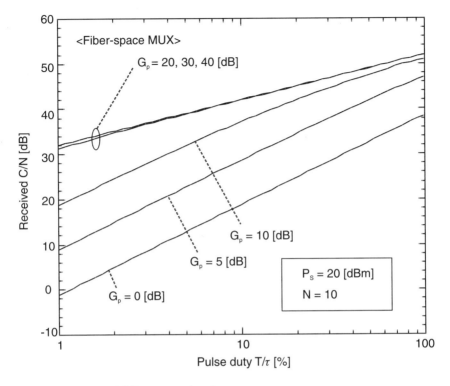

Figure 6.11 Received C/N versus pulse duty.

power because the thermal noise dominates the performance, while in the case for which G_p is large, C/N degradation is proportional to the pulse duty because the beat noise between amplified spontaneous emission and signal dominates the C/N performance. Consequently, the use of optical preamplification is also effective to reduce performance degradation due to the pulse width reduction.

6.3.2.4 Summary

This section described the photonic self-synchronous TDMA bus link system for fiber optic radio access networks, which need no time synchronization control among RBSs connected to a fiber optic bus link and can share a single laser source among RBSs. The configurations of the systems utilizing fiber-space multiplexing and polarization multiplexing were described, and the C/N performance of the radio signal detected at an RCS was theoretically analyzed. The splitting power ratio used at each RBS was optimized to balance the C/N performance of all RBSs.

The theoretical analysis clarifies the relationship among the C/N, number of connected RBSs, transmitting optical power, and preamplifier gain. We showed that the use of an optical preamplifier is effective for improving the received C/N performance, and a C/N of 40 dB can be attained when 12 RBSs are connected.

6.3.3 Asynchronous TDMA Highway

In Section 6.3.2, the self-synchronous TDMA bus link system was described, in which synchronization control for TDM of signals is automatically performed by means of a fiber optic delay line equipped in each RBS.

Another approach to solve the difficulty of synchronization control among RBSs is to adopt an asynchronous access technique. The asynchronous access system does not require any synchronization control among RBSs, and allows for the flexible construction of networks because the system can easily extend some RBSs connected to the fiber optic bus link without any changes in the system.

However, from the perspective of the future picocellular radio environment for multimedia data transmissions, such as wireless ATM or IP routing, it is expected that wideband radio packet communications will become popular. This means that each RBS receives radio bursts occasionally, and the probability that all RBSs are receiving any radio signal at the same time is not very high. In such an environment, the TDMA system in which all RBSs are always keeping their own optical channel does not seem to be effective from the viewpoint of the efficient use of the fiber optic link.

This section therefore describes an asynchronous TDMA bus link system and discusses its availability [11, 12]. The asynchronous TDMA system allows an RBS to access the fiber optic bus link only when it receives radio bursts. Such an asynchronous access system, however, causes collisions among radio bursts if several RBSs access the fiber optic link at the same time, and the collisions cause radio burst losses.

Thus, the proposed asynchronous TDMA system uses the natural bandpass sampling technique in order to convert received radio bursts into narrow pulsed optical bursts, which reduces the probability of the collision. The converted signal is referred to as pulse amplitude modulated/optical intensity modulation bursts, and the probability that a radio burst is lost due to a collision is referred to in this chapter as the *burst loss probability*. The burst loss probability depends on the number of RBSs connected to the bus link and the traffic intensity generated at each radio zone.

This section theoretically investigates the dependencies, proposes access control methods to reduce the burst loss probability, and investigates their effectiveness.

Section 6.3.3.1 describes the concept and the configuration of an asynchronous TDMA bus link system, and Section 6.3.3.2 shows the analytical model for the burst loss probability. Sections 6.3.3.3 and 6.3.3.4 theoretically analyze the burst loss probability and the C/N, respectively. Section 6.3.3.5 shows and discusses numerical results of the burst loss probability and the C/N.

6.3.3.1 Asynchronous TDMA Bus Link System

Figure 6.12 illustrates the configuration of the proposed asynchronous TDMA bus link system.

Many RBSs and RCSs are connected with a fiber optic bus link having an intermediate routing node. The function of the intermediate routing node is to demultiplex the transferred radio signals and switch them into the desired optical path if they have different destination RCSs.

Radio burst signals received at an RBS directly modulate a laser diode after a digital header signal is attached to recognize the start point of the burst frame, the base station number, and the destination RCS address. As a result, the intensity-modulated optical signal with an optical digital header

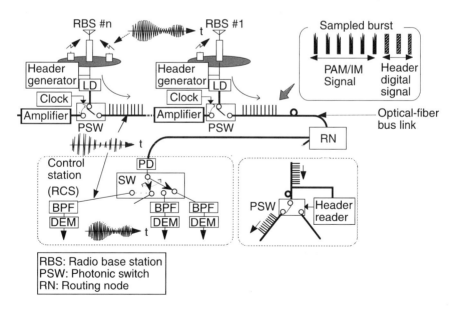

Figure 6.12 Configuration of asynchronous TDMA bus link system. (DEM: demodulator.)

signal is obtained. Then this intensity-modulated optical signal is asynchronously fed into the fiber optic bus link by means of a photonic switch. This photonic switch not only connects an RBS to the optic bus link but also performs the natural bandpass sampling of the optical intensity-modulated signal at the same time. Consequently, radio signals launched from many RBSs are asynchronously time-division multiplexed using the pulse amplitude modulated/intensity modulated signal format.

The asynchronous TDMA bus link system has the additional advantages of very flexible construction of the networks because the system can easily extend some RBSs connected to the fiber optic bus link without any changes to the system. Furthermore, fast and broadband operation of the routing node can be expected because the routing node is composed of a simple photonic switch. However, the routing node requires the subsystem to detect a digital header signal and recognize both the destination address and the start point of the pulse amplitude modulated/intensity modulated burst. It seems that the routing node and the RCS are able to handle more complicated configurations than RBSs. However, the asynchronous TDMA system has the problem that burst signals may be lost due to collisions among asynchronously transmitted pulse amplitude modulated/intensity modulated bursts. This problem is discussed in detail in the next section.

At the receiver in the RCS, radio signals can be reproduced by demultiplexing after photodetection, and bandpass filtering, if we use the appropriate sampling frequencies described in Section 6.3.1.1.

6.3.3.2 Radio Burst Collision

In the asynchronous TDMA bus link system, pulse amplitude modulated/intensity modulated bursts generated by radio signals at RBSs are asynchronously multiplexed into the fiber optic bus link. Thus the radio bursts may be lost due to collisions if several RBSs transmit their radio bursts at the same time. Figure 6.13 illustrates the mechanism of collision among pulse amplitude modulated/intensity modulated bursts.

A pulse amplitude modulated/intensity modulated burst is lost when its transmission is accidentally terminated at the photonic switch of an intermediate RBS that is transmitting a radio burst at the same time. The occurrence of a collision between two radio bursts drops the one transferred from the RBS that is located further from the RCS. In other words, the collision between the radio burst from jth RBS and that from kth RBS ($j < k$) causes the loss of the radio burst from kth RBS. Consequently, the throughput of radio bursts decreases, and the quality of the received radio signal deteriorates.

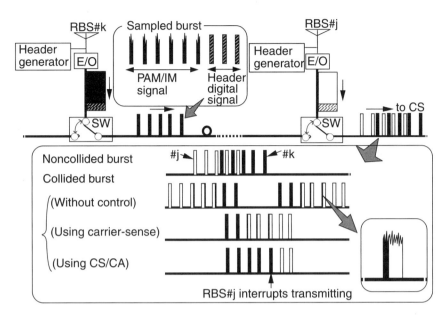

Figure 6.13 Mechanism of collision among pulse amplitude modulated/intensity modulated (PAM/IM) bursts.

One of the methods used to reduce the burst loss probability is to reduce the pulse width of pulse amplitude modulated/intensity modulated signals, because even if two RBSs transmit radio bursts at the same time, the burst loss does not occur unless their pulse transmitting time is just the same. However, the reduction in pulse width causes a decrease in the signal power detected at the RCS under the condition of the fixed optical peak power. The relationship between the pulse width reduction and the degradation of C/N performance is discussed in Section 6.3.3.5. Here we consider the use of some access controls in order to reduce the burst loss probability without the reduction of received signal power: carrier sense control, and carrier sense/collision avoidance control. In the theoretical analysis of burst loss probability, we consider the following three types of access control:

1. *No control:* Each RBS transmits the pulse amplitude modulated/ intensity modulated bursts without any attention to other busts on the fiber optic bus link. This type of system is the most fundamental and simplest one.

2. *With carrier sense control:* Each RBS is always monitoring the existence and the timing of optical pulses that are already fed into

the bus link from other RBSs. When an RBS tries to transmit its own signals, it can start its sampling and transmission at a different time than the monitored pulses and can avoid the collision. However, the optical pulses that the RBS can monitor are limited to those transmitted from RBSs located further away than that.

In other words, further RBSs cannot recognize whether the nearer RBSs are transmitting signals or not and, of course, the timing of the optical pulses. Therefore, they are unable to avoid the collision even if carrier sense control is applied. In such a case, if the further RBS starts the transmission of its optical pulses when a nearer RBS is already transmitting, the pulses may be lost due to the collision.

3. *With carrier sense and carrier sense/collision avoidance (CS/CA) control:* In the above, we mentioned that we cannot avoid collisions perfectly even if carrier sense control is applied. However, collisions can be detected at each RBS by the carrier sense. Then in this type of control, each RBS not only monitors transferred optical pulses, but also interrupts its own transmission of signals when the collision is detected. Hence, the further RBS on the optical bus link has the higher priority of transmitting the signals than the nearer RBS, in contrast to control types 1 and 2 listed above, in which the signals transmitted from further RBSs are lost when a collision occurs.

Another technique to reduce the probability of burst collision is pulse width control of the pulse amplitude modulated/intensity modulated bursts. The improvement due to pulse width control is based on the fact that the narrower pulse width is sufficient for the nearer RBSs to attain the required received signal power. The improvement on performance due to this technique is discussed in Section 6.3.3.5.

6.3.3.3 Theoretical Analysis of Burst Loss Probability

In this section, we theoretically analyze burst loss probability. RBSs are numbered from one, with RBS 1 being the RBS nearest the RCS.

Concerning the radio signal model, we assume that one radio carrier is used to multiplex all users with the TDMA format in a radio zone. We also assume that the length of the TDMA radio burst is a fixed value, H (in seconds), and that the interval between two successive bursts has an exponential distribution with mean a. According to these assumptions, mean traffic per RBS is given by

$$A_{\text{RBS}} = \frac{H}{a + H} \quad [\text{erl/RBS}] \tag{6.22}$$

Let A_{kj} denote the event that a pulse amplitude modulated/intensity modulated burst from the jth RBS is not lost due to a collision with bursts from the kth RBS ($k \neq j$) when the jth RBS receives a radio burst at time t_0 and transmits it into the fiber optic bus link. In this chapter, the term *collision* refers to the case in which two bursts overlap and the optical pulses included in them further overlap. Thus, the probability of the event A_{kj}, $P(A_{kj})$, is given by

$$P(A_{kj}) = 1 - P_{\text{over}} \cdot p_c \tag{6.23}$$

where P_{over} is the probability that bursts overlap and P_c is the probability that the included optical pulses overlap.

In the following analysis, $P(A_{kj})$ is derived first, and then we derive throughput, $P_{\text{through}j}$, which is the probability that a certain burst from the jth RBS successfully reaches the RCS without colliding with other bursts. Because traffic controls in all radio zones are mutually independent, $P(A_{kj})$ ($k = 1, 2, \ldots, M$) is mutually independent of the value of k. Therefore, $P_{\text{through}j}$ is given by

$$P_{\text{through}j} = \prod_{k=1}^{M} P(A_{kj}) \tag{6.24}$$

where M is the total number of RBSs connected to a bus link. By the use of $P_{\text{through}j}$, the burst loss probability of the jth RBS's burst, $P_{\text{loss}j}$, is written by

$$P_{\text{loss}j} = 1 - P_{\text{through}j} \tag{6.25}$$

Now, to derive $P(A_{kj})$ we consider two cases as shown in Figures 6.14(a) and 6.14(b), respectively: One is the case in which the jth RBS starts the transmission during the transmission of the kth RBS [Figure 6.14(a)]. For example, of $k < j$, the kth RBS is just transmitting a burst when the burst from the jth RBS reaches the kth RBS's photonic switch, and for $k > j$, the jth RBS starts to transmit a burst while the burst from kth RBS is passing through the jth RBS's photonic switch. Figure 6.14(b) shows the case in which the jth RBS starts its transmission in the presence of no other burst on the bus link.

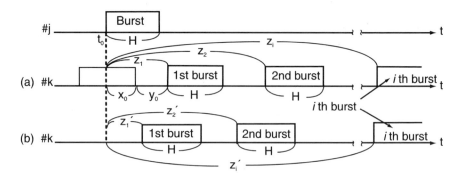

Figure 6.14 Collision among bursts: case that the #jth RBS starts the transmission (a) just during the transmission of the other #kth RBS and (b) in the presence of no other burst on the bus link.

For Figure 6.14(a), let z_i denote the time interval between the start point of the burst from the jth RBS, t_0, and that of ith burst from the kth RBS. Here z_1 is composed of two portions, x_0 and y_0, as shown in Figure 6.14.

For Figure 6.14(b), z_i' is defined in a manner similar to that for z_i in Figure 6.14(a). The probability that a state is represented by the case shown in Figure 6.14(a) or (b), respectively, P_a and P_b. These values are derived from the probability that a certain RBS is transmitting a burst at an arbitrary time. Because this probability is given by mean traffic per RBS, A_{RBS} (erlangs), P_a and P_b are given by $H/(a + H)$ and $a/(a + H)$, respectively.

Analysis of No Control Case

Let $P_{\mathrm{over}}(i)$ denote the probability that a burst from the jth RBS overlaps with i bursts from kth RBS. Taking into account that burst length is the fixed value H, $P_{\mathrm{over}}(i)$ is given by

$$P_{\mathrm{over}}(0) = p_b \int_H^\infty P_{z_1'}(z_1')dz_1' \tag{6.26}$$

$$P_{\mathrm{over}}(1) = p_a \int_H^\infty P_{z_1}(z_1)dz_1 + p_b \int_H^\infty P_{z_1'}(z_1')dz_1' \tag{6.27}$$

$$P_{\text{over}}(2) = p_a \int_0^H p_{z_1}(z_1) dz_1 \tag{6.28}$$

$$P_{\text{over}}(i) = 0 \qquad \text{for } i \geq 3 \tag{6.29}$$

where $p_{z_i}(z_i)$ and $p_{z_i'}(z_i')$ are the probability density function (pdf) of z_i and z_i', respectively. x_0 is a uniform distributed variable in the interval $0 \leq x_0 \leq H$, and y_0 is an exponential distributed variable with mean a. Because x_0 and y_0 are mutually independent, the pdf of $z_1 = x_0 + y_0$, $p_{z_1}(z_1)$, is derived from the convolution of $p_{x_0}(x_0)$ and $p_{y_0}(y_0)$ as

$$p_{z_1}(z_1) = \begin{cases} \dfrac{1}{H}(1 - e^{-z_1/a}) & 0 \leq z_i \leq H \\[2mm] \dfrac{1}{H} e^{-z_1/a}(e^{1/a} - 1) & z_i > H \\[2mm] 0 & z_i < 0 \end{cases} \tag{6.30}$$

However, z_1' is also an exponential distributed variable with mean a, and its pdf is given by

$$p_{z_1'}(z_1') = \begin{cases} \dfrac{1}{a} e^{z_1/a} & z_1' \geq 0 \\[2mm] 0 & \text{otherwise} \end{cases} \tag{6.31}$$

Substitution of (6.30) and (6.31) into (6.26)–(6.29) gives

$$P_{\text{over}}(0) = \frac{1}{1 + (H/a)} e^{-H/a} \tag{6.32}$$

$$P_{\text{over}}(1) = \frac{2}{1 + (H/a)} (1 - e^{-H/a}) \tag{6.33}$$

$$P_{\text{over}}(2) = \frac{(H/a)}{1 + (H/a)} \left\{ 1 + \frac{a}{H} (e^{-H/a} - 1) \right\} \tag{6.34}$$

Now, a burst is composed of optical pulse amplitude modulated/intensity modulated pulses with sampling interval T_s and pulse width τ_j.

Then, two bursts do not collide unless their optical pulses overlap with each other. Because the sampling time can be modeled as a uniform distributed random variable in the interval $\left[-\dfrac{T_s}{2}, \dfrac{T_s}{2}\right]$, the probability that a pulse in the burst from kth RBS overlaps with one from the jth RBS, p_{cjk}, is given by

$$p_{cjk} = \begin{cases} \dfrac{\tau_j + \tau_k}{T_s}; & \tau_j + \tau_k \leq T_s \\ 1; & \tau_j + \tau_k > T_s \end{cases} \qquad (6.35)$$

Therefore, when i bursts from the kth RBS overlap with a burst from the jth RBS, the probability that no optical pulse overlaps between the jth and kth RBS's bursts is given by $(1 - p_{cjk})^i$. By averaging it with respect to i, $P(A_{kj})$ is obtained as

$$P(A_{kj}) = \begin{cases} \displaystyle\sum_{i=0}^{\infty} (1 - P_{cjk})^i P_{\text{over}}(i) & k < j \\ 1 & k \geq j \end{cases} \qquad (6.36)$$

Finally substituting (6.36) into (6.24) gives

$$P_{\text{through}j} = \prod_{K=1}^{M} \sum_{i=0}^{\infty} (1 - p_{cjk})^i P_{\text{over}}(i) \qquad (6.37)$$

Analysis of the Case with Carrier Sense Control

As a result of carrier sense control use, a burst from the jth RBS is lost only when it reaches the photonic switch of the kth RBS, which is nearer the RCS, and the RBS is just transmitting its own burst [see Figure 6.14(a)]. Hence, $P(A_{kj})$ is given by

$$P(A_{kj}) = \begin{cases} (1 - p_{cjk})p_a + p_b & k < j \\ 1 & k \geq j \end{cases} \qquad (6.38)$$

and substituting (6.38) into (6.24) gives

$$P_{\text{through}j} = \prod_{k=1}^{j-1} [(1 - p_{cjk})p_a + p_b] \qquad (6.39)$$

Analysis of the Case with Carrier Sense/Collision Avoidance Control

In the case of carrier sense/collision avoidance use, the burst from the jth RBS is lost when it collides with bursts from the further kth RBSs ($k = j + 1, \ldots, M$), unlike the case of only carrier sense control use. The jth RBS interrupts to transmit its own burst in order to avoid the collision with bursts from the further RBSs if detecting the occurrence of a collision. Therefore, the burst signal of the jth RBS is lost when it collides with the first burst signal of the further kth RBS in the cases shown in Figures 6.14(a) and 6.14(b). Then $P(A_{kj})$ is given by

$$
P(A_{kj}) = \begin{cases} p_b \int_H^\infty P_{z_1'}(z_1')dz' + p_a \int_H^\infty P_{z_1}(z_1)dz_1 \\ \quad + (1 - p_{cjk})\left[p_a \int_0^H P_{z_1}(z_1)dz_1 + p_b \int_0^H P_{z_1'}(z_1')dz' \right] & k > j \\ 1 & k \leq j \end{cases}
$$

$$(6.40)$$

and substituting (6.40) into (6.24) gives

$$
P_{\text{through}j} = \prod_{k=j+1}^{M} P(A_{kj}) \tag{6.41}
$$

6.3.3.4 Theoretical Analysis of Received C/N

In this section, we theoretically analyze the C/N of a radio burst received at the RCS. We assume that an optical amplifier is equipped just before every photonic switch and before the RCS in order to compensate for fiber transmission loss and switch insertion losses. The power spectral density of the amplified spontaneous emission, N_{sp}, is given by

$$
N_{sp} = \frac{n_{sp}}{\eta_a} \frac{G_a - 1}{G_a} h\nu \tag{6.42}
$$

where G_a, n_{sp}, η_a, and $h\nu$ are the amplifier gain, the spontaneous emission factor, the quantum efficiency of the amplifier, and the photon energy, respectively. In the following analysis, we ignore the nonlinearity of laser diode and the noise generated in the radio link.

When the kth RBS receives a radio burst, $g_k(t)$, and the pulse amplitude modulated/intensity modulated burst generated from it are transferred to the RCS without any collision with other bursts, the photodetector output current, $i_{out}(t)$, at the RCS is written by

$$i_{out}(t) = \sum_k \sum_{l=-\infty}^{\infty} \alpha_o P_r (1 + g_k(t)) p(t - lT_s + t_k) + n(t) \quad (6.43)$$

where $p(t)$, t_k, P_r, and α_o are a rectangular pulse with the unit amplitude and the pulse width of τ_k, the start time of sampling, the average received optical power, and the responsivity of the photodetector, respectively. In this chapter, G_a (in decibels) is assumed to equal the transmission loss between two neighboring RBSs, L (in decibels), which includes the fiber loss and the insertion loss of the photonic switch. Thus, the received optical power from all RBSs, P_r, is equal to the transmitting optical power, P_t, and $1/T_s$ is the sampling frequency, which is given by the double of the bandwidth of radio signal, B_{RF}:

$$\frac{1}{T_s} = 2B_{RF} \quad (6.44)$$

The white noise photocurrent is given by

$$n(t) = i_{RIN}(t) + i_{shot}(t) + i_{th}(t) + i_{s-sp}(t) + i_{sp-sp}(t) \quad (6.45)$$

where $i_{RIN}(t)$, $i_{shot}(t)$, $i_{th}(t)$, $i_{s-sp}(t)$ and $i_{sp-sp}(t)$ are the relative intensity noise current, the shot noise current, the receiver thermal noise current, the beat noise current among amplitude spontaneous emissions and optical signal, and the beat noise current among amplitude spontaneous emissions, respectively. These noise powers in the bandwidth of the radio signal, B_{RF}, are given by

$$\langle i_{RIN}^2 \rangle = RIN(\alpha_o P_r)^2 B_{RF} \quad (6.46)$$

$$\langle i_{shot}^2 \rangle = 2e\alpha_o (P_r + m_k N_{sp} W) B_{RF} \quad (6.47)$$

$$\langle i_{th}^2 \rangle = \frac{4kT_{th}}{R} B_{RF} \quad (6.48)$$

$$\langle i_{s-sp}^2 \rangle = 4\alpha_o^2 m_k N_{sp} P_r B_{RF} \quad (6.49)$$

$$\langle i_{s-sp}^2 \rangle = 2\alpha_o^2 (m_k N_{sp})^2 (W - f_0) B_{RF} \qquad (6.50)$$

where RIN, m_k, W, k, T_{th}, and R are the power spectral density of the relative intensity noise of laser diodes, the number of optical amplifiers between the kth RBS and the RCS, the bandwidth of the optical filter at the RCS, the Boltzmann constant, the noise temperature, and the load resistance, respectively.

After demultiplexing of pulse amplitude modulated signals from the output current $I_{out}(t)$ by the distributor, and filtering by the ideal bandpass filter with the center frequency of f_0 and the bandwidth of B_{RF}, we can reproduce an original radio burst, $\hat{g}_k(t)$, which is given by

$$\hat{g}_k(t) = \frac{\tau_k}{T_s} \alpha_o P_r g_k(t) \qquad (6.51)$$

Therefore, the carrier power of $\hat{g}_k(t)$ is given by

$$\langle i_k^2 \rangle = \frac{1}{2} \left(\frac{\tau_k}{T_s} \right)^2 (\alpha_o P_r)^2 \qquad (6.52)$$

However, after the demultiplexing and filtering, the noise power $n_{PAM}(t)$ is equal to τ_k / T_s times the power of $n(t)$. From (6.46) through (6.50) and (6.52), consequently, the received C/N of the reproduced radio signal at the RCS is given by

$$\left(\frac{C}{N} \right)_{RCS} = \frac{\langle i_k^2 \rangle}{\frac{\tau_k}{T_s} \left[\langle i_{RIN}^2 \rangle + \langle i_{shot}^2 \rangle + \langle i_{th}^2 \rangle + \langle i_{s-sp}^2 \rangle + \langle i_{sp-sp}^2 \rangle \right]} \qquad (6.53)$$

$$= \frac{\frac{1}{2} \frac{\tau_k}{T_s} (\alpha_o P_r)^2}{\left[RIN(\alpha_o P_r)^2 + 2e\alpha_o \{ P_r + m_k N_{sp} W \} + \frac{4kT_{th}}{R} + 4\alpha_o^2 m_k N_{sp} P_r + 2\alpha_o^2 (m_k N_{sp})^2 (W - f_0) \right] \frac{1}{2T_s}}$$

6.3.3.5 Performance Evaluations

In this section, the burst loss probability and the C/N are calculated and compared among the three types of transmission control: no control, with carrier sense control, and with carrier sense/collision avoidance control. Parameters used for the calculations are shown in Table 6.3.

Performance in Case of No Control

Figure 6.15 shows burst loss probabilities versus pulse width of pulse amplitude modulated/intensity modulated bursts for different values of average traffic A_{RBS}. The abscissa is the pulse width normalized by the sampling interval T, that is, the pulse duty. In this figure, numerical results of P_{loss} for the tenth RBS's bursts are shown assuming that 10 RBSs are connected

Table 6.3
Parameters Used in Calculations

RIN	−152 dB/Hz	f_0	1 GHz
α_0	0.8 A/W	R	50Ω
n_{sp}	2.0	T_{th}	300K
η_a	0.5	L	5 dB
W	1 THz	P_f	10 dBm

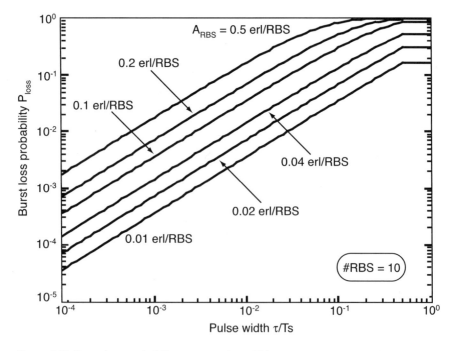

Figure 6.15 Burst loss probability versus pulse width.

in the fiber optic bus link. We can see that burst loss probability is much improved as pulse width is reduced for any values of A_{RBS}.

Figure 6.16 indicates required pulse width for the tenth RBS's bursts to satisfy $P_{loss} = 10^{-3}$ as a function of average traffic per RBS. The figure also shows the received C/N when the value of the required pulse width is used. The required pulse width and the received C/N in the synchronous TDMA (STDMA) method are also shown so as to compare system performance. We can see that the received C/N in the proposed system deteriorates in comparison with STDMA because we must reduce the pulse width into $1/100 \sim 1/1,000$ in comparison with that in the STDMA system in order to satisfy the quality of $P_{loss} = 10^{-3}$. However, even under the condition of $\tau/T_s = 10^{-4}$ and $A_{RBS} = 0.4$ (in erlangs/RBS), the received C/N of more than 40 dB is obtained at the RCS.

Figure 6.17 shows burst loss probability and received C/N versus the RBS number, which is numbered from RCS for the case of $\tau/T_s = 10^{-3}$. In case of no control, P_{loss} and C/N deteriorate as the RBS is located farther

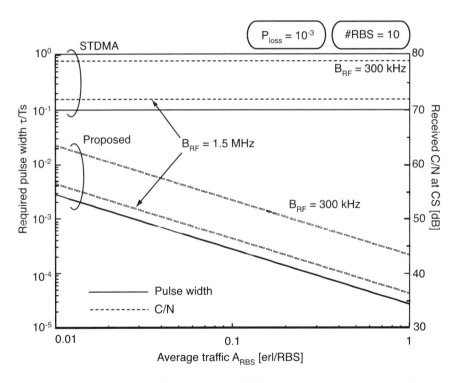

Figure 6.16 Required pulse width and received C/N as a function of average traffic per RBS.

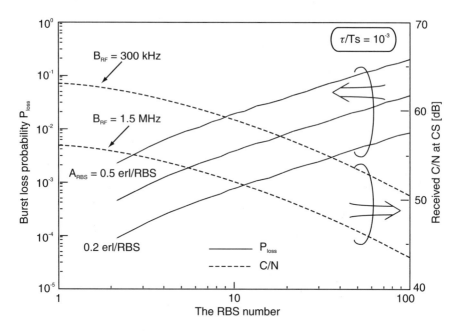

Figure 6.17 Burst loss probability and received C/N versus the RBS number.

from the RCS (i.e., as the RBS number increases), because of the use of a photonic switch to multiplex pulse amplitude modulated/intensity modulated bursts. However, we can easily add some RBSs without any troublesome changes to the whole system if we can put up with a little deterioration of burst loss probability and received C/N.

Performance Improvements as a Result of Carrier Sense, Carrier Sense/ Collision Avoidance, and Variable Pulse Width Controls

Figure 6.18 shows the relationship between burst loss probability and RBS number for three types of transmission control (no control, carrier sense control, and carrier sense/collision avoidance (CS/CA) control) when 10 RBSs are connected to the bus link. In this figure, the solid lines show P_{loss} for the case in which all RBSs use the same pulse width by which the C/N of the farthest RBS, that is, the most deteriorated C/N obtained in the system, can attain the required C/N of 40 dB. As shown in Figure 6.17, the burst from the RBS nearer to the RCS has a better-received C/N at the RCS; in other words, the narrower pulse width is sufficient for the nearer RBS to obtain the required C/N. Thus, it is expected that the burst loss probability would be reduced due to such pulse width reduction. Then, the dashed lines in this figure show the burst loss probability for the case in

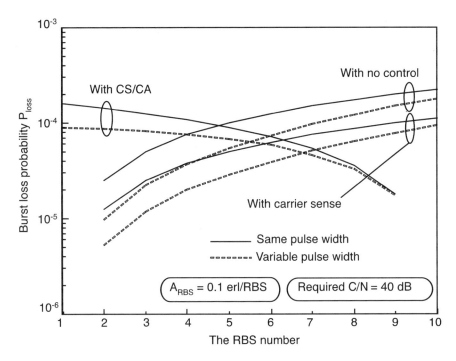

Figure 6.18 Burst loss probability and received C/N versus the RBS number with variable pulse width and control.

which each RBS uses an independent pulse width by which each RBS can attain the required C/N (variable pulse width).

We can also see from Figure 6.18 that using a variable pulse width is more effective to improve P_{loss} for the nearer RBS in any types of control. We can see from the figure that we can reduce burst loss probability to half of that with no control by using carrier sense control. However, the characteristic of P_{loss} in the system with carrier sense/collision avoidance is reverse to that of the systems with no control or with carrier sense because the RBS located the farthest away has the higher priority to transmit bursts. Comparing all types of control methods, we found that the carrier sense method with variable pulse width is the best choice for minimizing the worst P_{loss}. From another viewpoint, that of system simplicity, the method without any carrier sense and only with variable pulse width is effective.

Figure 6.19 shows the number of connected RBSs versus pulse width in the case of no control or carrier sense control. The number of RBSs connected is defined as the maximum number of RBSs that can satisfy a required burst loss probability of 10^{-3} when all RBSs use the same pulse

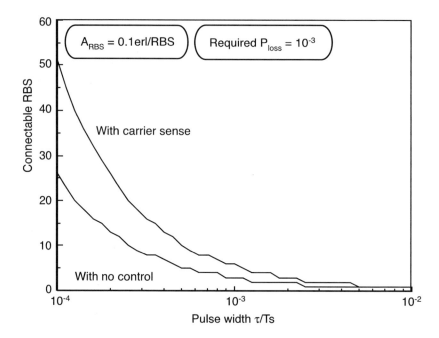

Figure 6.19 Maximum number of connectable RBSs versus pulse width in the network with carrier sense or with no control.

width. We can see that the system with carrier sense control can connect double numbers of RBSs in comparison with the system with no control for any values of pulse width. However, the received C/N decreases as the total number of RBSs increases. But even when 50 RBSs are connected to the bus link using a pulse duty of 10^{-4}, a C/N of more than 36 dB can be obtained by calculations under the condition of $B_{RF} = 1.5$ MHz.

6.3.3.6   Summary

This section describes the asynchronous TDMA bus link system for fiber optic radio access networks that can remove the difficulty of time synchronization control of time-division multiplexing of signals and can provide the capability of easy extension of networks. We showed that performance degradation due to collision among signals occurs. Then the carrier sense control and variable pulse width technique were proposed to improve the system performance.

We have theoretically analyzed the burst loss probability and the C/N performance of the proposed system.

- We can multiplex and transfer radio signals without any time synchronization among RBSs with a burst loss probability of 10^{-3} and received C/N of more than 40 dB by using a photonic natural bandpass sampling technique.

- The use of carrier sense control reduces the burst loss probability to half of that for the case of no control.

- Variable pulse width technique can improve the burst loss probability and it is more effective for the performance of the nearer RBS.

6.4 Photonic CDMA

Section 6.2 described the various types of multiple access methods for fiber optic radio networks and compared these methods. Among them, the CDMA method is suitable for the uplink of fiber optic radio networks. In conventional mobile communication using CDMA modulation, the radio signal is spread and despread at the electrical stage. So the radio highway networks of the conventional electrical CDMA method will use optical intensity modulation. However, the photonic CDMA method is more suitable than the electrical CDMA method because of the high process gain that is related to the high chip rate of photonic short pulses. Also, the routing and multiple access can be realized in the photonic domain without detecting the radio signal. In this section, conventional photonic CDMA methods are discussed, as is the necessity for a new type of optical CDMA method, direct optical switching (DOS-CDMA).

6.4.1 Conventional CDMA

Two CDMA methods are applied to fiber optic radio highway networks. One is the electrical CDMA and intensity modulation method, in which radio signals are spectrum spread in the electrical domain and converted into photonic intensity modulation signals at the laser diode, and their correlations are performed in the electrical domain after photodetection using the receiver site photodetector, as shown in Figure 6.20.

This method is the most conventional one and works for conventional photonic intensity modulation networks. However, it is difficult to use a chip rate of more than hundreds of megachips per second because of the band limitation of the electrical circuits. As a result, the coding gain is rather low and the number of multiple access RBSs is reduced.

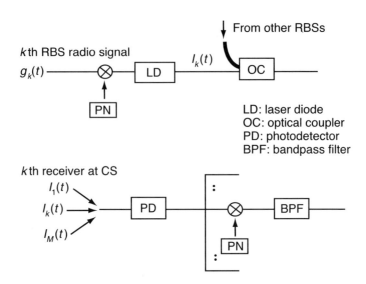

Figure 6.20 Configuration of the conventional electrical CDMA and photonic intensity modulation method.

Another candidate CDMA method is the photonic CDMA method, in which radio signals are spectrum spread and correlated in the optical domain. The optical CDMA method is more suitable than the conventional electrical CDMA method because radio signals are multiplexed in the optical domain, and high processing gains can be obtained by using the high chip rate of photonic short pulses and broad bandwidth of optical devices.

Up to this time, various optical CDMA methods have been studied mainly for digital optical LANs. Optical CDMA methods such as using fiber delay lines [13–17], optical phase masks [18, 19], and coherent optical phase modulations [20–22] have been proposed and studied.

Figure 6.21 illustrates examples of conventional optical CDMA methods. In the optical CDMA method using fiber delay lines, the encoding and the decoding are performed by the sum of delayed optical pulses in the time domain. In the optical CDMA method using optical phase masks, the spectral encoding and the decoding are performed by phase modulation in the optical. However, these methods have no flexibility in assigning code sequences, and these methods cannot be applied for multiple access for fiber optic radio highway networks. This optical phase CDMA method requires coherent optical phase modulators, and its correlators at the receiver site are very complicated and need very fine accuracy in the optical phase that is very sensitive to temperature changes.

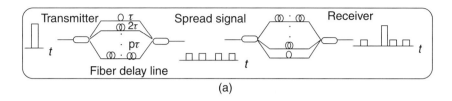

(a)

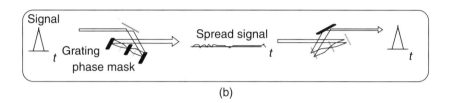

(b)

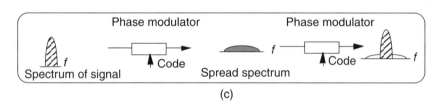

(c)

Figure 6.21 Conventional photonic CDMA methods: optical CDMA using (a) fiber delay lines in the time domain, (b) phase masks in the optical frequency domain, and (c) the coherent phase modulator.

However, an optical CDMA method that uses optical disk patterns has been proposed; however, in this method, the assignment of code sequences is not flexible.

Therefore, optical CDMA methods for fiber optic radio highway networks from the viewpoints of flexibility in assigning code sequences and the ability of optical CDMA to use optical intensity modulation is investigated for the solution of this code flexibility and simplicity.

6.4.2 DOS-CDMA Using the Gold Code

The remainder of this section is composed as follows: In Section 6.4.2.1, we describe the principle of the DOS-CDMA scheme. In Section 6.4.2.2, the two types of bus connection methods that can be used in the DOS-CDMA system are compared. The first type is the optical coupler connection type and the other is the optical switch connection type.

In the optical switch connection-type system, an optical switch is used not only to spread the spectrum of the optical signals but also to launch them into the fiber optic bus link. In Section 6.4.2.3, we theoretically analyze the carrier to interference-plus-noise ratio (CINR) of the radio signals regenerated at the RCS for two types of bus connection methods considering the chip pulse erasure at the optical switch. In Section 6.4.2.4, some numerical results are discussed and compared between two types of bus connection methods [23].

6.4.2.1 Principle Behind the DOS-CDMA Method

The DOS-CDMA scheme uses on-off type switching spectral spreading. Each radio signal received at each RBS is transmitted to an RCS by an analog-type optical pulse amplitude modulation scheme, and the multiple access among many RBSs is performed by the DOS-CDMA scheme. The regeneration of the radio signal is based on the bandpass natural sampling theory described in Section 6.3.1.1. Figure 6.22 shows the system model of the DOS-CDMA system, which illustrates the principle of the DOS-CDMA process at the RBS transmitter and the correlating process at the RCS receiver.

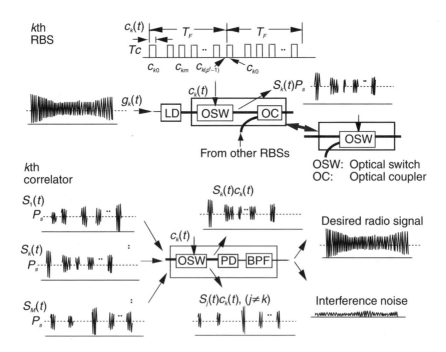

Figure 6.22 Principle of the DOS-CDMA process.

At an RBS, a received radio signal is converted into an optical intensity modulated signal by modulating the laser diode directly and is next sampled at an optical switch, which is driven with a certain code sequence, $c_k(t)$, and the output signal of the optical switch is transmitted to a receiver through optical fiber. At the output of the optical switch in the transmitter, we can obtain optical pulse amplitude modulated/intensity modulated signals whose pulses are positioned according to the code pattern of $c_k(t)$. At the receiver, many pulse amplitude modulated/intensity modulated signals from many RBSs are correlated with the code sequence, $c_k(t)$, at an optical switch, then directly detected at a photodetector and interpolated at a bandpass filter to regenerate the desired radio signal. It is assumed that the code sequence $c_k(t)$ matched with the one at the RBS is regenerated at the RCS and the synchronization between two code sequences is taken. The radio signal, which is contaminated by all other radio signals, is fed into a demodulator in order to obtain the information data.

The radio signal $g_k(t)$ at the kth RBS is represented by

$$g_k(t) = \text{Re}[a_k(t)e^{j2\pi f_{\text{RF}}t}] \qquad (6.54)$$

where f_{RF} is RF and $a_k(t)$ is the complex envelope with its bandwidth, B_{RF}. The optical on-off switching CDMA is performed at the optical switch driven by a code sequence, $c_k(t)$, whose frame period and chip width are T_F (sec) and T_c, respectively, and its pulse amplitude is 1 or 0. The intensity of the optical pulse amplitude modulated/intensity modulated signal at the output of the optical switch is given by

$$P_k(t) = P_S\{1 + g_k(t)\}e_k(t) \qquad (6.55)$$

where P_s is the average transmitted optical power before optical switching.

The $P_k(t)$ value is a bandpass natural sampled signal converted from a radio signal because optical switching is a window-type sampling. Therefore, a radio signal can be conveyed by the optical carrier with its signaling format kept and regenerated from the pulsed signals by the interpolation at a bandpass filter if we choose a sampling period of less than or equal to half of the inverse of the radio signal bandwidth, $\leq 1/(2B_{\text{RF}})$. In the proposed DOS-CDMA system, since a pseudorandom sequence is chosen as a code sequence which drives the optical switch at the transmitter, the durations between optical pulses become various values according to the type of code sequence used, but each pulse is surely repeated with its frame period of T_F. Therefore, in order to regenerate the radio signal after interpolation,

T_F of less than or equal to $1/(2B_{RF})$ should be chosen. From the viewpoint of simplicity, using T_F of much less than $1/(2B_{RF})$ is not effective, because a much faster speed for the optical switch is required. Hence, in this paper, T_F is set to be the maximum value, that is, $1/(2B_{RF})$.

To improve the quality of the regenerated radio signal in the DOS-CDMA system, a code sequence with the highest possible autocorrelation and the lowest possible cross-correlation has to be chosen. In the DOS-CDMA using an optical intensity modulated/direct detection scheme, a uniphase code has to be used as a code sequence, while PN codes like the maximum length or Gold code are used in conventional radio CDMA systems. References [13–17] have reported that a prime code sequence is the best code because a uniphase orthogonal code can provide the highest autocorrelation and the lowest cross-correlation of various orthogonal codes. So in the proposed DOS-CDMA system, the prime code sequence is employed as a spread spectrum (SS) code.

A set of prime codes has the preferable feature of an intensity modulated/direct detection CDMA system in which there are very few coincidences of 1's among code sequences. Prime codes with length p^2 are derived from prime sequences obtained from a Galois field, GF(p), where p is a prime number. Table 6.4 shows an example of prime sequences and prime code sequences for a prime number p of 7. Each prime sequence element s_{mn} is obtained by the product of the corresponding m and n modulo p. Letting $c_m = (c_{m0}, c_{m1}, \ldots, c_{mj}, \ldots, c_{m(p-1)})$ denote the mth prime code sequence, a jth code element, c_{mj}, is given by

$$c_{mj} = \begin{cases} 1 & j = s_{mn} + np \\ 0 & \text{otherwise} \end{cases} \tag{6.56}$$

In the DOS-CDMA scheme using prime codes, $T_F (= p^2 T_c ac)$ is set to $1/(2B_{RF})$ in order to gain the largest code length, p^2, at the same switching speed of the optical switch; thus, the chip width, T_c, is given by T_F/p^2. When an optical switch correlator is driven with the kth prime code sequence, $c_k(t)$, at the receiver, the optical pulse amplitude modulated/intensity modulated signal transmitted from the kth RBS is extracted from all CDMA signals. Then the output current of the photodetector is composed of a desired signal component, $S_k(t)$, interference components, $I(t)$, and additive noise components, $N(t)$. $S_k(t)$ and $I(t)$ are given, respectively, by

$$S_k(t) = \alpha P_r g_k(t) c_k(t) \tag{6.57}$$

Table 6.4
Prime Sequences and Prime Code Sequences for Prime Number

			Prime Sequences for $p = 7$				
m	s_{m0}	s_{m1}	s_{m2}	s_{m3}	s_{m4}	s_{m5}	s_{m6}
0	0	0	0	0	0	0	0
1	0	1	2	3	4	5	6
2	0	2	4	6	1	3	5
3	0	3	6	2	5	1	4
4	0	4	1	5	2	6	3
5	0	5	3	1	6	4	2
6	0	6	5	4	3	2	1

			Prime Code Sequences for $p = 7$				
m	c_{m0}	c_{m1}	c_{m2}	c_{m3}	c_{m4}	c_{m5}	c_{m6}
0	1000000	1000000	1000000	1000000	1000000	1000000	1000000
1	1000000	0100000	0010000	0001000	0000100	0000010	0000001
2	1000000	0010000	0000100	0000001	0100000	0001000	0000010
3	1000000	0001000	0000001	0010000	0000010	0100000	0000100
4	1000000	0000100	0100000	0000010	0010000	0000001	0001000
5	1000000	0000010	0001000	0100000	0000001	0000100	0010000
6	1000000	0000001	0000010	0000100	0001000	0010000	0100000

$$I(t) = \alpha P_r \sum_{j=1, j \neq k}^{M} g_j(t) c_j(t) c_k(t) \tag{6.58}$$

where α, P_r, and M are the responsivity of the photodetector, the average received optical power at the correlator, and the total number of connected RBSs, respectively. Here, we derive the carrier-to-interference power ratio (CIR) in the DOS-CDMA system. We derive the power spectral density of signal component, $S_k(f)$, by calculating the autocorrelation of $S_k(t)$ from (6.4). The power spectral density of $S_k(t)$, $S_k(f)$, is given by

$$S_k(f) = (\alpha P_r)^2 \frac{1}{p^2} \left(1 + \frac{1}{p} - \frac{1}{p^2} \right) \tag{6.59}$$

$$\left\{ G_k(f) + \frac{p-1}{p^2 + p - 1} \sum_{i=-\infty, i \neq 0}^{\infty} \text{sinc}^2 \left(\frac{\pi i}{p} \right) G_k(f - 2iB_{\text{RF}}) \right\}$$

where $G_k(f)$ is the power spectrum of $g_k(t)$ and $\text{sinc}(x)$ is $\sin(x)/x$.

Figure 6.23 shows the normalized both-sides power spectral density (psd) of the signal component. The first three terms of $S_k(f)$ are the desired signal component around f_{RF} and $-f_{RF}$, and the other terms are the frequency-shifted components caused by bandpass sampling. Images of these shifted components cause the distortion in the desired signal as the self-interference if they overlap over the signal components as shown in Figure 6.23(a). We can perfectly remove the self-interference components by setting the value of the RF f_{RF} at $(j + 1/2)B_{RF}$ or j/T_c (where j is an integer) as shown in Figure 6.23(b). Without such special values of f_{RF}, however, the self-interference component may not deteriorate the signal quality because its

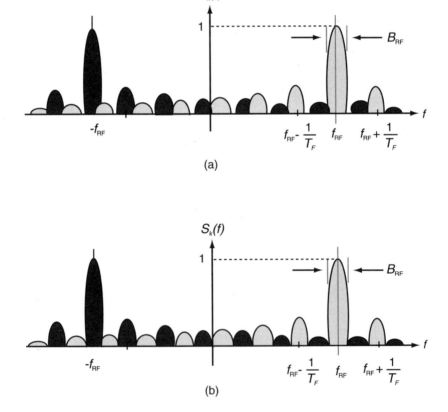

Figure 6.23 Normalized both-sides psd of the signal component: two components of desired signal and frequency shifted signal are (a) overlapped and (b) not overlapped.

power is much lower compared with that of the carrier signal component. We examine the carrier signal-to-self-interference power ratio (CSIR) and Figure 6.24 shows some numerical results for the f_{RF} of 1.93 GHz and B_{RF} of 900 kHz. In the small value of p, the sinc function causes the up and down movement in the CSIR, but as p increases, CSIR tends to be a saturated value of more than 30 dB, which is a good enough value to obtain the radio signal quality in DOS-CDMA. As p increases, the saturated value of CSIR is determined by the relation between f_{RF} and B_{RF}. Therefore, in the following analysis, we will ignore the self-interference components.

From (6.59), the carrier power of the regenerated radio signal, C_0, is given by

$$C_0 = (\alpha P_r)^2 \frac{1}{p^2} \left(1 + \frac{1}{p} - \frac{1}{p^2} \right) \tag{6.60}$$

Let σ_c^2 denote the average variance of the cross-correlation of the prime code. Then, CIR_0 is given by [14–18]

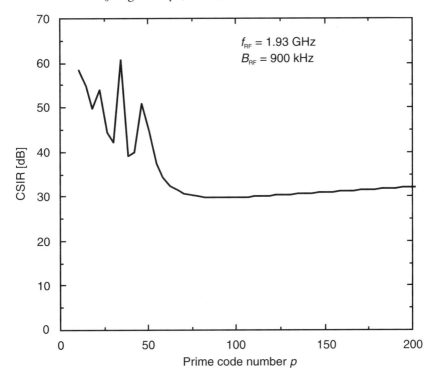

Figure 6.24 Relationship between CSIR and prime code number.

$$\mathrm{CIR}_0 = \frac{C_0}{I_0} = \frac{p^2}{\sigma_c^2 (M - 1)} \qquad (6.61)$$

Table 6.5 shows σ_c^2 for different values of prime number p calculated by using computer simulation. Table 6.5 shows that σ_c^2 is increased only slightly as p increases but has a saturated value of 0.329 for $p = 97$.

6.4.2.2   System Configuration

Figure 6.25 illustrates the configuration of the proposed DOS-CDMA cable-to-the-air system. From the viewpoint of a cost-effective configuration, we adopt a bus-type fiber optic link for the cable-to-the-air system. M radio zones are connected to the bus link, where the radio signals from radio terminals in each zone are multiplexed by the DOS-CDMA scheme and transmitted to the RCS. The RBS in each zone equips only a laser diode, an optical switch, and an automatic gain controller.

As mentioned in Section 6.4.2.1, after the direct intensity modulation of the laser diode, the optical on-off switching spread spectrum is performed at the optical switch, and in the bus-type fiber link, many DOS-CDMA signals are multiplexed by CDMA. At the receiver, received optical powers are different among the received DOS-CDMA signals from M RBSs because the optical loss between each RBS and RCS is different. Also, intensity modulated indices are different among the received CDMA signals because the radio signal received by the RBS has various amplitudes due to fading and the different distance between terminals and an RBS. These differences cause the near-far problem in a CDMA system. For this reason, RBSs are equipped with an automatic gain controller to control the amplitude of a

Table 6.5
Average Variance of the Cross-Correlation of the Prime Code

p	σ_c^2
7	0.272
11	0.298
13	0.303
23	0.318
31	0.322
47	0.326
71	0.328
97	0.329

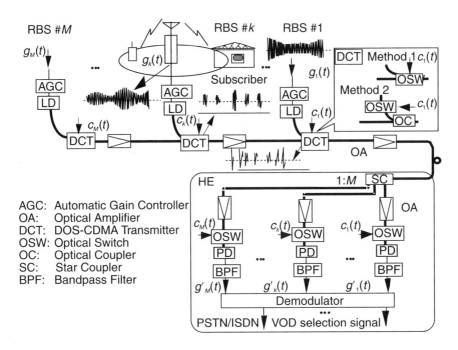

Figure 6.25 Configuration of the proposed DOS-CDMA system.

received radio signal in order to keep the optical modulation index constant at the laser diode. They are also equipped with an optical axis to compensate optical loss between two RBSs.

At the RCS, optical CDMA signals from RBSs are at first power-split into each of M receivers, then matched with one of different prime codes at each optical switch correlator and detected at the photodetector. Finally, the desired radio signal of each RBS is regenerated by the interpolation in BPF and then fed to a demodulator. In the conventional bus-type fiber optic link, each node is usually connected to a bus link with a passive optical coupler. In this case, insertion loss and coupling loss are seen in the coupler. In the proposed DOS-CDMA cable-to-the-air system where an optical coupler connection is used, the optical signal beat noise is caused by interference between two lights arriving at the photodetector at the same time. However, an optical switch is used not only to perform switching spread spectrum of the optical intensity modulated signal but also to launch them into the bus link.

In this section, we propose a configuration for the cable-to-the-air system that uses an optical coupler connection and an optical switch connection. These two connection methods at the RBSs are shown in Figure 6.25.

In the case of the optical switch connection, when an intensity modulated/CDMA signal from RBSs that are further away from the RCS arrives at the optical switch of those RBSs that are closer and transmitting their signals, the signal collision at the optical switch (OSW) of each RBS will cause the erasure of some chip pulses in the DOS-CDMA signal as shown in Figure 6.26. Details are given in the next section.

6.4.2.3 Theoretical Analysis of Optical Coupler Connection Bus

In this section, we theoretically analyze the CINRs of the regenerated radio signals at the RCS for an optical coupler connection bus.

At the kth correlator, the average received optical power, P_{r_k}, can be written

$$P_{r_k} = 10 \log_{10} P_s - k(L_f + L_{oc}) + kG - 10 \log_{10} M + G_M \quad \text{[dB]}$$

(6.62)

where L_{oc}, L_f, G, and G_M (all in decibels) are the coupling loss plus the insertion loss of an optical coupler, the fiber loss between RBSs, the gain of the optical axis at the RBS, and the gain of the optical axis at the output of a 1:M star coupler, respectively. We assume that an optical axis equipped at each RBS has gain G of $L_f + L_{oc}$ and G_M is equal to $10 \log_{10} M$. From (6.62),

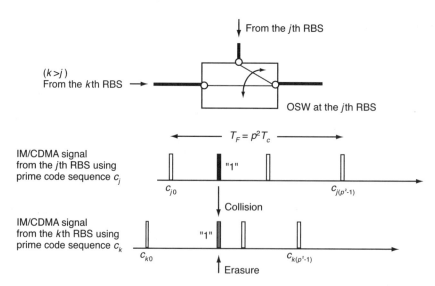

Figure 6.26 Collision between two intensity modulated (IM)/CDMA signals.

$$P_{r_k} = P_s (k = 1, 2, \ldots, M) \qquad (6.63)$$

At the RCS, each correlator receives M optical signals with the same modulation index of 1 and the same received power, $P_r = P_s$; thereby, the carrier power and the CINR of the regenerated radio signal are C_0 and CIR_0 given by (6.60) and (6.61) in the case of $P_r = P_s$, respectively. At the output of the bandpass filter, we consider additive noise currents composed of RIN, shot noise, receiver thermal noise, beat noise among amplified spontaneous emissions of an optical amplifier and optical signal, beat noise among amplified spontaneous emissions, and optical signal beat noise.

Considering that the average number of coincidences of 1's between any prime code sequence pair in the interval of the code frame period, T_F, is 1 (which is analyzed in Section 6.4.2.2), the total noise power, N_c, is written

$$N_c = N_{\text{RIN}} + N_{\text{shot}} + N_{th} + N_{s-sp} + N_{sp-sp} + \langle N_{s-s} \rangle \qquad (6.64)$$

and each power is given by

$$N_{\text{RIN}} = \zeta_{\text{RIN}} \left(\frac{\alpha P_s}{p^2} \right)^2 (p^2 + M - 1) B_{\text{RF}} \qquad (6.65)$$

$$N_{\text{shot}} = 2e \left\{ \left(\frac{\alpha P_s}{p^2} \right) (p + M - 1) + \alpha M (N_{sp} + N_{spM}) W \right\} B_{\text{RF}} \qquad (6.66)$$

$$N_{th} = \frac{4 k_B T}{R_L} B_{\text{RF}} \qquad (6.67)$$

$$N_{s-sp} = 4 \alpha M (N_{sp} + N_{spM}) \left(\frac{\alpha P_s}{p^2} \right) (p + M - 1) B_{\text{RF}} \qquad (6.68)$$

$$N_{sp-sp} = 2 \alpha^2 M^2 (N_{sp} + N_{spM})^2 (W - f_{\text{RF}}) \qquad (6.69)$$

where e, ζ_{RIN}, W, k_B, T, and R_L are the electric charge, the power spectral density of the RIN, the bandwidth of optical filter at the RCS, Boltzmann's constant, the noise temperature, and the load resistance, respectively. The power spectral densities of the amplified spontaneous emissions, N_{sp} and N_{sp_M}, are given by

$$N_{sp} = \frac{\eta_{sp}}{\eta_a} \cdot \frac{10^{G/10} - 1}{10^{G/10}} h\nu \qquad (6.70)$$

$$N_{sp_M} = \frac{\eta_{sp}}{\eta_a} \cdot \frac{M - 1}{M} h\nu \qquad (6.71)$$

respectively, where η_{sp}, η_a, and $h\nu$ are the spontaneous emission factor, the quantum efficiency of the optical axis, and the photon energy, respectively.

The optical signal beat noise, $\langle N_{s-s} \rangle$, is due to an interference between two optical carriers. In this analysis, we assume that the kth RBS uses a laser diode (LD) with its center frequency of f_k and its single-mode Gaussian-shaped spectrum, and we also assume that f_k is a random variable with a uniform probability density. The power spectral density of the optical signal-signal beat noise is given by

$$S_{s-s}(f) = \left(\frac{\alpha P_s}{p^2} \right)^2 \qquad (6.72)$$

$$\sum_{\substack{j=1 \\ k \neq j}}^{M} \sum_{k=1}^{M} \left\{ \frac{1}{2\sigma\sqrt{2\pi}} \exp\left[-\frac{(f - \Delta f_{jk})^2}{2\sigma^2} \right] + \frac{1}{2\sigma\sqrt{2\pi}} \exp\left[-\frac{(f + \Delta f_{jk})^2}{2\sigma^2} \right] \right\}$$

where

$$\Delta f_{jk} = f_j - f_k \qquad (6.73)$$

$$\sigma = \Delta\nu/(2\log 2) \qquad (6.74)$$

$$\Delta\nu = \sqrt{\Delta\nu_{LD} + (p^2 B_{RF})^2} \qquad (6.75)$$

where $\Delta\nu_{LD}$ is the full-width at half-maximum of the laser diode, and $\Delta\nu$ is the full-width at half-maximum after spread spectrum by prime codes with the prime number p. The power spectral density of the optical signal beat noise, $S_{s-s}(f)$, appears in the RF band after photodetection, but its frequency location depends on the frequency difference among laser diodes of M RBSs. So we treat its power N_{s-s} as a random variable and derive its average power $\langle N_{s-s} \rangle$. Assuming that optical carrier frequencies at RBSs, f_j ($j = 1, 2, \ldots, M$) are mutually independent random variables with mean of f_0 and a uniform pdf in the range of $|f_j - f_0| < \dfrac{\Delta F}{2}$, then the power of optical signal beat noise falling in bandwidth B_{RF}, N_{s-s}, is given by

$$
N_{s-s} = \left(\int\limits_{-f_{\mathrm{RF}} - B_{\mathrm{RF}}/2}^{-f_{\mathrm{RF}} + B_{\mathrm{RF}}/2} + \int\limits_{f_{\mathrm{RF}} - B_{\mathrm{RF}}/2}^{f_{\mathrm{RF}} + B_{\mathrm{RF}}/2} \right) S_{s-s}(f)\,df
$$

$$
= \left(\frac{\alpha P_s}{p^2} \right)^2 \sum_{j=1}^{M} \sum_{\substack{k=1, \\ k \neq j}}^{M} \left[erfc \left\{ \frac{-2(f_{\mathrm{RF}} + \Delta f_{jk}) - B_{\mathrm{RF}}}{2\sqrt{2}\sigma} \right\} \right.
$$

$$
- erfc \left\{ \frac{-2(f_{\mathrm{RF}} + \Delta f_{jk}) + B_{\mathrm{RF}}}{2\sqrt{2}\sigma} \right\} + erfc \left\{ \frac{-2(f_{\mathrm{RF}} - \Delta f_{jk}) - B_{\mathrm{RF}}}{2\sqrt{2}\sigma} \right\}
$$

$$
\left. - erfc \left\{ \frac{-2(f_{\mathrm{RF}} - \Delta f_{jk}) + B_{\mathrm{RF}}}{2\sqrt{2}\sigma} \right\} \right\}
$$

(6.76)

We can obtain the average power of optical signal beat noise falling in bandwidth B_{RF}, $\langle N_{s-s} \rangle$, by ensemble averaging N_{s-s} with the pdf of Δf_{jk} ($j, k = 1, 2, \ldots, M, j \neq k$).

Finally, the CINR of the regenerated radio signal in the optical coupler connection system is given by

$$
\mathrm{CINR}_c = \frac{C_0}{N_c + I_0}
$$

(6.77)

6.4.2.4 Theoretical Analysis of Optical Switch Connection Bus

In the case of the optical switch connection, some optical chip pulses in the CDMA signal from kth RBS may be lost at the optical switch of other RBSs that are located between the kth RBS and the RCS. Therefore, the chip pulse erasure has to be taken into consideration for the CINR analysis. First, we theoretically derive the average number of 1's in the prime code sequence successfully reaching the RCS and then analyze the CINR.

At the optical switch of each RBS, an optical intensity modulated signal converted from a radio signal is spectrum spread and simultaneously launched into the fiber optic bus link. If an intensity modulated/CDMA signal from the kth RBS arrives at the optical switch of the jth ($k > j$) RBS which is transmitting its own signal, a signal collision occurs and causes the erasure of some chip pulses from the kth RBS. Figure 6.26 illustrates the collision between two intensity modulated/CDMA signals. When the collision between two intensity modulated/CDMA signals occurs at the optical switch of the jth RBS, we lose the intensity modulated/CDMA signal from the kth RBS located farther than the jth RBS because the multiplexing of intensity modulated/CDMA signals is performed by using an optical switch.

Here, we examine the number of coincidences of 1's between any two prime code sequences in the code word period, T_F. For the 0th prime code sequence, c_0, the number of coincidences of 1's with any other sequences, c_n ($n \neq 0$), is always one in the code frame period T_F, and this property is kept for all shifted versions of c_n ($n \neq 0$). However, between any other two code sequences, c_m and c_n ($m \neq 0$, $n \neq 0$, $m \neq n$), the number of coincidences of 1's yields none, one, or two. In other words, the peak of the cross-correlation function is 1 between c_0 and c_n ($n \neq 0$) sequences, and 1 or 2 between c_m and c_n ($m \neq 0$, $n \neq 0$, $m \neq n$). Let N_{mni} be the number of shifted versions of c_m sequences that has i coincidences of 1's with c_n sequences. For a prime number $p = 2q - 1$, we can find that N_{mni} is

$$N_{0ni} = \begin{cases} 0 & i = 0, 2 \\ 4q^2 - 4q + 1 = p^2 & i = 1 \end{cases} \text{ for } n = 1, 2, \ldots, p - 1$$

(6.78)

$$N_{mni} = \begin{cases} q^2 - q & i = 0, 2 \\ 2q^2 - 2q + 1 & i = 1 \end{cases} \text{ for } m, n = 1, 2, \ldots, p - 1. \; m \neq n$$

(6.79)

In the code frame period, therefore, the average number of coincidences of 1's is one for any prime code sequence pair. In an actual DOS-CDMA cable-on-the-air system, a code sequence collides with another sequence asynchronously, so we have to consider the partial collision between two chip pulses. For the simplicity of analysis, however, we assume the full chip pulse is lost even in this case.

From the above results, when a DOS-CDMA signal from an RBS comes to another optical switch, one chip pulse is erased due to the collision on the average. Moreover, a DOS-CDMA signal transmitted from the kth RBS may lose from 1 to ($k - 1$) chip pulses because it passes through

$(k - 1)$ optical switches to the RCS. Hence, letting χ_k denote the average number of chips successfully reaching the RCS, χ_k is given by

$$\chi_k = p - (k - 1) \tag{6.80}$$

This is a worst-case estimation because the same chip pulse comes into collision with different chip pulses of more than two RBSs. The kth average received optical power, P_{r_k}, at the output of the star coupler is given by

$$P_{r_k} = 10 \log_{10} P_s + kG' - kL_f - 10 \log_{10} M + G_M \quad \text{[dB]} \tag{6.81}$$

$$k = 1, 2, \ldots, M$$

We assume that G_M is equal to $10 \log_{10} M$, as was done for the case of an optical coupler connection. For ease of discussion, we consider $G' = G_c + G_a$ (in decibels). G' is the gain of the optical axis at each RBS to compensate for the optical loss caused by the chip pulse erasure. We set the value of G_c to keep the carrier power of the radio signal from the farthest Mth RBS equal to that in the optical coupler connection at the correlator and G_a (in decibels) is an additional gain. The gain G_c satisfies the following equation:

$$P_{r_M} \frac{\chi_M}{p^2} = P_s \frac{1}{p} \tag{6.82}$$

Consequently, the desired G_c is derived as

$$G_c \text{ [dB]} = \frac{10}{M} \log_{10} \frac{p}{\chi_M} + L_f \tag{6.83}$$

and the average received optical power, P_{r_k}, at the kth correlator is given by

$$P_{r_k} = P_s \left(\frac{p}{\chi_M} \right)^{k/M} g_a^k \tag{6.84}$$

where $G_a = 10 \log_{10} g_a$.

The carrier power of the kth radio signal regenerated at the RCS can be written as

$$C_k = C_0 \left(\frac{\chi_k}{p}\right)^2 \left(\frac{p}{\chi_M}\right)^{2k/M} g_a^{2k} \quad k = 1, 2, \ldots, M \qquad (6.85)$$

However, intensity modulated CDMA signals from RBSs located farther than the kth RBS never reach the kth correlator because they are erased at the optical switch of the kth RBS during the interval of 1's in $c_k(t)$ as shown in Figure 6.26. Thus, we consider only the intensity modulated signals from RBSs nearer than the kth RBS as the interference. Then the interference power contaminating the kth radio signal regenerated at RCS can be written as

$$I_k = \frac{\sigma_c^2}{p^2} \sum_{j=1}^{k-1} C_j \qquad (6.86)$$

Hence, the CIR of the kth radio signal in the optical switch connection system is given by

$$\text{CIR}_k = \frac{C_k}{I_k} = \epsilon_k \text{CIR}_0 \quad k = 1, 2, \ldots, M \qquad (6.87)$$

$$\epsilon_k = \frac{M - 1}{\displaystyle\sum_{j=1}^{k-1} \left(\frac{p}{\chi_M}\right)^{2(j-k)/M} \left(\frac{\chi_j}{\chi_k}\right)^2 g_a^{2(j-k)}} \qquad (6.88)$$

where CIR_0 is the CIR obtained in the optical coupler connection system of (6.61).

In the optical switch connection system, since intensity modulated/CDMA signals from RBSs located farther than the kth RBS never reach the kth correlator, RIN, shot noise, beat noise among amplified spontaneous emissions of an optical axis, and optical signal and beat noise among amplified spontaneous emissions are different from those in the optical coupler connection system as follows:

$$N'_{\text{RIN}} = \zeta_{\text{RIN}} \left(\frac{\alpha P_{r_k}}{p^2}\right)^2 (\chi_k^2 + k - 1) B_{\text{RF}} \quad k = 1, 2, \ldots, M$$

$$(6.89)$$

$$N'_{shot} = 2e \left[\frac{\alpha P_{r_k}}{p^2} (\chi_k + k - 1) + \alpha \left\{ N'_{sp} \sum_{j=1}^{k} \left(\frac{g'}{l_f} \right)^j + MN_{spM} \right\} W \right] B_{RF}$$

$$k = 1, 2, \ldots, M$$

$$(6.90)$$

$$N'_{s-sp} = 4\alpha \left\{ N'_{sp} \sum_{j=1}^{k} \left(\frac{g'}{l_f} \right)^j + MN_{spM} \right\} \frac{\alpha P_{r_k}}{p^2} (\chi_k + k - 1) B_{RF}$$

$$k = 1, 2, \ldots, M$$

$$(6.91)$$

$$N'_{sp-sp} = 2\alpha^2 \left\{ N'_{sp} \sum_{j=1}^{k} \left(\frac{g'}{l_f} \right)^j + MN_{spM} \right\}^2 (W - f_{RF}) \qquad (6.92)$$

where $G_a = 10 \log_{10} g_a$, $L_f = 10 \log_{10} l_f$ (in decibels) and N'_{sp} is given by (6.70) substituted $G = G'$. Note that in optical switch connection, no optical signal beat noise occurs that is different from the optical coupler connection. Therefore, the total noise power at the kth correlator output, N_{swk}, is

$$N_{swk} = N'_{RIN} + N'_{shot} + N_{th} + N'_{s-sp} + N'_{sp-sp} \qquad k = 1, 2, \ldots, M$$

$$(6.93)$$

and the CINR of the kth regenerated radio signal, $CINR_k$, is given by

$$CINR_k = \frac{C_k}{I_k + N_{swk}} \qquad k = 1, 2, \ldots, M \qquad (6.94)$$

6.4.2.5 Numerical Results

In this section, some numerical results of CINR in the DOS-CDMA cable-on-the-air system for both the optical coupler and the optical switch connections are shown and discussed. Parameters used for calculation are shown in Table 6.6.

Figure 6.27 shows the CIR of the kth RBS for the prime number $p = 23$, and the number of connected RBSs is $M = 20$. In the case of $G_a = 0$ dB, the CIR in the optical switch connection system is degraded compared with the optical coupler connection system as k nears M because the coefficient ϵ_k [see (6.88)], which expresses the CIR reduction effect due to the erasure of chip pulse, becomes less than 1 as k nears M. However, the CIR of the nearer RBS to the RCS in the optical switch connection

Table 6.6
Parameters Used for Calculation

Responsivity of photodetector α	0.8 A/W
Power spectral density of the relative intensity noise ζ_{RIN}	−152 dB/Hz
Bandwidth of optical filter W	1 THz
Noise temperature T	300K
Load resistance R_L	50Ω
Spontaneous emission factor of optical axis η_{sp}	2.0
Quantum efficiency of the optical axis η_a	0.5
Radio signal frequency f_{RF}	1.9 GHz
Radio signal bandwidth B_{RF}	300 kHz
Coupling plus insertion loss of optical coupler L_{OC}	3 dB
Fiber loss between RBSs L_f	0.5 dB
Full-width half-maximum of the laser diode Δv_{LD}	10 MHz
Difference of center frequency of the laser diode ΔF	1 THz

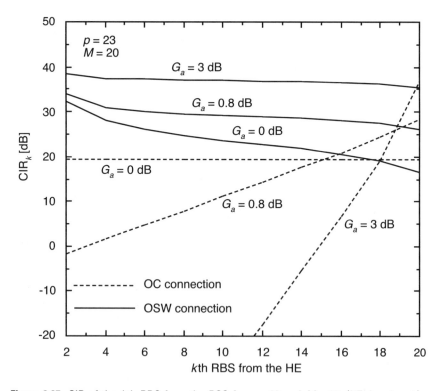

Figure 6.27 CIR of the kth RBS from the RCS for $p = 23$ and $M = 20$. (HE: head end.)

system is improved compared with the optical coupler connection system because intensity modulated/CDMA signals from RBSs located farther than the kth RBS never reach the kth correlator; therefore, the interference power is reduced. However, (6.87) and (6.88) show that by increasing the additional gain, G_a, the CIR_k can be improved for any kth radio signal. Figure 6.27 shows that as the G_a increases, the CIR_k is improved and we can obtain almost the same CIR for all radio signals from M RBSs. However, the additional gain is not effective for improving the CIR in the optical coupler connection system, because the radio signals from the nearer RBS to the RCS suffer interference with amplified large power while the CIR of the radio signal from the farther RBS can be improved as shown in Figure 6.27.

Figure 6.28 shows the CIR of the farthest Mth RBS as a function of prime code number p for an M of 20, 40, and 80. The upper abscissa is the optical switch speed. When G_a is 0 dB in the optical switch connection system, the coefficient ϵ_M is less than 1 as p nears M, thus the CIR_M of the farthest Mth RBS is degraded compared with the optical coupler connection system. This is due to the erasure of the chip pulse of the Mth radio signal. However, this penalty can be reduced and the CIR_M comes to the same as that in the optical coupler connection system as p increases more than M. It is also found from Figure 6.28 that by introducing additional gain ($G_a > 0$ dB), the CIR_M of the farthest Mth RBS is more improved than that in the optical coupler connection system as increases for any p and M. This is because as p increases more than M, (6.88) shows that ϵ_M nears $\dfrac{M-1}{\sum\limits_{j=1}^{M-1} g_a^{2(j-M)}}$, which is more than 1, and yields the same value regardless of M.

Here, we have to consider the limitations in G_a and p to realize the optical switch connection system. Regarding p, the achievable switching speed of the optical switch gives the limitations of the prime code number, p. If we use the optical switch with its speed of almost 10 GHz for the case of $B_{RF} = 900$ kHz, we can increase p up to about 80.

The possible additional gain, G_a, is limited by the optical power limitation in the optical fiber. Figure 6.29 shows the relationship between the average received optical power from the farthest Mth RBS, P_{r_M}, normalized by the average transmitted optical power, P_{r_M}/P_s, and the number of connected RBSs, M in the case of $p = 79$ ($1/T_c = 11.2$ GHz). For example, in the case for which the limitation of P_{r_M}/P_s is 20 dB, the upper limits in numbers of connected RBSs in the optical switch connection system are 63,

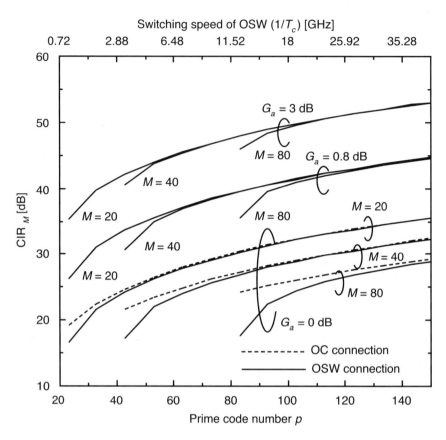

Figure 6.28 CIR of the farthest Mth RBS versus prime code number, p for M of 20, 40, and 80.

41, 30, 23, and 19 for G_a = 0.2, 0.4, 0.6, 0.8, and 1 dB, respectively. G_a can be increased up to 0.8 dB for P_{rM}/P_s = 20 dB and M = 20. Thus, it is seen from Figure 6.28 that for M = 20, the CIR in the optical switch connection system can be more improved by about 8 dB than that in the optical coupler connection system by introducing the G_a of 0.8 dB.

Figure 6.29 shows the C/N as a function of prime code number p for P_s of 0 dBm and M of 20 and 40. When G_a is 0 dB, C/Ns for both optical coupler and optical switch connection systems are dominated by the beat noise among amplified spontaneous emission of an optical amplifier and the optical signal, N_{s-sp} and N'_{s-sp}, respectively, for small p but affected by the receiver thermal noise, N_{th}, as p gets larger because the carrier power decreases in proportion to $1/p^2$. As p nears M, the C/N in the optical switch connection system is degraded by the erasure of chip pulse compared with the optical

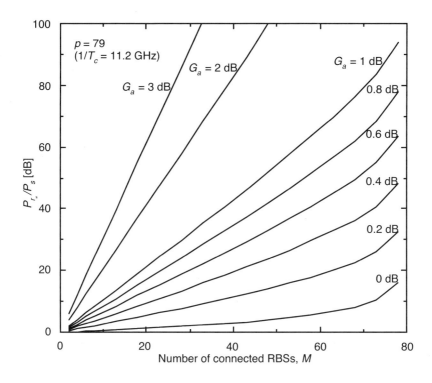

Figure 6.29 Relationship between P_{r_M}/P_S and M for $p = 79$ in the optical switch connection system.

coupler connection system. However, the C/Ns of both optical coupler and optical switch connection systems are similar as p increases more than M because the erasure of the chip pulse can be neglected and N_{s-sp} and N'_{s-sp} are dominated by the optical axis at the output of the star coupler. However, when G_a is more than 0 dB, the C/N in the optical switch connection system is dominated by N'_{s-sp} caused by the optical amplifier at the RBS; thus, the C/N is more improved than that in the optical coupler connection system as G_a increases. For example, we can see from Figure 6.30 that for $P_s = 0$ dBm, $M = 20$, and $p = 79$, the C/N in the optical switch connection system with G_a of 0.8 dB can be improved by 11 dB more than that in the optical coupler connection system.

We can see from Figures 6.28 and 6.29 that the CINR is dominated by the CIR for any p. By introducing additional gain, the CINR in the optical switch connection system can be more improved than that in the

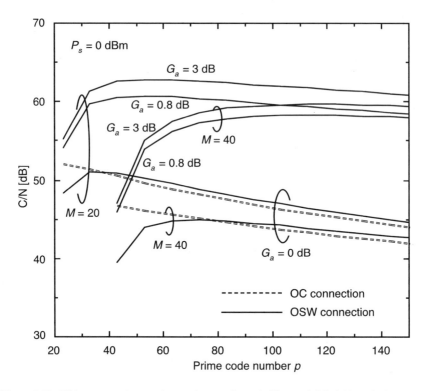

Figure 6.30 C/N versus prime code number, p, for = 0 dBm and M of 20 and 40.

optical coupler connection system. It is found that for $M = 20$, the CINR in the optical switch connection system can be improved by about 8 dB more than that in the optical coupler connection system by introducing the G_a of 0.8 dB.

Figure 6.31 shows the CINR as a function of the number of connected RBSs, M, for $p = 79$ ($1/T_c = 11.2$ GHz) and = 0 dBm. In the optical switch connection system, when $G_a = 0$ dB and M nears p, the CIR is worse than the optical coupler connection system. However, the CINR of the optical switch connection system is improved more than that in the optical coupler connection system as G_a increases. In the optical coupler connection system, the number of connected RBSs is determined by the required CINR. However, in the optical switch connection system, the number of connected RBSs is determined by both the required CINR and the optical power limitation. Figures 6.29 and 6.31 show that when the required CINR is 30 dB and the required P_{r_M}/P_s is 20 dB, the numbers of connected RBSs in the optical switch connection system are 63, 41, 30, and 23 for G_a =

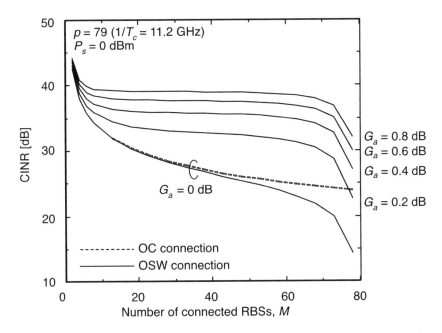

Figure 6.31 CINR versus number of connected RBSs M for $p = 79$ ($1/T_c = 11.2$ GHz).

0.2, 0.4, 0.6, and 0.8 dB, respectively, while the number of connected RBSs in the optical coupler connection system is 18. Thus it is found that in the optical switch connection system with $G_a = 0.2$ dB, three times the number of RBSs can be connected to the cable-on-the-air system compared with the optical coupler connection system.

6.4.2.6 Summary

In Section 6.4.2, the DOS-CDMA method using prime code sequences was described. The configuration of the system has been investigated by using optical coupler and optical switch connections. We have theoretically analyzed the CINR of regenerated radio signals at the RCS for two types of bus connection methods considering the chip pulse erasure at the optical switch. The following results are obtained:

- In the optical switch connection system, by introducing the additional optical gain at each RBS, the CIR for all RBSs can be almost the same when the prime code number exceeds the number of connected RBSs.

- In the optical switch connection system with additional gain, the CINRs for all RBSs and the number of connected RBSs can be improved compared with the optical coupler connection system. For example, by using the optical switch connection system with additional gain, $G_a = 0.2$ dB, three times the number of RBSs can be connected to the system with the CINR of 30 dB and the average received optical power from the farthest RBS normalized by the average transmitted optical power of 20 dB for prime number $p = 79$ compared with the optical coupler connection system. Thus, the optical switch connection is an effective DOS-CDMA system in which an optical switch is used not only to spread the spectrum of optical signals but also to launch them into the fiber optic bus link.

6.4.3 DOS-CDMA Using PN Codes

In the DOS-CDMA method described in Section 6.4.2, only optical uniphase orthogonal codes such as prime codes are applied to obtain a desired process gain because an optical switch is used to despread DOS-CDMA signals at each correlator.

However, the prime code suffers a limit in the number of distinct code sequences, which results in the limitation of the number of RBSs connected to fiber optic radio networks. In addition, in fiber optic radio networks using the DOS-CDMA scheme with prime codes, the optical power efficiency is low because the pulse duty of a prime code is quite low. Therefore, we should consider a new type of the correlator for the DOS-CDMA method to which conventional PN codes such as maximal length codes and Gold codes can be applied, because those codes are usually used in radio systems and are generally superior in the number of distinct code sequences compared with prime codes.

For digital networks using the optical CDMA method, the sequence inversion keyed direct sequence CDMA methods that have been proposed [24, 25] require specially balanced PN codes. To allow the use of any unbalanced PN codes, the power splitting ratio of the power divider at the optical correlator has been controlled [26], and the transmission of two channels using two wavelengths or two orthogonal polarizations has been proposed [27]. In sequence inversion keyed CDMA methods, however, binary digital data are encoded and transmitted with the positive polarity and the negative polarity of bipolar codes; therefore, their correlators at the receiver cannot be applied to DOS-CDMA signals that are converted from radio signals by the on-off switching CDMA method. In this section, the optical polarity-reversing correlator for a DOS-CDMA radio highway network using PN codes is described [28–32].

6.4.3.1 Principle of OPRC

The optical polarity-reversing correlator (OPRC) can be realized with various optical devices such as optical switches, matched filters, and Mach-Zehnder interferometers. We propose the OPRC with a Mach-Zehnder interferometer or two optical switches. Figure 6.32 illustrates the configuration of the transmitter and the OPRC for the optical CDMA using the PN code $c_k(t)$ having the value of 1 or −1. The $c_k(t)$ has the frame period, T_F, and the chip width, T_c. In the interval of T_F, $c_k(t)$ has the code length L, which

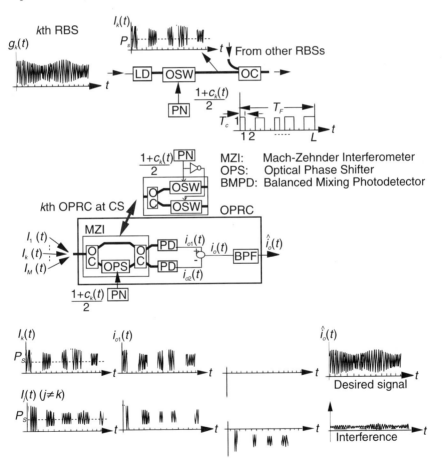

Figure 6.32 Configuration of the transmitter and the OPRC for optical CDMA using PN codes.

is the number of chips, and a number of 1's of $(L + 1)/2$. At the kth RBS, the laser diode is directly modulated by the radio signal with an optical modulation index of 1. DOS-CDMA is performed at an optical switch driven with the biased PN code $[1 + c_k(t)]/2$. Then, the intensity of the optical signal from the kth RBS is written by

$$I_k(t) = P_s\{1 + g_k(t)\} \frac{1 + c_k(t)}{2} \tag{6.95}$$

where $g_k(t)$ is the radio signal with the bandwidth B_{RF} and the carrier frequency f_{RF}, and P_s is the average transmitted optical power before the optical switching.

At the kth OPRC, many intensity modulated/CDMA signals are received and split into two signals at the Mach-Zehnder interferometer or the optical coupler. We assume that the biased PN code, $[1 + c_k(t)]/2$, matched with the one at the RBS, is regenerated at the RCS by using the retiming code generator such as the retiming block and the synchronization between two code sequences is taken. The OPRC is driven by the biased PN code, $[1 + c_k(t)]/2$, matched with the one at the RBS. In the case of the OPRC realized with a Mach-Zehnder interferometer, the phase of the optical phase shifter is shifted according to the biased PN code, $[1 + c_k(t)]/2$. When $[1 + c_k(t)]/2$ is 1, the phase difference between both arms of the Mach-Zehnder interferometer is zero; thus, intensity-modulated/CDMA signals are output to the upper port of the Mach-Zehnder interferometer through the second optical coupler. When $[1 + c_k(t)]/2$ is 0, the phase difference between both arms of the Mach-Zehnder interferometer is set to π; thus, the lower port of the Mach-Zehnder interferometer outputs intensity modulated/CDMA signals through the second optical coupler. In the case of OPRC realized with two optical switches, when $[1 + c_k(t)]/2$ is 1 the upper optical switch is set to on and the lower optical switch is set to on when $[1 + c_k(t)]/2$ is 0.

Thus, output currents of the balanced mixing photodetector are expressed as

$$i_{o1}(t) = \alpha \sum_{j=1}^{M} P_{r_k} g_j(t) \frac{1 + c_j(t)}{2} \cdot \frac{1 + c_k(t)}{2} + i_{n1}(t) \tag{6.96}$$

$$i_{o2}(t) = \alpha \sum_{j=1}^{M} P_{r_k} g_j(t) \frac{1 + c_j(t)}{2} \cdot \frac{1 - c_k(t)}{2} + i_{n2}(t) \tag{6.97}$$

where α is the responsivity of the photodetector and $i_{n1}(t)$ and $i_{n2}(t)$ are additive noise currents, respectively. Equations (6.96) and (6.97) show that the positive polarity of the desired kth code $c_k(t)$ matches with the kth one at the RBS at the upper port of the Mach-Zehnder interferometer or the upper optical switch. However, it is obstructed at the lower port of the Mach-Zehnder interferometer or the lower optical switch. The desired kth signal at the input to the bandpass filter is a bandpass natural sampled signal converted from the kth radio signal, $g_k(t)$, because optical switching is a window-type sampling. Therefore, a radio signal is conveyed by the optical carrier with its signaling format kept and can be regenerated from the pulsed signals by the interpolation at the bandpass filter whose principle is discussed in Section 6.3.1.

However, for the interference signal $I_j(t)$ $(j \neq k)$, both ports of the Mach-Zehnder interferometer or both optical switches generate interference only when the positive polarity of $c_j(t)$ coincides with the positive or negative polarity of $C_k(t)$. Two interferences of balanced mixing photodetector are subtracted and the interference is suppressed at the bandpass filter. However, in the unipolar-type correlator for optical CDMA, PN codes cannot be applied because the interference is not suppressed though the code sequence length increases.

6.4.3.2 System Configuration

Figure 6.33 illustrates the configuration of the optical CDMA radio highway network using the OPRC.

Each RBS is equipped with an automatic gain control to control the amplitude of a received radio signal in order to keep the optical modulation index constant at the laser diode. After the direct intensity modulation at the laser diode, DOS-CDMA is performed by on-off switching at the optical switch driven by a biased PN code. In the bus link, many intensity modulated/CDMA signals from M RBSs are multiplexed by the DOS-CDMA method and transmitted to the carrier sense. When any biased PN code sequence coincides with another biased PN code sequence, the numbers of coincidences of chips is $(L+1)^2/4L$ out of chips of L. In the case of the fiber optic bus link using the optical switch connection, the signal collision at the optical switch of each RBS will cause the erasure of enormous chip pulses in DOS-CDMA signals. Because enough additional gain cannot be inserted at each RBS owing to the optical power limitation, the far-near problem will happen in the fiber optic link. Thus we consider a system configuration that uses only the optical coupler connection.

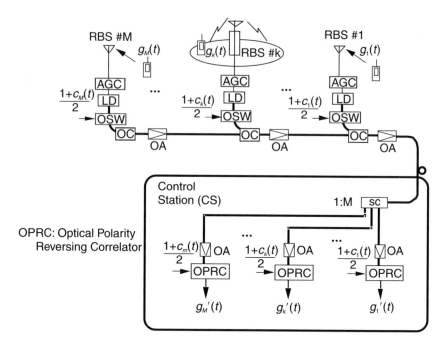

Figure 6.33 Configuration of the optical CDMA radio highway network using the OPRC.

At the carrier sense, optical CDMA signals are at first power split to each of M receivers. Intensity modulated/CDMA signals are correlated at the OPRC driven by a biased PN code matched with the RBS. To compensate for the optical loss between RBSs and the receiver, each RBS and RCS is equipped with optical axes. At the kth OPRC, the average received optical power, P_{r_k}, can be written as

$$P_{r_k} = 10 \log_{10} P_s - k(L_f + L_{oc}) + kG - 10 \log_{10} M + G_M - 2L_{oc} \quad [\text{dB}]$$
$$(6.98)$$

where L_{oc}, L_f, G, and G_M (all in decibels) are the coupling loss plus the insertion loss of an optical coupler, the fiber loss between RBSs, the gain of the optical axes at the RBS, and the gain of the optical axes at the output of 1:M SC, respectively. The last term, $2L_{oc}$, is the coupling loss caused by the two optical couplers at the Mach-Zehnder interferometer in Figure 6.32.

6.4.3.3 Theoretical Analysis of CINR Performance

The input current to the bandpass filter is given by

$$i_o(t) = i_{o1}(t) - i_{o2}(t) \quad (6.99)$$
$$= i_{S_k}(t) + i_I(t) + i_N(t)$$

where $i_{S_k}(t)$, $i_I(t)$, and $i_N(t)$ are the desired signal, the interference, and the additive noise, respectively. To obtain the same optical received power at the OPRC, we assume that an optical axis equipped at each RBS has gain G of $L_f + L_{oc}$ and the gain of optical axis at the output of the star coupler, G_M, is equal to $10 \log_{10} M + 2L_{oc}$. Thus, the same received power $P_{r_k} = P_s$ ($k = 1, 2, \ldots, M$) is obtained from (6.98). We also assume that each Mach-Zehnder interferometer receives M optical signals with the same modulation index of 1. At the input of the bandpass filter, $i_{S_k}(t)$ and $i_I(t)$ are expressed as

$$i_{S_k}(t) = \alpha P_s g_k(t) \frac{1 + c_k(t)}{2} \tag{6.100}$$

$$i_I(t) = \sum_{j=1, j \neq k}^{M} \alpha P_s g_j(t) \frac{1 + c_j(t)}{2} c_k(t) \tag{6.101}$$

respectively.

We consider the carrier power of the regenerated radio signal and the interference power at the bandpass filter from the psds of the desired signal and the interference. The autocorrelation of desired signal, $i_{S_k}(t)$, is expressed as

$$R_S(\tau) = \left(\frac{\alpha P_s}{2}\right)^2 R_g(\tau) \left\{ 1 + \frac{2}{L} + R_c(\tau) \right\} \tag{6.102}$$

where $R_g(\tau)$ and $R_c(\tau)$ are the autocorrelations of $g_k(t)$ and $c_k(t)$, respectively. Though we assume that radio signals, $g_j(t)$ ($j = 1, 2, \ldots, M$), are unmodulated carriers, the calculated carrier power of regenerated radio signals can be applied irrespective of the modulation method of radio signals. Thus, we assume that radio signals, $g_j(t)$ ($j = 1, 2, \ldots, M$), are unmodulated carriers that have the autocorrelation function, $R_g(\tau)$:

$$R_g(\tau) = \frac{1}{2} \cos(2\pi f_{RF} \tau) \tag{6.103}$$

The autocorrelation of $c_k(t)$ is given by

$$R_c(\tau) = \frac{1}{L^2} + \frac{L+1}{L^2} \sum_{k=-\infty, k \neq 0}^{\infty} \text{sinc}^2\left(\frac{\pi k T_c}{T_F}\right) \cos\left(\frac{2\pi k \tau}{T_F}\right) \tag{6.104}$$

Thus, the power spectral density of $i_{S_k}(t)$, $S_S(f)$, is given by

$$S_S(f) = \left(\frac{\alpha P_s}{2}\right)^2 \left\{ S_g(f)\left(1 + \frac{2}{L}\right) + S_g(f) \otimes S_c(f) \right\}$$

$$= \left(\frac{\alpha P_s}{2}\right)^2 \left\{ \left(\frac{L+1}{L}\right)^2 S_g(f) \right. \tag{6.105}$$

$$\left. + \frac{L+1}{L^2} \sum_{k=-\infty, k \neq 0}^{\infty} \text{sinc}^2\left(\frac{\pi k T_c}{T_F}\right) S_g(f - k/T_F) \right\}$$

where $S_g(f)$ and $S_c(f)$ are the power spectral densities of $g_k(t)$ and $c_k(t)$, respectively. The first term of $S_S(f)$ is the desired signal component around f_{RF} and $-f_{RF}$, and the second terms are the frequency-shifted components caused by bandpass sampling. Images of these shifted components will cause the distortion in the desired signal as the self-interference if they overlap over the signal. However, we can perfectly remove the self-interference components by setting the value of the RF f_{RF} at $(j + 1/2)B_{RF}$ or j/T_c (j is an integer). Without such special values of f_{RF}, we can ignore the self-interference component because it may not deteriorate the signal quality [29]. Since the biased PN code sequence drives the optical switch periodically with the sampling period T_F at the RBS transmitter, each signal pulse in one code sequence period has the same period T_F. Then, if the periodic frequency $1/T_F$ is set to be at least twice the radio signal bandwidth (= $2B_{RF}$), the radio signal can be regenerated from only one pulse train in the code sequence by the interpolation at the bandpass filter.

From (6.105), C can be obtained as

$$C = \frac{1}{2}\left(\frac{\alpha P_s(L+1)}{2L}\right)^2 \tag{6.106}$$

The autocorrelation of $i_I(t)$, $R_I(\tau)$, is derived as

$$R_I(\tau) = (M - 1)\left(\frac{\alpha P_s}{2}\right)^2 R_g(\tau)\left\{ R_c(\tau)\left(1 + \frac{2}{L}\right) + R_\gamma(\tau) \right\} \tag{6.107}$$

where $R_\gamma(\tau)$ represents the autocorrelation of $c_i(t) \cdot c_k(t)$. $R_\gamma(\tau)$ is given by [34]

$$R_\gamma(\tau) = \overline{a_0^2} + \sum_{k=-\infty, k\neq 0}^{\infty} \frac{1}{L} \frac{1}{1 + (k\pi/L)} \cos\left(\frac{2\pi k\tau}{T_F}\right) \quad (6.108)$$

where the value of $\overline{a_0^2}$ depends on the kind of code sequence used. The maximal length code is usually used in radio system and [14, 34, 35] have reported that the Gold code is excellent as a PN code can provide the required small cross-correlation. Thus, Gold codes and maximal length codes are used for the optical CDMA using the OPRC. The $\overline{a_0^2}$'s for Gold codes and maximal length codes are given by [33, 34]

$$\overline{a_0^2} = \begin{cases} \dfrac{L^2 + L - 1}{4L^3} & \text{for Gold codes} \\[3mm] \dfrac{L^2 + L - 1}{L^3} & \text{for maximal length codes} \end{cases} \quad (6.109)$$

The power spectral density of $i_I(t)$, $S_I(f)$, is given by

$$S_I(f) = (M-1)\left(\frac{\alpha P_s}{2}\right)^2 S_g(f) \otimes \left\{ S_c(f)\left(1 + \frac{2}{L}\right) + S_\gamma(f) \right\}$$

$$= (M-1)\left(\frac{\alpha P_s}{2}\right)^2 \left[\left(\frac{L+2}{L^3} + \overline{a_0^2}\right) S_g(f) \right. \quad (6.110)$$

$$+ \sum_{k=-\infty, k\neq 0}^{\infty} \left\{ \frac{(L+1)(L+2)}{L^3} \operatorname{sinc}^2\left(\frac{\pi k T_c}{T_F}\right) \right.$$

$$+ \left. \frac{1}{L}\frac{1}{1 + (k\pi/L)} \right\} S_g\left(f - \frac{k}{T_F}\right) \right]$$

where $S_\gamma(f)$ is the power spectral density of $c_i(t) \cdot c_k(t)$. The interference power at the output of the bandpass filter, I, can be obtained as

$$I = \begin{cases} \dfrac{M-1}{2}\left(\dfrac{\alpha P_s}{2}\right)^2 \dfrac{L^2 + 5L + 7}{4L^3} & \text{for Gold codes} \\[4mm] \dfrac{M-1}{2}\left(\dfrac{\alpha P_s}{2}\right)^2 \dfrac{(L+1)^2}{L^3} & \text{for maximal length codes} \end{cases}$$

$$(6.111)$$

Hence, from (6.106) and (6.111), the CIRs for Gold codes and maximal length codes can be written as follows:

$$
\mathrm{CIR} = \begin{cases} \dfrac{4}{M-1} \dfrac{L(L+1)^2}{L^2 + 5L + 7} & \text{for Gold codes} \\[2em] \dfrac{L}{M-1} & \text{for maximal length codes} \end{cases}
$$

$$(6.112)$$

For discussion, we consider the CIR for the case of a unipolar-type correlator for optical CDMA as shown in Figure 6.32. When prime codes are applied, the CIR is given by [28, 29]

$$
\mathrm{CIR}_{uc,p} = \frac{1}{(M-1)} \frac{p^2}{\sigma_c^2} \qquad \text{for prime codes} \qquad (6.113)
$$

where σ_c^2 denotes the average variance of the cross-correlation of the prime code, whose length, L, is p^2. However, when PN codes are applied to the unipolar-type correlator as has been proposed for the OPRC, from (6.96), the interference current at the input of the bandpass filter is given by

$$
i_{I,uc}(t) = \sum_{j=1, j \neq k}^{M} \frac{\alpha P_s}{4} g_j(t) \{1 + c_j(t)\}\{1 + c_k(t)\} \qquad (6.114)
$$

The autocorrelation function of $i_{I,uc}(t)$, $R_{I,uc}(\tau)$, is given by

$$
R_{I,uc}(\tau) = (M-1)\left(\frac{\alpha P_s}{2}\right)^2 R_g(\tau) \qquad (6.115)
$$

$$
\left\{ 1 + \frac{4}{L} + \frac{4}{L^2} + 2\left(1 + \frac{2}{L}\right) R_c(\tau) + R_\gamma(\tau) \right\}
$$

The power spectral density of $i_{I,uc}(t)$, $S_{I,uc}(f)$, is given by

$$S_{I,uc}(f) = (M-1)\left(\frac{\alpha P_s}{4}\right)^2 \left[S_g(f)\left(1 + \frac{4}{L} + \frac{4}{L^2}\right) + S_g(f) \right.$$

$$\otimes \left. \left\{2\left(1 + \frac{2}{L}\right)S_c(f) + S_\gamma(f)\right\}\right]$$

$$= (M-1)\left(\frac{\alpha P_s}{4}\right)^2 \left[\left\{1 + \frac{4}{L} + \frac{4}{L^2} + \frac{2(L+2)}{L^3} + \overline{a_0^2}\right\} S_g(f) \right.$$

$$+ \sum_{k=-\infty, k\neq 0}^{\infty} \left\{ \frac{2(L+1)(L+2)}{L^3} \operatorname{sinc}^2\left(\frac{\pi k T_c}{T_F}\right) \right.$$

$$\left. \left. + \frac{1}{L}\frac{1}{1+(k\pi/L)}\right\} S_g\left(f - \frac{k}{T_F}\right)\right] \tag{6.116}$$

The interference power at the output of the bandpass filter can be obtained as

$$I_{uc} = \begin{cases} \dfrac{M-1}{2}\left(\dfrac{\alpha P_s}{4}\right)^2\left(1 + \dfrac{17L^2 + 25L + 18}{4L^3}\right) & \text{for Gold codes} \\[3ex] \dfrac{M-1}{2}\left(\dfrac{\alpha P_s}{4}\right)^2\left(1 + \dfrac{5L^2 + 7L + 3}{L^3}\right) & \text{for maximal length codes} \end{cases} \tag{6.117}$$

Hence, in the case of a unipolar-type correlator using PN codes, the CIR is written

$$CIR_{uc} = \begin{cases} \dfrac{4}{M-1}\dfrac{4L(L+1)^2}{4L^3 + 17L^2 + 25L + 15} & \text{for Gold codes} \\[3ex] \dfrac{4}{M-1}\dfrac{L(L+1)^2}{L^3 + 5L^2 + 7L + 3} & \text{for maximal length codes} \end{cases} \tag{6.118}$$

We analyze the noise power at the output of the bandpass filter. We consider additive noise currents composed of the RIN, shot noise, receiver thermal noise, beat noise between optical signal and amplified spontaneous emission, beat noise between amplified spontaneous emissions, and optical signal beat noise. The total noise power, N, is written

$$N = N_{\text{RIN}} + N_{\text{shot}} + N_{th} + N_{s-sp} + N_{sp-sp} + \langle N_{s-s} \rangle \quad (6.119)$$

When any biased PN code sequence coincides with the positive and the negative polarity of any PN code sequence, the numbers of coincidences of 1's in the interval of T_F are $(L + 1)^2/4L$ and $(L^2 - 1)/4L$, respectively. Each noise power is given by

$$N_{\text{RIN}} = \zeta_{\text{RIN}} \left(\frac{\alpha P_s}{L} \right)^2 \left[\left(\frac{L + 1}{2} \right)^2 \right. \quad (6.120)$$

$$\left. + (M - 1) \left\{ \left(\frac{(L + 1)^2}{4L} \right)^2 + \left(\frac{L^2 - 1}{4L} \right)^2 \right\} \right] B_{\text{RF}}$$

$$N_{\text{shot}} = 2e \left\{ \left(\frac{\alpha P_s}{L} \right) \frac{M(L + 1)}{2} + \alpha M (N_{sp} + N_{spM}) W \right\} B_{\text{RF}} \quad (6.121)$$

$$N_{th} = \frac{8 k_B T}{R_L} B_{\text{RF}} \quad (6.122)$$

$$N_{s-sp} = 4 \alpha M (N_{sp} + N_{spM}) \left(\frac{\alpha P_s}{L} \right) \frac{M(L + 1)}{2} B_{\text{RF}} \quad (6.123)$$

$$N_{sp-sp} = 2 \alpha^2 M^2 (N_{sp} + N_{spM})^2 (W - f_{\text{RF}}) \quad (6.124)$$

where e, ζ_{RIN}, W, k_B, T, and R_L are the electric charge, the power spectral densities of the RIN, the bandwidth of optical filter at the RCS, Boltzmann's constant, noise temperature, and load resistance, respectively. The power spectral densities of the amplified spontaneous emission, N_{sp} and N_{spM}, are given by

$$N_{sp} = \frac{\eta_{sp}}{\eta_a} \cdot \frac{10^{G/10} - 1}{10^{G/10}} h\nu \quad (6.125)$$

$$N_{spM} = \frac{\eta_{sp}}{\eta_a} \cdot \frac{10^{G_M/10} - 1}{10^{G_M/10}} h\nu \quad (6.126)$$

respectively, where η_{sp}, η_a, and $h\nu$ are the spontaneous emission factor, the quantum efficiency of the optical axis, and the photon energy, respectively. The optical signal beat noise, $\langle N_{s-s} \rangle$, is the result of an interference between two optical signals. $\langle N_{s-s} \rangle$ is composed of two interference components: One is an interference between optical signals of a desired signal and an interference, $\langle N_{s_s-s_i} \rangle$; another is one between optical signals of two interferences, $\langle N_{s_i-s_i} \rangle$.

Therefore, the CINR of the regenerated radio signal is given by

$$\text{CINR} = \frac{C}{I + N} \qquad (6.127)$$

6.4.3.4 Numerical Results and Discussions

In this section, we provide some numerical results and discussions. The parameters used for calculation are those shown in Table 6.6. Figure 6.34 shows the relationship between the code length L and the CIR for the

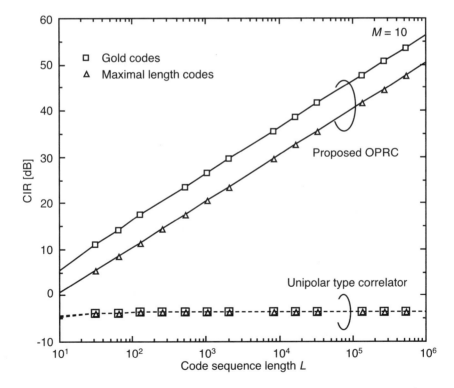

Figure 6.34 Relationship between the code sequence length, L, and the CIR.

proposed OPRC and the unipolar type correlator shown in Figure 6.33 for the optical CDMA radio highway in the case of using PN codes for $M = 10$.

In the case of the unipolar-type correlator using maximal length codes and Gold codes, there is no improvement in the CIR even though L increases. In this case, we can see from (6.118) that the CIR almost equals $4/(M-1)$ regardless of L. However, we see from (6.112) that in the proposed OPRC, the CIR is improved as the L increases.

The results show that PN codes such as maximal length codes and Gold codes can be applied to the optical CDMA radio highway using the proposed OPRC. However, in the case of the optical CDMA radio highway using the unipolar-type correlator, we see that PN codes cannot be applied and only unipolar orthogonal codes can be applied. References [9, 35, 36] have reported that the prime code is the best unipolar orthogonal code that can provide the highest CIR. So in the next figure, we compare the number of distinct code sequences that result in a limitation of the number of RBSs connected to the radio highway for maximal length codes, Gold codes, and prime codes.

Figure 6.35 shows the number of distinct code sequences versus the code sequence length L. Gold code sequences are generated by combining a pair of preferred maximal length sequences using modulo-2 addition if the number of preferred maximal length sequences is at least two [33, 34]. However, in the case of prime codes, the number of distinct code sequences is equal to the prime number p for the code sequence length of p^2 [9].

The numbers of distinct code sequences for maximal length codes and Gold codes are larger than that for prime codes. For example, comparing maximal length codes and Gold codes of $L = 32,767$ with prime codes of $p^2 = 32,041$, the numbers of distinct code sequences for maximal length codes and Gold codes are 10 times and 183 times larger than that for prime codes, respectively. Therefore, using the proposed OPRC can assign a larger number of distinct code sequences to RBSs in the optical CDMA radio highway than does using the unipolar type correlator with prime codes.

Figures 6.36(a) and 6.36(b) show the relationship between the code length L and the CINR in the case of $M = 30$ for P_s of 0 dBm and -10 dBm, respectively. We calculate the CINR versus L where the Gold code sequence is defined. A little better CIR can be obtained by the proposed OPRC using Gold codes than the unipolar-type correlator using prime codes.

In the case of $P_s = 0$ dBm, the C/N in the proposed OPRC using Gold codes is dominated by the optical signal beat noise, $\langle N_{s-s} \rangle$. However, in the case of the unipolar-type correlator using prime codes, the C/N is

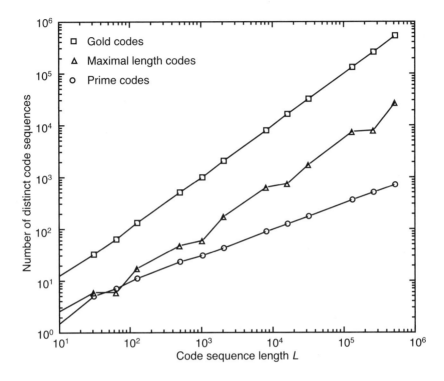

Figure 6.35 Number of distinct code sequences versus the code sequence length, L.

dominated by the beat noise among the amplified spontaneous emission of an optical axis and the optical signal, N_{s-sp}, and deteriorated by the receiver thermal noise, N_{th}, as L becomes larger because the carrier power decreases in proportion to $1/L$. Thus, when radio signal bandwidth, B_{RF} is increased to 900 kHz, the C/N is deteriorated by the increase in B_{RF}. When P_s is 0 dBm as shown in Figure 6.37(a), the CINR for the proposed OPRC using Gold codes is dominated by CIR. In the case of the unipolar-type correlator using prime codes, the CINR is dominated by CIR until L comes to 262,143, from which the CINR is dominated by the C/N of N_{th}. Thus, when P_s is 0 dBm, the CINR is slightly improved by the proposed OPRC using Gold codes compared with the unipolar-type correlator using prime codes because the CIR is determined by the kind of used code sequence.

In the case of $P_s = -10$ dBm as shown in Figure 6.36(b), the C/N in the proposed OPRC using Gold codes is dominated by the N_{s-sp}, while in the case of the unipolar-type correlator using prime codes, the C/N is dominated by the N_{th} and deteriorated severely as L becomes larger.

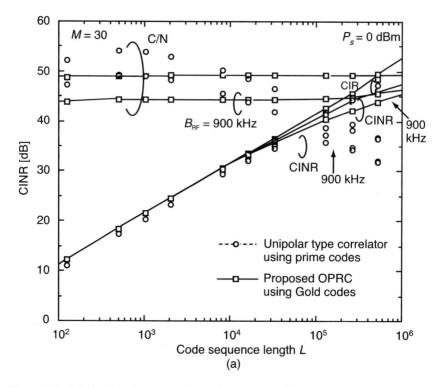

Figure 6.36 Relationship between the code sequence length, L, and the CINR for (a) 0 dBm and (b) –10 dBm.

In the case of the conventional correlator using prime codes, as L increases, the CINR is improved by the CIR but deteriorated by the C/N of N_{th} as L increases to greater than 16,383. Thus, in this case, the maximal CINR is determined by the CIR and the C/N. In the case of the small average transmitted optical power such as P_s = –10 dBm for the unipolar-type correlator using prime codes, we can see from Figures 6.35 and 6.36(b) that L should be increased to assign a larger number of distinct code sequences to RBSs, but the CINR is deteriorated as L gets larger. However, in the case of the proposed OPRC using Gold codes, because the C/N is almost constant regardless of L, the CINR is improved and not deteriorated as L increases.

Figure 6.37 shows the relationship between the average transmitted optical power and the CINR in the case of M = 40 for different switching speeds in the DOS-CDMA, $1/T_c = 2B_{RF} \cdot L$. In the unipolar-type correlator using prime codes, when $1/T_c$ is large, the C/N dominated by N_{th} severely deteriorates the CINR as P_s decreases because the signal power is in proportion to P_s^2/L. However, in the proposed OPRC using Gold codes, the CINR is

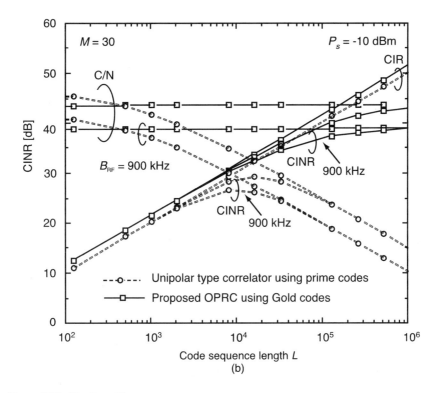

Figure 6.36 (Continued.)

not so abruptly deteriorated with the C/N as P_s decreases because the signal power is nearly in proportion to $P_s^2/4$ from (6.106). Thus, the CINR is more improved by the proposed OPRC using Gold codes compared with the unipolar-type correlator using prime codes as P_s decreases.

Figure 6.38 shows the relationship between the switching speed, $1/T_c$, and the number of maximum connected RBSs, M_{max}, in the case for which CINR is 30 dB.

Figure 6.38 shows that with the proposed OPRC using Gold codes, the number of maximum connected RBSs can be much improved for small P_s such as -10 dBm compared with the conventional correlator using prime codes. For example, in the case of $1/T_c = 9.8$ GHz by the proposed OPRC using Gold codes, 1.3 and 2.5 times the RBSs can be accessed to the radio highway with the CINR of 30 dB compared with the unipolar-type correlator using prime codes for P_s of 0 dBm and -10 dBm, respectively.

6.4.3.5 Summary

In this section, we described an optical polarity-reversing correlator for the DOS-CDMA radio highway network using PN codes such as maximal length

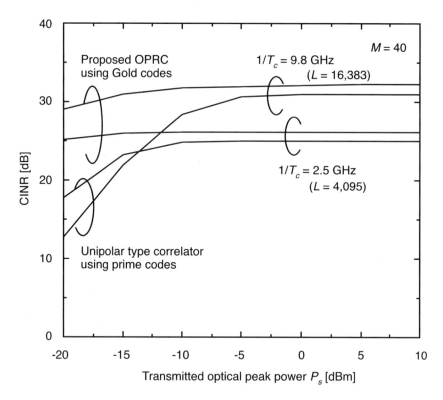

Figure 6.37 Relationship between the average transmitted optical power, P_s, and the CINR.

codes and Gold codes, which are usually used in radio systems. The CINR of the regenerated radio signal at the control station has been theoretically analyzed. The following results were obtained:

- By using the OPRC with PN codes, we can assign more distinct code sequences to radio base stations connected to the DOS-CDMA radio highway network than can be assigned with the unipolar prime code type correlator. For example, comparing maximal length codes and Gold codes of code length $L = 32,767$ with prime codes of code length $p^2 = 32,041$, the numbers of distinct code sequences for maximal length codes and Gold codes are 10 times and 183 times larger than that for prime codes, respectively.

- For small average transmitted optical powers, the proposed optical polarity reversing correlator using Gold codes can much more improve the number of maximum connected RBSs over the unipolar

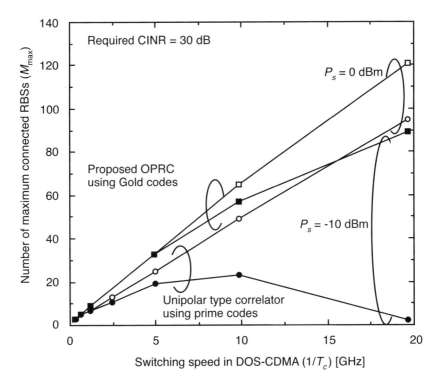

Figure 6.38 Relationship between the switching speed in the DOS-CDMA, $1/T_c$, and the number of maximum connected RBSs, M_{max}.

prime codes type correlator. For example, in the case of the switching speed in DOS-CDMA of 9.8 GHz with an OPRC using Gold codes, 1.3 and 2.5 times the number of RBSs can be accessed to the DOS-CDMA radio highway network with a CINR of 30 dB compared with the unipolar prime codes type correlator for average transmitted optical powers of 0 dBm and −10 dBm, respectively.

6.4.4 DOS-CDMA Routing Switch

In the DOS-CDMA methods described in Sections 6.4.2 and 6.4.3, because the optical power is zero during the interval of zero parts in prime codes or negative polarity parts in PN codes, the received optical power is less than the transmitted laser power at the RBS. Also, to enjoy the flexibility of conventional CDMA radio systems using all intervals of PN codes, we propose the reversing optical intensity CDMA (ROI-CDMA) method, in which the spectrum spreading is performed in the optical domain for all intervals of a PN code [37]. We apply the ROI-CDMA routing scheme to conventional CDMA radio systems in order to route CDM signals to the RCSs, which provide different types of radio services. At the desired RCS of this network, the despreading for two-layered spectrum spreading is performed at the OPRC by using the code sequence of two-layered spectrum spreading at one time.

6.4.4.1   ROI-CDMA Routing Method for CDMA Radio Systems

Figure 6.39 illustrates the configuration of the radio highway using the ROI-CDMA routing method. RBSs are connected to the fiber optic bus link, where CDM signals from radio terminals are multiplexed by the ROI-CDMA scheme. At the ROI-CDMA transmitter, the spreading is performed with code sequences such as $b_r(t)$ and $b_q(t)$ used as the routing data of destination RCSs according to services or operators. At the bus-type fiber link, many

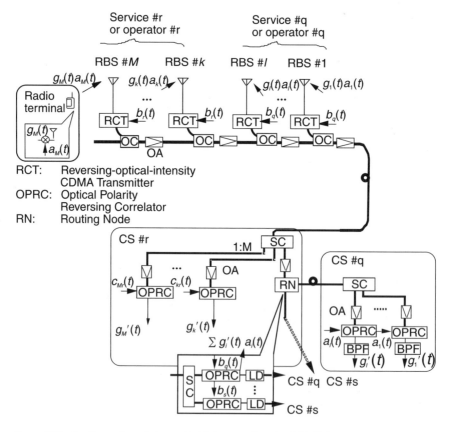

Figure 6.39 Configuration of the radio highway using the ROI-CDMA routing method.

intensity modulated/CDMA signals are transmitted to the RCS. Figure 6.40 illustrates the configurations of the ROI-CDMA transmitter and the OPRC.

At the receiver of the RCS, ROI-CDMA signals from RBSs are first power split into OPRCs and the routing node. To compensate for the optical loss between RBS and the receiver, each RBS and RCS is equipped with optical axes. At the kth OPRC, the average received optical power, P_{r_k}, can be written as shown in (6.98). We assume that the code sequence used in

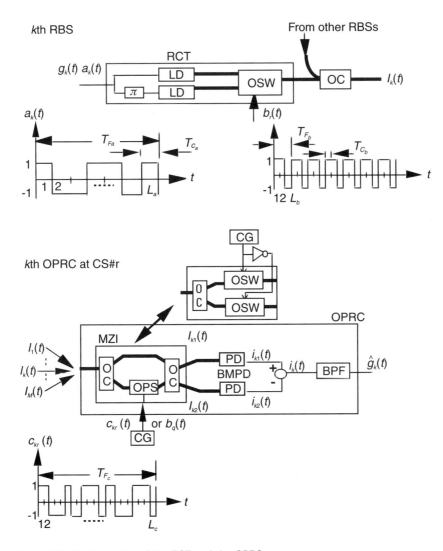

Figure 6.40 Configuration of the RCT and the OPRC.

spreading at the ROI-CDMA transmitter and the code sequence of two-layered spectrum spreading are regenerated at the RCS by using a retiming code generator such as the retiming block. When the RCS is the desired destination RCS, the code sequence $b_r(t)$ is regenerated at the RCS#r, the despreading for two-layered spectrum spreading is performed at the OPRC at one time and radio signals are regenerated. However, in the case where routing is to other RCSs, for example, when $b_q(t)$ is regenerated at the RCS#r, the despreading is performed at the OPRC by using the code $b_r(t)$ used in spreading at the RCT; thus, CDM signals to the RCS#q are regenerated.

Regenerated CDM signals are converted into intensity modulated signals at the laser diode and routed to the routing destination RCS#q.

6.4.4.2 Theoretical Analysis of DOS-CDMA Routing

Theoretical analyses of CINR performance are performed in this section. The spread-spectrum signal received at the kth RBS, $g_{a_k}(t)$, is written by

$$g_{a_k}(t) = g_k(t)a_k(t) \qquad (6.128)$$

where $g_k(t)$ is the radio signal with its bandwidth B_{RF} and carrier frequency f_{RF}, and $a_k(t)$ is the PN code with the frame period of T_{F_a}, the chip width of T_{c_a}, and the code length of L_a, which is the number of chips. To simplify the discussion, each RBS receives CDM signals from a radio terminal. $g_{a_k}(t)$ is split into two signals and one of them is phase shifted by π. Next, two laser diodes are directly intensity modulated by them with an optical modulation index of 1. Optical spectrum spreading is performed by switching outputs of two laser diodes at the optical switch according to the polarity of a PN code sequence, $b_r(t)$, having frame period T_{F_b}, chip width T_{c_b}, and code length L_b. Code sequences such as $b_r(t)$ and $b_q(t)$ are used as the routing data of destination RCSs.

We assume that chip synchronization occurs between $a_k(t)$ and $b_r(t)$. T_{c_a} is the integer multiple of T_{c_b} and L_b is less than L_a to simplify the discussion. Thus, $a_k(t)b_r(t)$ can be expressed as $c_{kr}(t)$ having frame period T_{F_a}, chip width T_{c_b}, and code length L_aL_b. The ROI-CDMA can be expressed as

$$I_k(t) = P_s\{1 + g_k(t)a_k(t)b_r(t)\} = P_s\{1 + g_k(t)c_{kr}(t)\} \qquad (6.129)$$

At the RCS, an OPRC consists of a Mach-Zehnder interferometer or two optical switches, a balanced mixing photodetector, and a bandpass filter.

Many intensity modulated/CDMA signals from many RBSs are received at the Mach-Zehnder interferometer, which is composed of two optical couplers and an optical phase shifter or two optical switches. It is assumed that an optical axis equipped at each RBS has gain G of $L_f + L_{oc}$ and the gain of the optical axis at the output of the star coupler, G_M, is equal to $10 \log_{10} M + 2L_{oc}$. Thus, each OPRC receives optical signals with the same received power $P_{r_k} = P_s$ $(k = 1, 2, \ldots, M)$ from (6.98).

6.4.4.3 The Desired Destination Control Station

When the RCS is the desired destination RCS, code sequence $b_r(t)$ is regenerated at the kth receiver of the RCS#r, and the kth OPRC is driven by the PN code, $c_{kr}(t)$. During the positive polarity interval of $c_{kr}(t)$, the phase shift of optical phase shifter is set to zero or the upper optical switch is set to on. Thus, intensity modulated/CDMA signals are output to the upper port of the balanced mixing photodetector. During the negative polarity interval of $c_{kr}(t)$, the phase shift of optical phase shifter is set to π or the lower optical switch is set to on. Thus, intensity modulated/CDMA signals are output to the lower port of the balanced mixing photodetector. Output currents of the balanced mixing photodetector are expressed as

$$i_{k1}(t) = \alpha \sum_{\substack{j=1 \\ i=q\,\text{or}\,r}}^{M} P_s g_j(t) c_{ji}(t) \cdot \frac{1 + c_{kr}(t)}{2} + i_{n1}(t) \qquad (6.130)$$

$$i_{k2}(t) = \alpha \sum_{\substack{j=1 \\ i=q\,\text{or}\,r}}^{M} P_s g_j(t) c_{ji}(t) \cdot \frac{1 + c_{kr}(t)}{2} + i_{n2}(t) \qquad (6.131)$$

where α is the responsivity of the balanced mixing photodetector, and $i_{n1}(t)$ and $i_{n2}(t)$ are additive noise currents, respectively. The input current to the bandpass filter is given by

$$i_k(t) = i_{k1}(t) - i_{k2}(t) = i_{S_k}(t) + i_I(t) + i_N(t) \qquad (6.132)$$

where $i_{S_k}(t)$, $i_I(t)$, and $i_N(t)$ are the desired signal, the interference, and the additive noise, respectively. The terms $i_{S_k}(t)$ and $i_I(t)$ are expressed as

$$i_{S_k}(t) = \alpha P_s g_k(t) \qquad (6.133)$$

$$i_I(t) = \sum_{\substack{j=1,\,j\neq k \\ i=q \text{ or } r}}^{M} \alpha P_s g_j(t) c_{ji}(t) c_{kr}(t) \tag{6.134}$$

The power spectrum of $i_{S_k}(t)$ is given by

$$S_S(f) = (\alpha P_s)^2 S_g(f) \tag{6.135}$$

where $S_g(f)$ is the power spectrum of $g_k(t)$. We assume that radio signals, $g_j(t)$ ($j = 1, 2, \ldots, M$), are unmodulated carriers that have $S_g(f)$ of

$$S_g(f) = \frac{1}{4}\{d(f + f_{RF}) + d(f - f_{RF})\} \tag{6.136}$$

The carrier power at the output of the bandpass filter, C, can be obtained as

$$C = \int_{f_{RF}-B_{RF}/2}^{f_{RF}+B_{RF}/2} S_S(f)\,df + \int_{-f_{RF}-B_{RF}/2}^{-f_{RF}+B_{RF}/2} S_S(f)\,df = \frac{1}{2}(\alpha P_s)^2 \tag{6.137}$$

The autocorrelation of interference is expressed as

$$R_I(\tau) = (M-1)(\alpha P_s)^2 R_g(\tau) R_\gamma(\tau) \tag{6.138}$$

where $R_\gamma(\tau)$ represents the autocorrelation of a function $\gamma_{jk}(t)$ that is defined by $c_{ji}(t) c_{kr}(t)$ ($i = q$ or r). $R_\gamma(\tau)$ is given by

$$R_\gamma(\tau) = \overline{a_0^2} + \sum_{k=-\infty,\,k\neq 0}^{\infty} \frac{1}{L_a L_b} \frac{1}{1+(k\pi/L_a L_b)} \cos\left(\frac{2\pi k\tau}{T_{F_a}}\right) \tag{6.139}$$

where the value of $\overline{a_0^2}$ depends on the kind of used code sequences. The power spectrum of $i_I(t)$ is given by

$$S_I(f) = (M-1)\left(\frac{\alpha P_s}{2}\right)^2 S_g(f) \otimes S_g(f) \tag{6.140}$$

At the output of the bandpass filter, I is obtained as

$$I = \int_{f_{RF}-B_{RF}/2}^{f_{RF}+B_{RF}/2} S_I(f)df + \int_{-f_{RF}-B_{RF}/2}^{-f_{RF}+B_{RF}/2} S_I(f)df \qquad (6.141)$$

$$= \begin{cases} \dfrac{M-1}{2}(\alpha P_s)^2 \dfrac{(L_a L_b)^2 + (L_a L_b) - 1}{4(L_a L_b)^3} & \text{for Gold code} \\[3mm] \dfrac{M-1}{2}(\alpha P_s)^2 \dfrac{(L_a L_b)^2 + (L_a L_b) - 1}{(L_a L_b)^3} & \text{for maximal length code} \end{cases}$$

Hence, the CIR is written

$$\text{CIR} = \begin{cases} \dfrac{4(L_a L_b)^3}{(M-1)\{(L_a L_b)^2 + (L_a L_b) - 1\}} & \text{for Gold codes} \\[3mm] \dfrac{(L_a L_b)^3}{(M-1)\{(L_a L_b)^2 + (L_a L_b) - 1\}} & \text{for maximal length codes} \end{cases}$$

$$(6.142)$$

At the output of the bandpass filter, we consider additive noise currents composed of RIN, shot noise, receiver thermal noise, beat noise between optical signal and amplified spontaneous emission, beat noise between amplified spontaneous emissions, and optical signal beat noise. The total noise power, N, is written

$$N = N_{RIN} + N_{shot} + N_{th} + N_{s-sp} + N_{sp-sp} + \langle N_{s-s} \rangle \qquad (6.143)$$

Each noise power is given by

$$N_{RIN} = \zeta_{RIN}(\alpha P_s)^2 M B_{RF} \qquad (6.144)$$

$$N_{shot} = 2e\alpha\{P_s M + M(N_{sp} + N_{sp_M})W\}B_{RF} \qquad (6.145)$$

$$N_{th} = \frac{8k_B T}{R_L}B_{RF} \qquad (6.146)$$

$$N_{s-sp} = 4\alpha^2 M^2(N_{sp} + N_{sp_M})P_s B_{RF} \qquad (6.147)$$

$$N_{sp-sp} = 2\alpha^2 M^2 (N_{sp} + N_{spM})^2 (W - f_{\mathrm{RF}}) \qquad (6.148)$$

where e, ζ_{RIN}, W, k_B, T, and R_L are the electric charge, the power spectral density of the relative intensity noise, the bandwidth of optical filter at the RCS, Boltzmann's constant, noise temperature, and load resistance, respectively. The power spectral densities of the amplified spontaneous emissions, N_{sp} and N_{spM}, are given by (6.125) and (6.126), respectively.

6.4.4.4 Routing to Other RCSs

In the case of routing to other RCSs, for example, when $b_q(t)$ is regenerated at the RCS#r, at the routing node the despreading is performed at the OPRC by using the code $b_q(t)$ used in spreading at the ROI-CDMA transmitter. During the positive polarity interval of $b_q(t)$, the phase shift of the optical phase shifter is set to zero or the upper optical switch is set to on; thus, intensity modulated/CDMA signals are output to the upper port of the balanced mixing photodetector. During the negative polarity interval of $b_q(t)$, the phase shift of the optical phase shifter is set to π or the lower optical switch is set to on; thus, intensity modulated/CDMA signals are output to the lower port of the balanced mixing photodetector. Output currents of the balanced mixing photodetector are expressed as:

$$i_{l1}(t) = \alpha \sum_{\substack{j=1 \\ i=q\ or\ r}}^{M} P_s g_j(t) a_j(t) b_i(t) \frac{1 + b_q(t)}{2} + i_{n1}(t) \qquad (6.149)$$

$$i_{l2}(t) = \alpha \sum_{\substack{j=1 \\ i=q\ or\ r}}^{M} P_s g_j(t) a_j(t) b_i(t) \frac{1 + b_q(t)}{2} + i_{n2}(t) \qquad (6.150)$$

The input current to the bandpass filter is given by

$$i_l(t) = i_{l1}(t) - i_{l2}(t) \qquad (6.151)$$

$$= \alpha P_s \sum_{j=1}^{R} g_j(t) a_j(t) + \alpha P_s \sum_{j=R+1}^{M} g_j(t) a_j(t) b_r(t) b_q(t) + i_N$$

where R is the RBS number routed to RCS#q out of M RBSs.

6.4.4.5 Numerical Results and Discussions

Some numerical results are discussed here with parameters indicated in Table 6.4. Figure 6.41 shows the relationship between the CIR and the code

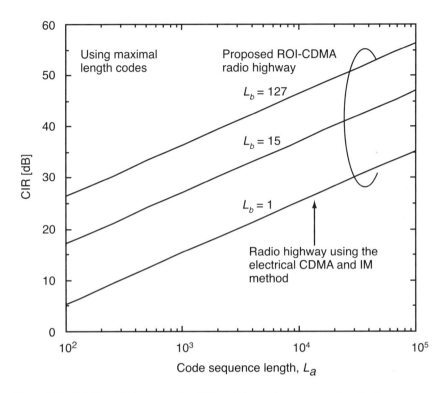

Figure 6.41 Relationship between the CIR and the code sequence length, L_a.

sequence length L_a in the domain of electrical signals for CDMA radios in the case for which 30 RBSs are used along with maximal codes. In the proposed ROI-CDMA radio highway, the process gain can be accomplished by increasing the code sequence length L_b at the ROI-CDMA transmitter compared with the radio highway using the electrical CDMA and intensity modulation method. Thus, we can see from this figure that in the proposed ROI-CDMA radio highway, the CIR is improved more by increasing the code sequence length L_b at the ROI-CDMA transmitter than by the radio highway using the electrical CDMA and intensity modulated method.

Figure 6.42 shows the relationship between the C/N and the code sequence length L_a in the domain of electrical signals for CDMA radios when 30 RBSs are being used. In the radio highway using the electrical CDMA and intensity modulated method, the C/N is dominated by the optical signal beat noise, $\langle N_{s-s} \rangle$. However, in the case of the ROI-CDMA radio highway, $\langle N_{s-s} \rangle$ is reduced by the subtraction process and the bandpass filter at the OPRC and the C/N is dominated by the beat noise among the

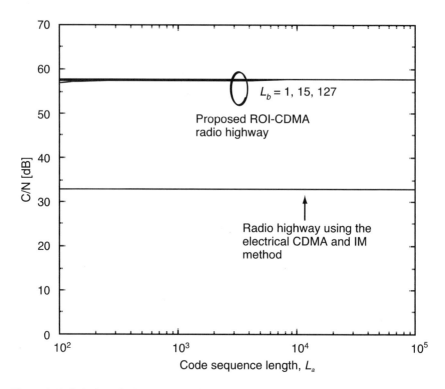

Figure 6.42 Relationship between the C/N and the code sequence length, L_a.

amplified spontaneous emissions of an optical axis and the optical signal, N_{s-sp}. In the case of the CDMA radio highway with the optical coupler bus connection, the C/N is deteriorated more by $\langle N_{s-s} \rangle$ than N_{s-sp}. Therefore, we see from this figure that the C/N in the ROI-CDMA radio highway is improved compared with the radio highway using the electrical CDMA.

Figure 6.43 shows the relationship between the CINR and the code sequence length L_a in the domain of electrical signals for CDMA radios in the case of 30 RBSs and maximal codes. In the radio highway using the electrical CDMA and intensity modulation method, the CINR is dominated by the CIR for the small value of L_a and by the C/N as L_a increases. However, in the proposed ROI-CDMA radio highway, the CINR is dominated by the CIR. Thus, we see from this figure that in the proposed ROI-CDMA radio highway, the CINR is improved more by increasing the code sequence length L_b at the RCT than in the radio highway using the electrical CDMA and intensity modulation method.

6.4.4.6   Summary

In this section, we have proposed the ROI-CDMA method in which spectrum spreading is performed during all intervals of a PN code in order to increase

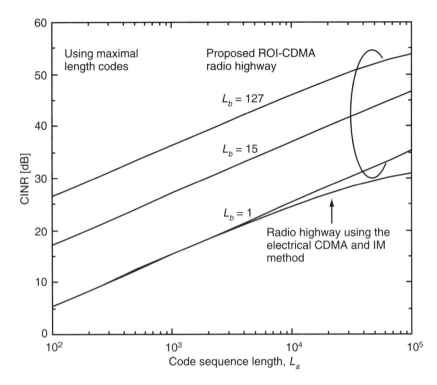

Figure 6.43 Relationship between the CINR and the code sequence length, L_a.

the received optical power at the receiver and have the flexibility required for conventional CDMA radio systems. We have applied the ROI-CDMA routing scheme to conventional CDMA radio systems in order to route CDM signals to the control station according to services or operators.

In the proposed network, at the desired control station the despreading for two-layered spectrum spreading is performed at the OPRC by using the code sequence of two-layered spectrum spreading at one time. It is found that the CINR power ratio can be improved more by increasing the code sequence length at the ROI-CDMA transmitter than that in the radio highway using the electrical CDMA and intensity modulation method.

6.5 Photonic Chirp Multiple Access

6.5.1 Requirements for a Multiband Highway

The fiber optic radio access networks have the potential to be universally used among different kinds of radio services or different providers. The realization of such universal radio access networks can reduce the construction time for and investment in infrastructures, thus retaining the advantages of microcellular systems. Different kinds of radio service are ordinarily operated under different frequency bands, and their RCSs are probably located at different locations. To handle such multiple radio services universally, the fiber optic radio access networks need a routing node in the networks, which distinguishes types of radio service and then switches each service signal into its desired RCS.

The conventional fiber optic link to transfer radio signals is SCM/ optical intensity modulation/direct detection link, which has the advantages of simplicity and cost efficiency in terms of the RBS configuration. However, from the viewpoint of the feasibility of the routing node, the SCM/optical intensity modulation/direct detection link has the primary disadvantage of it being impossible to distinguish among the types of radio service unless the optical signals are photodetected, because optical intensity includes the multiple radio services in subcarrier multiplexing format.

To realize seamless routing of radio service, this section treats the new application of a chirp Fourier transformer [38] to the fiber optic radio access networks. We refer to the system as the chirp multiple access system [39–41] and such a transform is referred to as the chirp multiplexing transform, where a chirp multiplexing transform–equipped RBS converts multiple service radio signals operated under the different frequency bands into ones operated in the TDMA format before E/O conversion. Therefore, the routing node composed of a simple photonic switch can distinguish and switch radio service in the optical stage. The chirp multiplexing transform signals can be demodulated with a correlation demodulator or inverse chirp multiplexing transform.

As for the photonic time switching, techniques such as self-routing photonic switching are progressing, and we can expect fast switch operation, high cost efficiency, and a large capacity. As the alternative to chirp Fourier transform, you may think of the FDMA-to-TDMA conversion scheme using digital signal processing. This system, however, is not suitable for our purposes because it spoils the flexibility of RBSs due to the requirement of digital processing and cannot handle broadband radio signals. However, the chirp Fourier transform realized by surface acoustic wave devices is suitable for our purposes because of its broadband characteristics and its function as a real-time analog Fourier transformer.

The chirp Fourier transform has been studied in many applications such as a spectrum analyzer, for time compression of signal and as a multiplexer and demultiplexer in satellite communications. A group demodulator using the chirp Fourier transform has been proposed in [42, 43], where the chirp Fourier transform transforms FDM radio signals into TDM signals, and digital processing can demodulate the transformed TDM signal. The BER performance has been clarified by using simulation analysis.

This section describes a new configuration for the chirp multiple access fiber optic radio access system that utilizes the chirp multiplexing transform/ optical intensity modulation/direct detection methods. It also describes the theoretical analyses of the S/N performance on the uplink considering the nonlinearity of laser diode, the receiver noises in the optical intensity modulation/direct detection link, and interchannel interference and intersymbol interferences caused by the chirp multiplexing transform. Furthermore, this section shows that the parallel use of chirp Fourier transform is also effective at improving the S/N performance of the chirp multiple access systems.

This section is organized as follows. Section 6.5.2 describes the principle of chirp multiplexing transform. Section 6.5.3.1 clarifies the characteristic of the chirp multiplexing transform signal. Section 6.5.3.2 describes the direct demodulation scheme for the chirp multiplexing transform signal and theoretically analyzes the signal-to-interchannel interference power ratio (S/ICI) and signal-to-intersymbol interference power ratio (S/ISI). Next, Section 6.5.3.3 proposes the parallel use of two chirp multiplexing transforms to reduce ICI and ISI. Section 6.5.3.4 describes theoretical analyses of S/N performance taking into account the laser diode nonlinearity and overall performance including S/N, S/ICI, and S/ISI. From the results of analysis, the performance of the proposed system is compared to the conventional SCM system.

6.5.2 Principle of Chirp Multiplexing Transform

Figure 6.44 illustrates the network configuration of the chirp multiplexing transform fiber optic radio access system and the principle of its operation for multiple radio services. The networks consist of RBSs for the universal use among multiple radio services, RCSs prepared for each radio service, a routing node, and fiber optic links to connect them. The main functions of RBS are chirp multiplexing transform and E/O conversion. The figure is illustrating the example case in which the RBS receives two service frequency bands, A and B. The chirp multiplexing transform converts such multiple radio services with an FDM format into ones with a TDM format in electric stage. Afterward, the converted signals are transmitted into a fiber optic link with E/O conversion toward a routing node. The routing node distinguishes

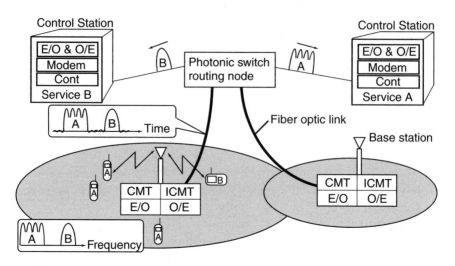

Figure 6.44 Network configuration of photonic chirp multiplexing transform.

the kind of radio service by its transmitted time and switches each service signal into the desired RCS by the use of simple photonic switch.

To investigate the operation of chirp multiplexing transform, we assume the band-limited signal $r(t)$.

$$r(t) = \text{Re}[u(t)e^{j2\pi f_0 t}] \tag{6.152}$$

where $u(t)$ is the complex envelope whose spectrum is band-limited into $|f| \leq (B_{\text{total}}/2) \cdot f_0$, which is the center frequency of radio. The configuration of chirp multiplexing transform is shown in Figure 6.45. The fundamental process of chirp multiplexing transform consists of two multiplications with a chirp signal and a filtering with a chirp filter. However, one process can convert only T_c time duration signals at intervals of two times T_c. Therefore, two identical processes (upper and lower processes in Figure 6.45) have to be operated in parallel in order to convert continuous signals that are longer than T_c without any omission of signal.

The first process in chirp multiplexing transform is the premultiplication of $r(t)$ with an up-chirp signal, $C_a(t)$. The term *up chirp* means that its frequency increases as time passes, while *down chirp* means that its frequency decreases. An up-chirp signal can be obtained as an impulse response of an up-chirp filter in which the higher frequency component is delayed longer. Introducing a complex lowpass up-chirp signal $c(t)$, the up-chirp signal with its center frequency f_a, $C_a(t)$, is given by

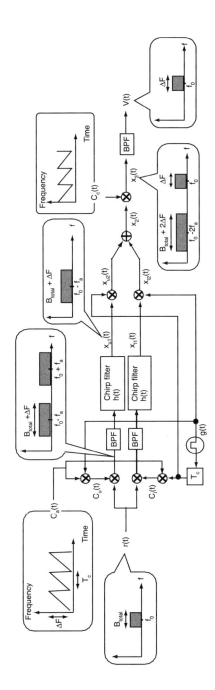

Figure 6.45 Configuration of chirp multiplexing transform.

$$C_a(t) = \text{Re}[\sqrt{2c(t)}e^{j2\pi f_0 t}] \tag{6.153}$$

$$c(t) = \begin{cases} \exp(j2\pi\beta t^2/2); & -\dfrac{T_c}{2} \le t \le \dfrac{T_c}{2} \\ 0; & \text{otherwise} \end{cases} \tag{6.154}$$

where β is the chirp slope (in hertz per second) and T_c is the chirping dispersion time. Then we can define the frequency range of sweep, ΔF, as

$$\Delta F = \beta T_c \tag{6.155}$$

For the upper and the lower processes, we need two up-chirp signals that are alternately gated at the intervals of $2T_c$, $C_u(t)$, and $C_l(t)$. They are obtained by multiplying an up-chirp signal with its period T_c, $\sum_{m=-\infty}^{\infty} C_a(t - 2kT_c)$, by two alternate gating functions, $G(t)$ and $G(t - T_c)$, with their period $2T_c$ and duration T_c:

$$G(t) = \sum_{k=-\infty}^{\infty} g(t - 2kT_c) \tag{6.156}$$

$$g(t) = \begin{cases} 1; & -\dfrac{T_c}{2} \le t \le \dfrac{T_c}{2} \\ 0; & \text{otherwise} \end{cases} \tag{6.157}$$

Then $C_u(t)$ and $C_l(t)$ are written, respectively,

$$C_u(t) = \sum_{m=-\infty}^{\infty} Ca(t - 2mT_c) \tag{6.158}$$

$$C_l(t) = \sum_{m=-\infty}^{\infty} Ca(t - (2m + 1)T_c) \tag{6.159}$$

The radio signal $r(t)$ is multiplied by the gated up-chirp signals, $C_u(t)$ and $C_l(t)$, respectively, on the upper path and the lower path as shown in Figure 6.45. The multiplication generates two components of $(f_0 - f_a)$ frequency band signals and $(f_0 + f_a)$ frequency band signals. Any image frequency component and any overlap of the two frequency band signals is not generated in the case of $|f_0 + f_a| \gg (\Delta F + B_{total})/2$ and $f_a \gg (\Delta F + B_{total})/2$. Therefore, we can use either of the frequency band signals for the next filtering process with a down-chirp filter. However, because the carrier frequency of radio signal f_0 is in the microwave or millimeter-wave band in recent trends, it seems comparatively difficult to realize a chirp filter with such a high center frequency, $(f_0 + f_a)$. Consequently, it is reasonable to use the up-chirp filter for the lower frequency band $(f_0 - f_a)$ in the filtering process. Hence,

$$r(t) \cdot C_u(t) = \text{Re}\left[\sum_{m=-\infty}^{\infty} \frac{\sqrt{2}}{2} u(t)c^*(t - 2mT_c)e^{j2\pi(f_0 - f_a)t} \right]$$

(6.160)

$$r(t) \cdot C_l(t) = \text{Re}\left[\sum_{m=-\infty}^{\infty} \frac{\sqrt{2}}{2} u(t)c^*(t - (2m + 1)T_c)e^{j2\pi(f_0 - f_a)t} \right]$$

(6.161)

and the impulse response of the up-chirp filter, $h(t)$, is given by

$$h(t) = \begin{cases} \text{Re}\left[\sqrt{\beta}e\exp\left[j2\pi\left(\frac{\beta}{2}t^2 - \Delta Ft\right) \right] \cdot e^{j2\pi(f_0 - f_a)t} \right] & 0 \le t \le 2T_c \\ 0 & \text{otherwise} \end{cases}$$

(6.162)

Therefore, the outputs after the filtering process on each path, $x_{u1}(t)$ and $x_{u2}(t)$, are written, respectively,

$$x_{u1}(t) = \{r(t)C_u(t)\} \otimes h(t) \tag{6.163}$$

$$= \text{Re}\left[\sum_{m=-\infty}^{\infty} U^{(2m)}(t)C_{if}^{(2m)}(t) \right]$$

$$x_{l1}(t) = \{r(t)C_l(t)\} \otimes h(t) \tag{6.164}$$

$$= \text{Re}\left[\sum_{m=-\infty}^{\infty} U^{(2m+1)}(t)C_{if}^{(2m+1)}(t) \right]$$

where $\otimes$ denotes convolution, and $U^{(k)}(t)$ and $C_{if}^{(k)}(t)$ are given by

$$U^{(k)}(t) = \begin{cases} \int_{\frac{T_c}{2}}^{t} u(\tau + kT_c)e^{-j2\pi\beta[t-(k+1)T_c]\tau}d\tau; & -\frac{T_c}{2} \le t - kT_c \le \frac{T_c}{2} \\[2mm] \int_{-\frac{T_c}{2}}^{\frac{T_c}{2}} u(\tau + kT_c)e^{-j2\pi\beta[t-(k+1)T_c]\tau}d\tau; & \frac{T_c}{2} \le t - kT_c \le \frac{3T_c}{2} \\[2mm] \int_{t-2T_c}^{\frac{T_c}{2}} u(\tau)e^{-j2\pi\beta[t-(k+1)T_c]\tau}d\tau; & \frac{3T_c}{2} \le t - kT_c \le \frac{5T_c}{2} \\[2mm] 0; & \text{otherwise} \end{cases}$$

(6.165)

$$C_{if}^{(k)}(t) = \frac{\sqrt{2\beta}}{4} e \exp\left[j2\pi\beta\frac{[t-(k+1)T_c]^2}{2} \right] \cdot e^{j2\pi(f_0 - f_a)t}$$

(6.166)

Figure 6.46 illustrates the relationship between the input and the output signals of the chirp filter in the time domain. The part of the output in the range of $T_c/2 \le t - kT_c \le 3T_c/2$ in (6.165), that is, $U^{(k)}(t)g(t - (k+1)T_c)$, is illustrated as the shadowed parts in Figure 6.46. The part corresponds to the Fourier transform of $u(t)$ gated in the time range T_c, and is necessary for our purposes. Then, the necessary parts from each path $x_{u2}(t)$ and $x_{l2}(t)$ are extracted by gating functions and summed. The result is written $x_2(t)$.

$$x_2(t) = \text{Re}\left[\sum_{k=\infty}^{\infty} U^{(k)}(t) C_{if}^{(k)}(t) g(t - (k+1)T_c) \right]$$

(6.167)

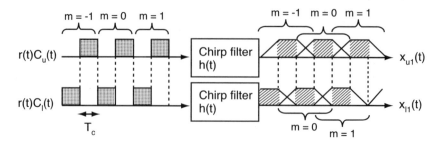

Figure 6.46 Input and output signals of the chirp filter.

We find from (6.167) that the center frequency of $x_2(t)$ increases periodically due to $C_{if}^{(k)}(t) \cdot x_3(t)$ in which the effect of $C_{if}^{(k)}(t)$ is eliminated and can be obtained by multiplying $x_2(t)$ by a down-chirp signal $C_c(t)$, which is given by

$$C_c(t) = \text{Re}[\sqrt{2}c^*(t)e^{j2\pi f_a t}] \tag{6.168}$$

After bandpass filtering $x_3(t)$ $(= x_2(t) \cdot C_c(t))$ by the use of a bandpass filter with center frequency f_0, we finally obtain the desired chirp multiplexing transform signal, $V(t)$:

$$V(t) = \text{Re}[v(t)e^{j2\pi f_0 t}] \tag{6.169}$$

$$v(t) = \sum_{k=\infty}^{\infty} U^{(k)}(t)g(t - (k+1)T_c) \tag{6.170}$$

where the amplitude constant $\sqrt{2\beta}/8$ and the phase constant are dropped. $U^{(k)}(t) \cdot g(t - (k+1)T_c)$ is referred to as the time-limited Fourier transform signal in this section because it is the Fourier transform output of $u(t + kT_c) \cdot g(t)$ where the conversion of the frequency axis to the time axis, that is, the f axis to the βt axis, is executed. The time waveform of the time-limited Fourier transform signal represents the spectrum information of $u(t + kT_c) \cdot g(t)$, but it is further limited in the time range of T_c. Consequently, it is necessary for βT_c to cover the spectrum range of radio signals, B_{total}.

$$\Delta F = \beta T_c \geq B_{\text{total}} \tag{6.171}$$

Under this condition, the chirp multiplexing transform can successfully perform FDM-to-TDM conversion of radio signals. However, we suppose that the spectrum shape of the time-limited Fourier transform signal represents the time waveform of $u(t + kT_c) \cdot g(t)$, and is also band-limited in the range of $\beta T_c = \Delta F$.

6.5.3 Chirp Multiple Access Architecture and Performance

6.5.3.1 Chirp Multiple Access System for FDM-Phase Shift Keying Radio Signals

In this section, we discuss the characteristics of the chirp multiple access system by assuming that FDM-phase shift keying radio signals are applied to the chirp multiplexing transform/optical intensity modulation/direct detection system. Figure 6.47 shows the analysis models of chirp multiplexing transform/optical intensity modulation/direct detection system and SCM/optical intensity modulation/direct detection system for a comparable system. In the chirp multiplexing transform/optical intensity modulation/direct detection system, the chirp multiplexing transform signal whose lowpass equivalent is represented by (6.170) directly modulates the laser diode and is transferred via a fiber optic link. The receiver at the RCS detects the optical signal and directly demodulates the information using the proposed demodulator described in Section 6.5.3.2. However, in the SCM/optical intensity modulation/direct detection system, the FDM-phase shift keying signal directly modulates the laser diode and is similarly transferred via an optic link and their information is demodulated by the optimum demodulation using a matched filter at the RCS.

The lowpass equivalent of FDM-phase shift keying radio signals is written

$$u(t) = \sum_{l=-M}^{M} \sum_{n=-\infty}^{\infty} f(t - nT)e^{j[2\pi l \Delta f t + \theta_{ln}]} \tag{6.172}$$

where θ_{ln} is the nth modulation phase of the lth FDM channel and $(2M + 1)$ is the number of FDM channels. T, Δf, and $f(t)$ are the symbol duration, the channel separation, and the pulse waveform band-limited into B, respectively. Then the total bandwidth, B_{total}, is written

$$B_{total} = (2M + 1)\Delta f \tag{6.173}$$

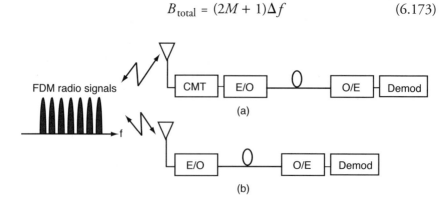

Figure 6.47 Performance analysis models of (a) the chirp multiplexing transform (CMT)/optical intensity modulation/direct detection system and (b) the SCM system.

We investigate the $k = 0$th time-limited Fourier transform signal in $\nu(t)$ when (6.172) is substituted into (6.170). We represent it as $\nu_0(t)$.

$$\nu_0(t) = \int_{-\frac{T_c}{2}}^{\frac{T_c}{2}} u(\tau)e^{-j2\pi\beta[t-T_c]\tau}d\tau; \qquad \frac{T_c}{2} \leq t \leq \frac{3}{2}T_c \qquad (6.174)$$

Assuming the perfect synchronization between the symbol timing of $u(t)$ and the chirping dispersion time T_c, $\nu_0(t)$ becomes

$$\nu_0(t) = \sum_{l=-M}^{M} \sum_{n=-\infty}^{\infty} S_{ln}(t)e^{j\theta_{ln}} \qquad \frac{T_c}{2} \leq t \leq \frac{3}{2}T_c \qquad (6.175)$$

$$S_{ln}(t) = \beta G(\beta t) \otimes [F[\beta(t - T_c) - l\Delta f]e^{-j2\pi[\beta(t-T_c)-l\Delta f]\tau T}] \qquad (6.176)$$

where $G(f)$ and $F(f)$ are the Fourier transforms of $g(t)$ and $f(t)$, respectively. $S_{ln}(t)$ corresponds to the pulse waveform of the nth symbol of the lth channel. Note that the words *channel* and *symbol number* described in this section consistently indicate those in the original FDM-phase shift keying radio signals defined in (6.172).

Equations (6.175) and (6.176) can be generally applied to any kind of pulse waveform, $f(t)$, and any value of Δf. However, we focus our discussion on the case for which $f(t)$ is pulse shaped by the root Nyquist filter. Then $F(f)$ is given by

$$F(f) = \begin{cases} 1 & 0 < |f| \leq \dfrac{1-\alpha}{2T} \\ \cos\left[\dfrac{\pi T}{2\alpha}\left(|f| - \dfrac{1-\alpha}{2T}\right)\right] & \dfrac{1-\alpha}{2T} < |f| \leq \dfrac{1+\alpha}{2T} \\ 0 & \text{otherwise} \end{cases}$$

$$(6.177)$$

where α is the roll-off factor. Figure 6.48 illustrates the relationship between the original radio signal $u(t)$ and its spectrum $U(f)$ and the resultant chirp multiplexing transform signal $\nu_0(t)$ and its spectrum $\nu_0(f)$. We can see from (6.176) and Figure 6.48 that the chirp multiplexing transform signal consists of multicarrier phase shift keying signals with carrier spacing βT,

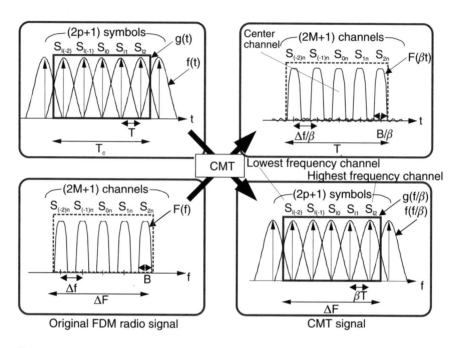

Figure 6.48 Radio signal and chirp multiplexing transform (CMT) signal.

and pulse duration $\Delta f / \beta$. The different symbol in a certain channel is converted to the different carrier frequency in the same time slot, while the different channel's symbol is converted to a different time slot. Furthermore, the chirp multiplexing transform signal is rectangularly band-limited into the frequency range of Δf, and its time waveform is limited in the time range of T_c, as described in Figure 6.48.

If the chirping dispersion time, T_c, is infinite ($T_c \rightarrow \infty$), we can ignore the convolution with $G(\beta t)$ in (6.176); therefore, the set of $S_{ln}(t)$ ($l = -M$, ..., M; $n = -\infty$, ..., ∞) becomes an orthogonal set and no interchannel interference or intersymbol interference occurs. However, we can see from (6.176) and Figure 6.48 that in the resultant time domain, each pulse waveform corresponding to a channel, $F(\beta t - l\Delta f) \otimes G(\beta t)$, has a spreading sidelobe even if $F(f)$ is perfectly band-limited. This spread causes interchannel interference. However, in the resultant frequency domain, each spectrum shape corresponding to a symbol, $f(f / \beta - nT) \cdot g(f / \beta)$, is partially truncated. This truncation causes intersymbol interference because of the imperfection of the orthogonality between different symbols even if the set of $f(t - nT)$ ($n = -\infty$, ..., ∞) is an orthogonal set.

Here, in order to estimate the number of carriers to be considered in the following analysis, we examine the symbol energy of $S_{ln}(t)$. Figure 6.49

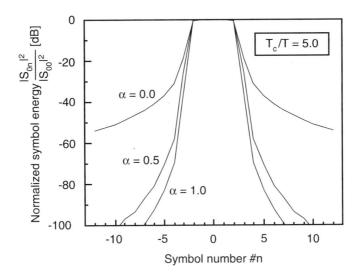

Figure 6.49 Normalized symbol energy versus symbol number.

shows the symbol energy of $S_{0n}(t)$ versus the symbol number n in the case of $T_c/T = 5.0$. Each symbol energy is normalized by the center symbol energy, S_{00}. We can see that the energy of the symbols out of center T_c/T symbols, that is, the energy of S_{0-3} and S_{03} in this example, suddenly decreases, and it is less than -36 dB of the energy of S_{00} for the case of $\alpha = 1.0$. Consequently, we approximate the number of carriers to be considered in the following discussion as T_c/T. Then when T_c/T is represented by $2p + 1$ (p is a positive integer), $\nu_0(t)$ can be rewritten as

$$\nu_0(t) \approx \sum_{l=-M}^{M} \sum_{n=-p}^{p} S_{ln}(t) e^{j\theta_{ln}} \qquad (6.178)$$

6.5.3.2 Theoretical Interchannel Interference and Intersymbol Interference Performance Analyses

Figure 6.50 shows the configuration of the proposed suboptimal correlation receiver. The term *suboptimal* indicates that the waveform correlated with the received chirp multiplexing transform (CMT) signal at the receiver is the waveform omitting the convolution term, $G(\beta t)$, in (6.176). The demodulator performs the multiplications of the received chirp multiplexing transform signal with $\nu_0(t)G(\beta t)F(\beta t - l\Delta f)$ and $e^{j2\pi[\beta nT]^t}$, which are followed by an integrator and a decision circuit. In the demodulation, the perfect synchronization of frequency and phase, and the condition of the

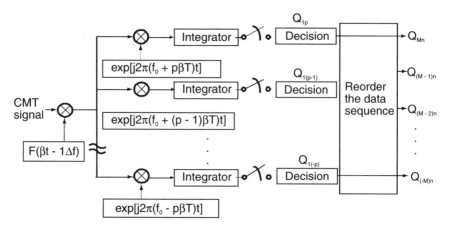

Figure 6.50 Configuration of suboptimal correlation receiver.

ΔfT = integer are assumed. The latter condition is desirable to make the configuration of the receiver simple. Under the condition of the $\Delta ft \neq$ integer, the carrier phase in $S_{ln}(t)$ (6.176) changes according to channel number l; therefore, the receiver needs to control the phase of multiplying carriers according to the value of l.

Furthermore, when the proposed receiver is applied to the case of asynchronous symbol timing between channels, both the phase and the frequency of the multiplying carriers have to be changed according to the channel number l.

The decision sample of the demodulated nth symbol of the lth channel, Q_{in}, is written by

$$Q_{in} = \int_{t_{la}}^{t_{lb}} \nu_0(t) S'_{ln}(t) dt \tag{6.179}$$

$$= \int_{t_{la}}^{t_{lb}} [S_{ln}(t) e^{j\theta_{in}}](S'_{ln}(t) dt) + \sum_{\substack{l=-M \\ k \neq n}}^{M} \sum_{\substack{s=-p \\ s \neq n}}^{p} \int_{t_{la}}^{t_{lb}} [S_{ks}(t) e^{j\theta_{ks}}] S'_{ln}(t) dt$$

$$S'_{ln}(t) = F[\beta(t - T_c) - l\Delta f] e^{2\pi[\beta(t-T_c)]\tau T} \tag{6.180}$$

$$t_{la} = T_c + \frac{1}{\beta}\left(l\Delta f - \frac{B}{2}\right) \tag{6.181}$$

$$t_{lb} = T_c + \frac{1}{\beta}\left(l\Delta f + \frac{B}{2}\right) \tag{6.182}$$

The first term of (6.179) represents the desired signal and the second term represents interchannel interference and intersymbol interference. Consequently, when we demodulate the nth symbol of the lth channel, the S/ICI and S/ISI are derived as follows:

$$\left(\frac{S}{ICI}\right)_{ln} = \frac{\left(\mathrm{Re}\left[\int_{t_{la}}^{t_{lb}} S_{ln}(t)S'_{ln}(t)dt\right]\right)^2}{\left(\sum_{\substack{k=-M\\k\neq l}}^{M}\mathrm{Re}\left[\int_{t_{la}}^{t_{lb}} S_{kn}(t)S'_{ln}(t)dt\right]\right)} \tag{6.183}$$

$$\left(\frac{S}{ISI}\right)_{ln} = \frac{\left(\mathrm{Re}\left[\int_{t_{la}}^{t_{lb}} S_{ln}(t)S'_{ln}(t)dt\right]\right)^2}{\left(\sum_{\substack{s=-p\\s\neq n}}^{M}\mathrm{Re}\left[\int_{t_{la}}^{t_{lb}} S_{ls}(t)S'_{ln}(t)dt\right]\right)} \tag{6.184}$$

Figures 6.51 and 6.52 show the S/ICI and the S/ISI performances of the center channel versus chirping dispersion time normalized by symbol duration (T_c/T). The total number of channels, $2M + 1$, is 3, and the roll-off factor α and the symbol number n are parameters. In the figures, $n = p$ indicates the highest frequency symbol in the chirp multiplexing transform signal.

As described in Section 6.5.3.1, the convolution of $G(\beta t)$ with $F(\beta t)$ causes spreading sidelobes of each channel's pulse; therefore, the interchannel interference is produced. Figure 6.51 shows that the performance of the highest frequency symbol deteriorates the most and its S/ICI does not change even if the T_c becomes large, though the S/ICI of the center frequency symbol or the identical frequency symbol is improved. The reason is that the larger spreading of the sidelobe in the time domain is caused by the

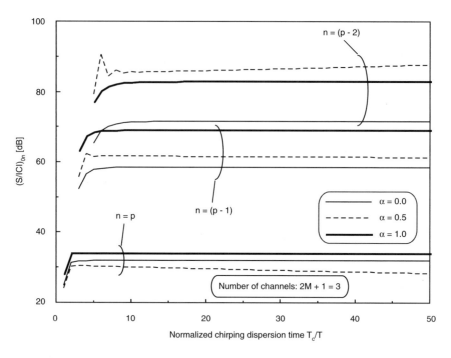

Figure 6.51 S/ICI versus chirping dispersion time.

larger truncation of the spreading sidelobe in the frequency domain, and the highest frequency symbol in the chirp multiplexing transform signal suffers from the largest truncation (see the chirp multiplexing transform signal in Figure 6.48).

However, we can see from Figure 6.52 that the difference of the roll-off factor influences the S/ISI more than that of the symbol number, and $\alpha = 1$ gives the maximum S/ISI. The reason is that the spreading of each symbol's spectrum shape in the chirp multiplexing transform signal becomes narrowest when $\alpha = 1$, resulting in the smallest truncated part and the smallest distortion.

6.5.3.3 Double Chirp Multiplexing Transform System

As shown in Section 6.5.3.2, the S/ICI and the S/ISI performances of the highest and lowest frequency symbols in the chirp multiplexing transform signal deteriorate the most and dominate total system performance. Figure 6.51 shows that a 34-dB difference exists between the S/ICI of $n = p$ and that of $n = p - 1$ when $T_c/T = 5.0$ and $\alpha = 1.0$.

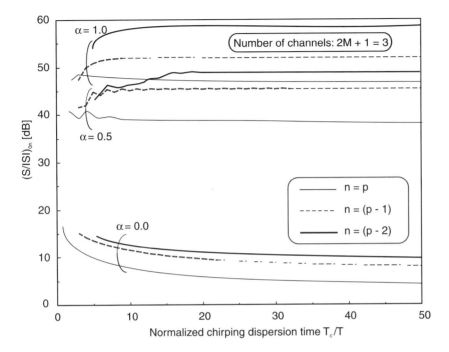

Figure 6.52 S/ISI versus chirping dispersion time.

A similar problem occurs in the group demodulation system using the chirp Fourier transform for FDM signals. To solve this problem, the dual-parallel use of the chirp Fourier transform has been proposed in [44]. In the system, the chirp sweep timings of the two transformers are offset by half of the chirping dispersion time from each other, and only the center $k/2$ symbols ($k = T_c/T$) from each transformer output are used for demodulation.

The parallel use of a chirp Fourier transformer is also effective for our systems, and we call the system a *double chirp multiplexing transform system*. However, we are faced with the problem that two fiber optic links are required in order to transfer each of the output signals from two chirp multiplexing transforms and it increases the complexity of the network configuration. This problem is inherent to our applications where the chirp multiplexing transforms and the demodulation functions are separately equipped in the networks. Therefore, we propose to multiplex the two parallel chirp multiplexing transform outputs into a single fiber optic link with TDM format.

Figure 6.53 illustrates the configuration of the double chirp multiplexing transform systems. It consists of two chirp multiplexing transforms

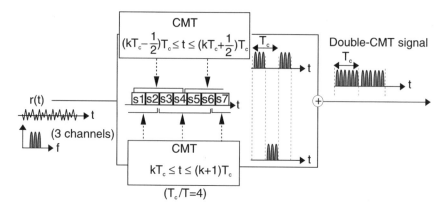

Figure 6.53 Configuration of double chirp multiplexing transform (CMT).

whose chirp sweep timings are offset from each other by half of the chirping dispersion time, $T_c/2$, in the case of T_c/T = even integer, otherwise by $(T_c + T)/2$ in the case of the T_c/T = odd integer. Equation (6.170) shows that when we enlarge the frequency range of the sweep, ΔF, into more than double that of the total radio bandwidth B_{total}, we can reduce the time width of the time-limited Fourier transform signal, $U^{(k)}(t)g(t - (k + 1)T_c)$, to less than $T_c/2$. Consequently two chirp multiplexing transform outputs can be time-division multiplexed by simply adding the two outputs.

Figure 6.54 illustrates the configuration of the receiver for the double chirp multiplexing transform system in the case of $T_c/T = 4.0$. At the receiver, we use only center $T_c/2T$ symbols in the frequency domain to demodulate, and it is sufficient to prepare only $T_c/(2T)$ demodulation circuits in the case of the T_c/T = even integer. The sampling process is executed at the interval of the half of $\Delta f/\beta$, and the demultiplexing for two parallel chirp

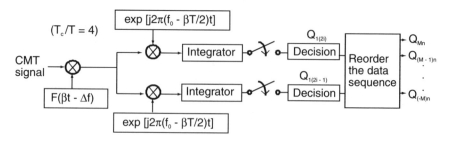

Figure 6.54 Configuration of the receiver for a double chirp multiplexing transform (CMT) system.

multiplexing transform outputs is not required. In the case of the $T_c/T =$ odd integer, we need $T_c + T/(2T)$ demodulation circuits.

Another important merit of the double chirp multiplexing transform system is robustness against the imperfect synchronization between the chirping dispersion time period and the symbol duration assumed in Section 6.5.3.2. The reasons are that the imperfection generates a more truncated part and performance degradation mainly to the highest or the lowest frequency symbol in the chirp multiplexing transform signal. Note that the double chirp multiplexing transform system does not use the damaged symbol for demodulation.

6.5.3.4 Theoretical Analyses of S/N and Overall Performances

In this section, we theoretically analyze the S/N obtained by the suboptimal correlation detection in the chirp multiplexing transform/optical intensity modulation/direct detection system. Here we assume that T_c is larger than T, and we neglect the effect of the convolution with $G(\beta t)$ in (6.174). Under this assumption, suboptimal correlation detection is considered to be the optimum detection scheme for maximizing the S/N because the signal for correlation at the receiver is matched to the waveform of the received signal.

In the chirp multiplexing transform/optical intensity modulation/direct detection systems, the laser diode is modulated by the chirp multiplexing transform signals. The output optical intensity of the laser diode, $P(t)$, can be characterized by the use of a polynomial equation when the nonlinear characteristics of the laser diode are taken into consideration. The chirp multiplexing transform–modulated signal is given by

$$P(t) = P_0[1 + mV(t) + a_2\{mV(t)\}^2 + a_3\{mV(t)\}^3 + \ldots]$$

$$(6.185)$$

where P_0 is the average transmitting optical power and m is the modulation index. When this optical signal is received at the RCS, the output current of the photodetector, $I_0(t)$, is given by

$$I_0(t) = \alpha_0 P_r(1 + mV(t)) + I_{\text{RIN}}(t) + I_{\text{shot}}(t) + I_{th}(t) + I_{im3}(t)$$

$$(6.186)$$

where P_r and α_0 are the average received optical power and the responsivity of the photodetector, respectively. $I_{\text{RIN}}(t)$, $I_{\text{shot}}(t)$, and $I_{th}(t)$ are the RIN current, the shot noise current, and the thermal noise current, respectively.

$I_{im3}(t)$ is the third-order intermodulation (IM3) distortion current. Assuming that each noise current is white noise current, the power spectral density level of each noise current is given by

$$n_{RIN} = RIN(\alpha_0 P_r)^2 \qquad (6.187)$$

$$n_{shot} = 2e\alpha_0 P_r \qquad (6.188)$$

$$n_{th} = \frac{4k_b T_{th}}{R} \qquad (6.189)$$

where RIN, k_B, and R are the power spectral densities of the RIN of the laser diode, Boltzmann's constant, noise temperature, and load resistance, respectively. When N unmodulated carriers with equal amplitude and equal frequency spacing are applied as $V(t)$, the IM3 power in the kth carrier band, $\langle I_{im3}^2 \rangle_{(N,k)}$, is given by

$$\langle I_{im3}^2 \rangle_{(N,k)} = \begin{cases} \frac{1}{2} \left(\frac{3}{4} a_3 m^3 D_{(N,k)}^{(2)} + \frac{3}{2} a_3 m^3 D_{(N,k)}^{(3)} \right)^2 (\alpha_0 P_r)^2 & N \geq 3 \\ 0 & N < 3 \end{cases}$$

$$\qquad (6.190)$$

$$D_{(N,k)}^{(2)} = \frac{1}{2} \left[N - 2 - \frac{1}{2} \{1 - (-1)^N\}(-1)^k \right] \qquad (6.191)$$

$$D_{(N,k)}^{(3)} = \frac{k}{2}(N - k + 1) + \frac{1}{4}\{(N - 3)^2 - 5\} - \frac{1}{8}\{1 - (-1)^N\}(-1)^{N+k}$$

$$\qquad (6.192)$$

The $\langle I_{im3}^2 \rangle_{(N,k)}$ has the largest value at the center carrier frequency band. We take the IM3 currents into consideration for the S/N analysis by translating the IM3 current into the band-limited white noise current with its power spectral density level of n_{im3}

$$n_{im3} = \begin{cases} \dfrac{(2p + 1) \cdot \langle I_{im3}^2 \rangle_{(2p+1,p)}}{\Delta F} & 2p + 1 \geq 3 \\ 0 & 2p + 1 < 3 \end{cases} \qquad (6.193)$$

where $2p + 1$ is the number of carriers represented by $2p + 1 = T_c/T$. The symbol energy E_s is written

$$E_s = \frac{1}{2}(\alpha_0 P_r)^2 m^2 \int_{-\infty}^{\infty} F^2(\beta t)\,dt \qquad (6.194)$$

$$= \frac{1}{2}(\alpha_0 P_r)^2 m^2 \frac{T_c}{T\Delta F}$$

where the modulation index m needs to satisfy the following condition in order not to cause overmodulation distortion at the laser diode in (6.185):

$$m \le \frac{1}{2p + 1} = \frac{T}{T_c} \qquad (6.195)$$

We can see from (6.185) and (6.195) that the symbol energy E_s decreases as ΔF increases because the pulse waveform of the chirp multiplexing transform (CMT) signal becomes narrower, and that E_s also decreases as T_c becomes large because the number of carriers, that is, T_c/T, increases and m should be proportional to the inverse of it.

The S/N attained by the correlation detection, that is, the energy contrast, is given by

$$\left(\frac{S}{N}\right)_{CMT} = \frac{E_s}{n_{RIN} + n_{shot} + n_{th} + n_{im3}} \qquad (6.196)$$

There exists an optimum modulation index m_{opt} which gives the maximum S/N because n_{RIN}, n_{shot}, and n_{th} are independent of the modulation index m, while E_s increases proportional to m^2, and n_{im3} increases proportional to m^6. The value of m_{opt} is given by

$$m_{opt} = \left(\frac{n_{RIN} + n_{shot} + n_{th}}{2n_{im3}|_{m=1}}\right)^{1/6} \qquad (6.197)$$

where m_{opt} needs to satisfy the condition of (6.195) at the same time. Regarding the frequency range of sweep, ΔF, we can choose any value of ΔF more than B_{total} [see (6.171)], but since E_s is proportional to the inverse of ΔF, it should be the following value to attain the maximum S/N:

$$\Delta F = \begin{cases} B_{total} & \text{single CMT} \\ 2B_{total} & \text{double CMT} \end{cases} \qquad (6.198)$$

Figure 6.55 shows the $(S/N)_{CMT}$ versus the number of FDM radio channels for both the single chirp multiplexing transform scheme and the double chirp multiplexing transform scheme, and for different values of the normalized chirping dispersion time, T_c/T. In the calculation, the optimum modulation index m_{opt} and parameters shown in Table 6.7 are used.

We can see from the Figure 6.55 that the S/N decreases as the number of channels increases and that the S/N of the double chirp multiplexing transform deteriorates more than that of the single chirp multiplexing

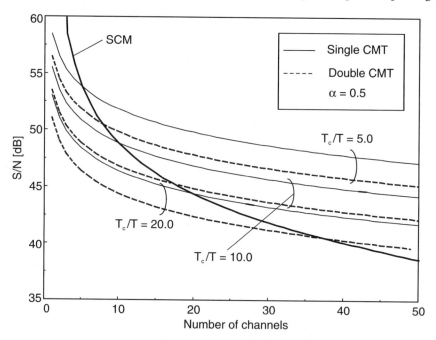

Figure 6.55 S/N versus the number of radio channels.

Table 6.7
Parameters Used in Calculations

RIN	−150 dBc/Hz	T_{th}	300K
α_0	0.8 A/W	R	50Ω
α_s	0.01	P_r	−10 dBm
$1/T$	192 symbols	Δf	2/T

transform scheme because the increase in the frequency range of the sweep, ΔF, causes a decrease in the symbol energy. Note also that the S/N for large T_c/T is degraded because of the increase of the intermodulation distortion power. Therefore, a small value for T_c/T is desirable.

In Figure 6.55, the S/N performance of the conventional SCM scheme is also shown. The S/N of the SCM system more rapidly decreases in comparison with the proposed system as the number of channels increases. The reason is that the IM3 power in the SCM system rapidly increases as the number of channels increases, whereas the IM3 power in the chirp multiplexing transform (CMT)/optical intensity modulation/direct detection system is independent of the number of channels, but dependent on the T_c/T. As a result, the S/N performance of the proposed system becomes superior to the SCM system in terms of the large number of channels in spite of the value T_c/T. In the case of $T_c/T = 5.0$, the S/N of the single chirp multiplexing transform system is superior if the number of channels is more than six, and that of the double chirp multiplexing transform system is superior if the number of channels is more than eight.

Figure 6.56 shows overall S/N performance of the proposed system in which the S/ICI, the S/ISI, and the S/N are taken into consideration. The overall S/N is given by

$$\left(\frac{S}{N+1}\right) = \left[\left(\frac{S}{ICI}\right)^{-1} + \left(\frac{S}{ISI}\right)^{-1} + \left(\frac{S}{N}\right)^{-1}\right]^{-1} \quad (6.199)$$

In the calculation of the S/ICI and the S/ISI, $\alpha = 0.5$ is assumed, and we examine the performances of the most degraded symbol, that is, the highest frequency symbol in the chirp multiplexing transform signal. We can see that the overall performance of the single chirp multiplexing transform system is always inferior to the SCM system in spite of the value of T_c/T because it is dominated by the low S/ICI performance.

However, the double chirp multiplexing transform system can much improve the overall performance because it is dominated by the S/N performance shown in Figure 6.55 and it is superior to the SCM system at $T_c/T = 5.0$ and the number of radio channels is more than 26. In the figure, a 3-dB improvement in the overall performance is obtained when the number of channels is 50 and $T_c/T = 5.0$. This superiority is caused by the fact the chirp multiplexing transform system can decrease the number of carriers and is tolerable against the laser-diode nonlinearity. Consequently, the performance of the chirp multiplexing transform system is superior to the SCM

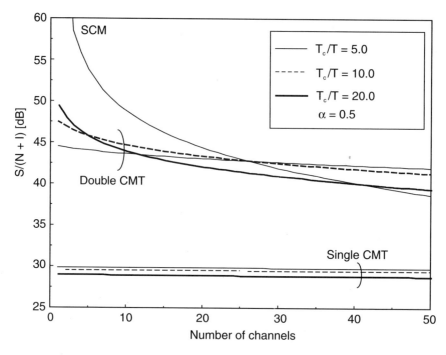

Figure 6.56 S/(N + I) versus the number of radio channels.

system if the condition holds that the laser-diode nonlinearity dominates the signal performance.

6.5.3.5 Summary

This section proposed a chirp multiplexing transform fiber optic radio access system suitable for fiber optic radio access networks at the point of seamless routing of radio services, and theoretically analyzed S/N performance on the uplink, taking into consideration interchannel interference, intersymbol interference, laser-diode nonlinearity, and the receiver noise inherent to IM/DD fiber optic links.

Furthermore, in order to decrease the interchannel interference and the intersymbol interference, a double chirp multiplexing transform system was proposed. The theoretical analysis investigated the relationship between S/N performance and parameters inherent to the chirp multiplexing transform systems and compared S/N performance to the conventional SCM radio access system. The following results were obtained:

- The chirp multiplexing transform can successfully transform multiple radio services with FDM formats into ones with TDM formats.

Then the chirp multiplexing transform radio access system can provide more simple and seamless radio access networks because a routing node composed of a photonic time switch can distinguish among types of radio service and switch them at the optical stage in the networks.

- By means of a suboptimal correlation demodulator, the chirp multiplexing transform signal can be directly demodulated with high S/ISI and high S/ICI performances. However, such performances are different according to the symbol number and the roll-off factor of the pulse shaping. In particular, the S/ICI performance of the highest and the lowest frequency symbol in the chirp multiplexing transform signal deteriorates the most, and the use of the largest value of roll-off factor gives the highest S/ISI performance.

- The double chirp multiplexing transform system can improve S/ICI and S/ISI performance, although S/N performance is deteriorated. The double chirp multiplexing transform system results in a 32-dB improvement of the S/ICI for the case of $T_c/T = 5.0$ and $\alpha = 1.0$.

- The S/N performance of the chirp multiplexing transform radio access system is superior to the conventional SCM system in spite of the value T_c/T when the number of radio channels becomes large.

- The overall S/N performance of the chirp multiplexing transform radio access system is superior to the conventional SCM system when the number of channels is large and the double chirp multiplexing transform is used. When the number of channels is 50, $T_c/T = 5.0$, and $\alpha = 0.5$, the improvement of the overall S/N performance is 3 dB.

6.5.4 Chirp Multiple Access Routing Networks

An asynchronous TDMA bus link system was discussed in Section 6.3.3. The section introduced the use of asynchronous TDMA bus link for fiber optic radio access networks, which results in simple construction and easy extension of the networks, because the difficulty of time synchronization among RBSs can be obviated, while the optical pulse loss occurs due to the asynchronous access of signals.

In that section, the asynchronous TDMA system was realized by transforming radio signals into narrow optical pulse amplitude modulated/optical intensity modulated pulses by the use of a photonic natural bandpass sampling technique.

This section describes the asynchronous TDMA bus link system using chirp access, in which radio signals are also transformed into narrow optical pulse amplitude modulated/optical intensity modulated pulses in order to allow the asynchronous access of radio signals with TDMA format. However, such a transformation is achieved by means of the chirp multiplexing transform system described in Sections 6.5.2 and 6.5.3, not by the use of the photonic natural sampling technique.

As described in Sections 6.5.2 and 6.5.3, the chirp multiplexing transform can transform radio signals with an FDMA format into ones with a TDMA format. Therefore, the fiber optic radio access networks using chirp multiplex access provide the possibility of universal use of the networks among different types of radio service because chirp multiple access is able to distinguish and switch radio service signals in the optical stage by means of a photonic switch at the routing node in the network.

Besides this advantage of chirp multiple access, if we can perform the time compression of output signals from the chirp multiplexing transform, a configuration that uses an asynchronous TDMA bus link combined with chirp multiplexing transform system is possible. Such a combination of asynchronous TDMA and chirp multiplexing transform provides much universal and flexible fiber optic radio access networks because the advantages of each system are also combined. Fortunately, the chirp multiplexing transform has the ability to compress the output at the same time if the chirp sweep bandwidth can be enlarged more than the bandwidth of the radio signal. Therefore, in this section, an asynchronous TDMA bus link system using chirp multiplexing transform is described and discussed.

Another topic dealt with in this section is the effect of traffic distribution on the pulse loss probability and the call blocking probability. The results obtained can be also applied to the performance of the asynchronous TDMA bus link systems using the photonic natural sampling techniques that were discussed in Section 6.3.1.

The remainder of this section is organized as follows: Section 6.5.4.1 describes the configuration of the asynchronous TDMA bus link systems using chirp multiple access. Section 6.5.4.2 theoretically analyzes the pulse loss probability and the call blocking probability, considering traffic distribution in the area covered by a bus link. Section 6.5.4.3 discusses the effect of traffic distribution on the pulse loss probability and the call blocking probability, and the relationship between them is shown by means of some numerical results. Section 6.5.4.4 introduces the optimum design method for determining the area size covered by a bus link and the number of RBSs connected to the bus link.

6.5.4.1 Configuration of Asynchronous TDMA Bus Link Using Chirp Multiple Access

Figure 6.57 illustrates the configuration of asynchronous TDMA bus link systems using chirp multiple access. Many RBSs and several RCSs are connected with a fiber optic bus link, and each RBS is equipped with the chirp multiplexing transform, which was described in Section 6.5.3.

The chirp multiplexing transform can perform the conversion of several radio service signals received under different frequency bands into ones with a TDMA format as described in Section 6.5.3. Therefore, the routing node included in the network can distinguish and switch radio service into the

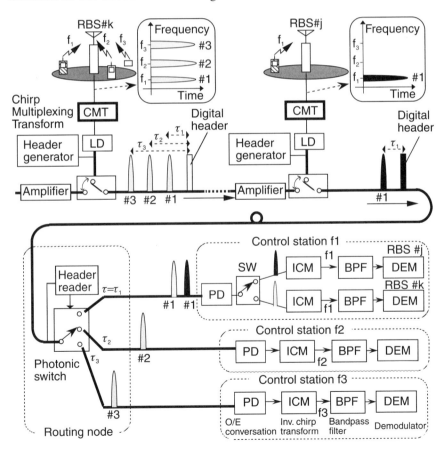

Figure 6.57 Asynchronous TDMA bus link systems using chirp multiple access.

appropriate RCS by means of a simple photonic switch. Besides such utilization of the chirp multiple access as an FDMA-TDMA converter, the system proposed in this section utilizes the chirp multiplexing transform as a time compressor for radio signals in the narrow pulse format. The time compression of radio signals allows asynchronous TDMA of radio signals into a fiber optic bus link as introduced in Section 6.5.3.

The RBS receives several radio services and feeds them into the chirp multiplexing transform. Figure 6.58 illustrates the configuration of the chirp multiplexing transform and the relationship between the received radio signals and resultant chirp multiplexing transform signal. In the chirp multiplexing transform, input radio signals, $r(t)$, are premultiplied with a periodic chirp signal, passed through a chirp filter, and postmultiplied with a chirp signal. The output of chirp multiplexing transform, $v(t)$, is represented by the following as described in Section 6.5.3:

$$V(t) = \text{Re}\left[\sum_{l=-\infty}^{\infty} U^{(l)}(t)e^{j2\pi f_0 t} \right] \tag{6.200}$$

$$U^{(l)}(t) = \begin{cases} \int_{-\frac{T_c}{2}}^{\frac{T_c}{2}} u(\tau + lT_c)e^{-j2\pi\beta[t-lT_c]\tau} d\tau & -\frac{T_c}{2} \le t - lT_c \le \frac{T_c}{2} \\ 0 & \text{otherwise} \end{cases}$$

$$\tag{6.201}$$

where $u(t)$ is the complex envelope of input radio signals. T_c and ΔF are the chirping dispersion time and the frequency sweep range of the chirp

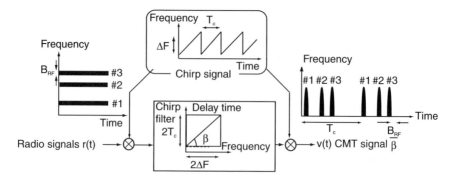

Figure 6.58 Configuration of chirp multiplexing transform (CMT).

multiplexing transform, respectively. β is the chirping angle given by $\Delta F/T_c$, l in the equation is an integer, and f_0 is a center frequency of the chirp filter. You can understand from (6.201) that the chirp multiplexing transform executes the Fourier transformation of original radio signals with every time width of T_c, and outputs the results as a function of time. Consequently, the one radio service with its bandwidth B_{RF} is transformed into periodical pulses with its pulse width B_{RF}/β and its interval T_c as illustrated in Figure 6.58. Therefore, if we enlarge the frequency sweep range, ΔF, then the bandwidth of radio signal B_{RF}, we can compress its pulse width at the same time. The relationship between the pulse duty of obtained pulse trains and the radio signal is given by

$$\frac{\tau}{T_c} = \frac{B_{\mathrm{RF}}}{\Delta F} \tag{6.202}$$

Figure 6.57 illustrates the example case in which the kth RBS is receiving three radio carriers, $f_1, f_2,$ and f_3, which correspond, respectively, to radio service 1, service 2, and service 3. The chirp multiplexing transform transforms them into TDM pulses whose time positions are different from each other ($\tau_1, \tau_2,$ and τ_3 in the figure). Then the TDM pulses are transformed into optical pulse amplitude modulation/optical intensity modulation pulses by a directly modulating laser diode and fed into a fiber optic bus link using the photonic switch after attaching a digital header for recognition of the start point of the TDM frame and the destination address of included radio service signal. Then, each RBS transmits the generated optical pulse amplitude modulation/optical intensity modulation pulse trains without any synchronization control with the other RBSs. In other words, the pulse amplitude modulation/optical intensity modulation pulses from many RBSs are asynchronously multiplexed in fiber optic bus link with TDMA format.

Such an asynchronous TDMA bus link system has the advantage of flexibility and simplicity because the network needs no synchronization control among the RBSs and the networks can be easily extended. Furthermore, such a system allows the unified transmission of multiple radio services in the TDMA format; therefore, switching of radio services at the optical stage can be executed. We consider the capability of photonic routing of radio service to provide the possibility for universal use of fiber optic networks among different types of radio service. However, one of the problems inherent to the asynchronous TDMA bus link systems is that a pulse amplitude modulation/optical intensity modulation pulse is lost if several RBSs transmit into a bus link at just the same time. For such a case, the pulse loss probability is discussed in Section 6.5.4.2.

Another problem to be studied is the degradation of received signal power due to pulse width narrowing and many optical devices passed in the transmission. Therefore, in the asynchronous TDMA bus link systems, we must compensate for such degradation of the received signal power by using an optical amplifier with an in-line type of insertion as illustrated in Figure 6.57, and the amplifier can also equalize the signal power received at the RCS from every RBS. The topic of the C/N performance of the asynchronous TDMA bus link and chirp multiplexing transform systems was discussed in Sections 6.5.2 and 6.5.3 in detail.

At the RCS, optical pulse amplitude modulation/optical intensity modulation pulses received from several RBSs are detected by a photodetector and demultiplexed into the signal from each RBS by means of a switch. Inverse chirp multiplexing transform with the same configuration as the transmitting chirp multiplexing transform regenerates the original radio signals, or the information data conveyed on the radio signal can be directly demodulated by means of a correlation demodulator as described in Section 6.5.3.

6.5.4.2 Theoretical Analyses of Pulse Loss and Call Blocking Probabilities

Since the pulse amplitude modulation/optical intensity modulation pulses from many RBSs are multiplexed asynchronously in the fiber optic bus link, some of the pulses may not be able to reach the RCS and become lost when several RBSs transmit at the same time. The pulse loss probability of asynchronous TDMA systems was studied in Section 6.5.2, but only the uniform traffic distribution model was considered in the analysis. Therefore, in this section, we theoretically analyze the pulse loss probability and the call blocking probability, taking the spatial traffic inequality into consideration.

As for the traffic distribution model, a Gaussian model is the most simple and popular one because traffic partiality can be simulated by means of only two parameters, its mean and variance, and it is also suitable for us to investigate the fundamental effect of traffic partiality on call blocking probability or on pulse loss probability.

Figure 6.59 illustrates the one-dimensional Gaussian traffic model used in our analysis. The fiber optic bus link covers the service area with its size X, and it is equally divided into N radio zones. The RBSs are numbered consecutively, starting with the one nearest the RCS. The traffic density at the distance x from the center of the service area, $f(x)$, is given by

$$f(x) = \frac{A}{\sigma\sqrt{2\pi}} \exp\left[-\frac{(x-m)^2}{2\sigma^2}\right] \qquad (6.203)$$

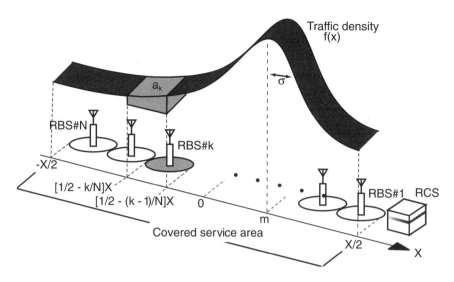

Figure 6.59 Traffic density distribution.

where m is the position with the peak intensity and σ is the standard deviation. By modifying these values, the various states of traffic partiality can be simulated. Furthermore, to fairly study the effect of traffic partiality on call blocking probability or on pulse loss probability, we determined that the total traffic covered by the service area, $[-X/2, X/2]$, is fixed to a_t. Thus, by solving the equation of $\int_{-2/X}^{2/X} f(x)dx = a_t$, A is given by

$$A = \frac{a_t}{\frac{1}{2}\left[\operatorname{erf}\left(\dfrac{X/2 - m}{\sigma\sqrt{2}}\right) - \operatorname{erf}\left(\dfrac{-X/2 - m}{\sigma\sqrt{2}}\right)\right]} \tag{6.204}$$

Consequently, the traffic generated in the kth zone, a_k, is given by

$$a_k = \int_{\left[-\frac{1}{2}+\frac{k-1}{N}\right]X}^{\left[-\frac{1}{2}+\frac{k}{N}\right]X} f(x)dx \tag{6.205}$$

$$= \frac{a_t\left[\operatorname{erf}\left(\dfrac{\left[-\dfrac{1}{2}+\dfrac{k}{N}\right]X - m}{\sigma\sqrt{2}}\right) - \operatorname{erf}\left(\dfrac{\left[-\dfrac{1}{2}+\dfrac{k-1}{N}\right]X - m}{\sigma\sqrt{2}}\right)\right]}{\left[\operatorname{erf}\left(\dfrac{X/2 - m}{\sigma\sqrt{2}}\right) - \operatorname{erf}\left(\dfrac{-X/2 - m}{\sigma\sqrt{2}}\right)\right]}$$

Figure 6.60 shows some example states of traffic distribution for the case in which 10 RBSs are connected and $a_t = 3.0$. The parameters σ and m are normalized by the service area size X. For example, $\sigma/X = 0.15$ and $m/X = 0$ is the state in which the center 30% of the service zone area includes 70% of the total traffic. However, the difference in the value of m causes a difference in the peak intensity even if the same value of σ/X is used, because of the assumption of fixed total traffic in the service area.

We assume that the capacity of every RBS is fixed to n channels, and the radio channels available in each radio zone are multiplexed in the FDMA format. The initial call blocking occurs if call connections over the channel capacity n are requested at the same time in the radio zone. The call blocking probability in the kth zone, R_k, is given by means of the Erlang B formula as follows:

$$R_k = \frac{\dfrac{a_l^n}{n!}}{\displaystyle\sum_{i=0}^{n} \dfrac{a_k^n}{i!}} \qquad (6.206)$$

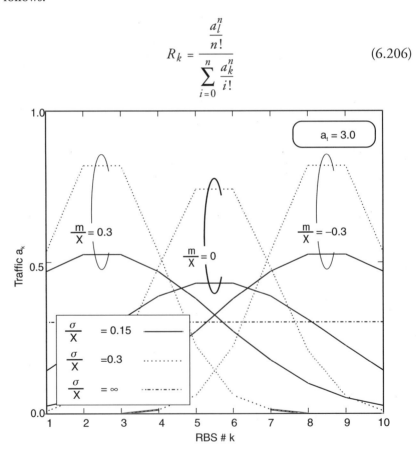

Figure 6.60 Traffic versus the RBS number.

Therefore, the average traffic virtually accepted by RBS can be estimated as $a_t(1 - R_k)$. We assume that the traffic accepted by RBSs is equally assigned to each channel on the average. Then, the average traffic per channel at kth RBS, a_{k_c}, is given by

$$a_{k_c} = \frac{a_k(1 - R_k)}{n} \qquad (6.207)$$

Figure 6.61 illustrates the mechanism of pulse loss. The pulse loss occurs at the stage of photonic switch multiplexing. If the jth RBS transmits its pulse amplitude modulation/optical intensity modulation pulse when the kth RBS's pulse is passing the photonic switch, the transmission of kth RBS's pulse is terminated and the pulse is lost. Therefore, the RBS that is nearer the RCS has priority in the transmission of pulses, and the pulse loss probability performance of the RBS that is further away is more deteriorated.

As described in Section 6.5.4.1, the chirp multiplexing transform at an RBS transforms one radio carrier into compressed pulse trains with its pulse duty $B_{RF}/\Delta F$ and its repetition interval T_c. When the RBS is receiving several carriers, such compressed pulse trains are generated as much as the number of receiving carriers. Under the condition that the jth RBS is receiving only one carrier when the kth RBS's pulse is passing the photonic switch, the probability that the pulse loss occurs is given by $2\tau/T_c$ because the occurrence time difference between the jth and the kth RBSs' pulses can be modeled as a uniform distributed random variable in the interval $[-X/2, X/2]$ as described in Section 6.5.2.

The probability that the jth RBS is receiving a carrier on a channel at an arbitrary time is given by a_{j_c}. Let A_{k_j} denote the event that one pulse amplitude modulated/optical intensity modulated pulse transmitted from

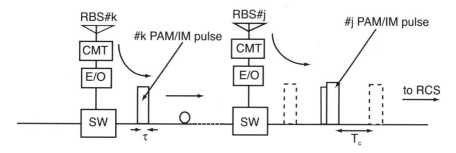

Figure 6.61 Mechanism of pulse amplitude modulated/optical intensity modulated pulse loss.

the kth RBS is not lost at the stage of photonic switching of the jth RBS ($j < k$). Considering that n channels are independently operated in the radio zone, the probability of event A_{k_j} occurring is given by

$$P(A_{k_j}) = \left[1 - \frac{2\tau}{T_c} a_{k_c} \right]^n \ (j < k)$$

(6.208)

Furthermore, because the traffic in all radio zones is mutually independent, the probability P_{th_k} that the pulse amplitude modulated/optical intensity modulated pulse from kth RBS will successfully reach the RCS without any pulse loss is

$$P_{th_k} = \prod_{j=1}^{k-1} P(A_{kj})$$

(6.209)

$$= \prod_{j=1}^{k-1} \left[1 - \frac{2\tau}{T_c} a_{k_c} \right]^n$$

Therefore, the pulse loss probability in the fiber optic bus link for the kth RBS's radio signal, P_{loss_k}, is given by

$$P_{\mathrm{loss}_k} = 1 - P_{th_k}$$

(6.210)

6.5.4.3 Numerical Results of Pulse Loss and Call Blocking Probabilities

This section shows some numerical results of the pulse loss probability and the call blocking probability. It is consistently assumed that radio bandwidth per carrier, B_{RF}, is 300 kHz and the maximum number of carriers received at an RBS, that is, the channel capacity n, is 4.

Figures 6.62 and 6.63 show the pulse loss probability versus the RBS number for different values of σ/X and m/X, respectively. An RBS transmits pulse amplitude modulated/optical intensity modulated pulses using a photonic switch, and the pulse loss occurs when a pulse amplitude modulated/optical intensity modulated pulse from any other RBS further from the RCS tries to pass the switch of the RBS that is transmitting at the same time. Therefore, the pulse amplitude modulated/optical intensity modulated pulse of the nearer RBS has priority over the transmission and, in particular, the pulse amplitude modulated/optical intensity modulated pulses of the first RBS are never lost.

The figures show that the pulse loss probability is degraded as the RBS number increases. Its degradation curve is different according to the value

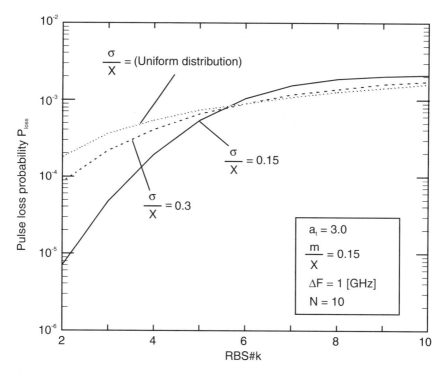

Figure 6.62 Pulse loss probability versus the RBS number when σ is changed.

of σ/X or m/X. The pulse loss probability of the kth RBS depends only on the traffic covered by RBSs nearer the RCS than that one. Therefore, in Figure 6.62, the pulse loss probability of the RBS located nearer to the RCS than the position of the peak intensity ($m/X = 0$) is much improved as σ/X decreases. Also in Figure 6.63, the pulse loss probability of the nearer RBS is improved as m/X decreases. However, the performance of the RBS furthest from the RCS, that is, the worst performance in this system, has little dependence on the parameters σ/X and m/X.

Figure 6.64 shows the pulse loss probability and the call blocking probability versus the number of connected RBSs. The increase of RBSs refers to the radio zone size reduction and the decrease of traffic per RBS. The performance of both depends on the number of RBSs. Thus, this figure is indicating the worst performance that would be obtained in all RBSs. The total traffic in the service area, a_t, is the parameter.

If more than four RBSs are connected, the pulse loss probability has little dependency on the number of RBSs. However, the call blocking probability is much improved by increasing the number of RBSs. As a result, lines of the two characteristics cross at the number of 14 connected RBSs at

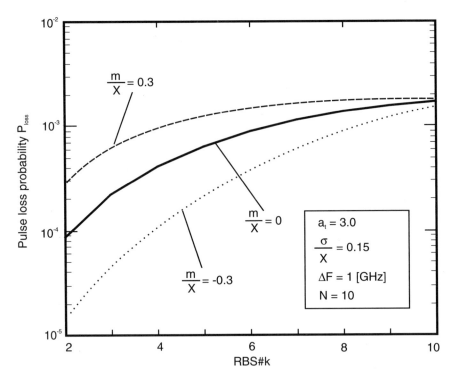

Figure 6.63 Pulse loss probability versus the RBS number when m is changed.

$a_t = 3.0$, and the number of 20 connected RBSs at $a_t = 5.0$. From our point of view, the pulse loss may cause interception of the communications, which seems to be a more serious problem than the initial call blocking. Therefore, the pulse loss probability has to be smaller or much smaller than the call blocking probability. In the case of Figure 6.64, more than 10 RBSs are allowed to be connected if $a_t = 5.0$.

Consequently, we should determine the network size covered by a bus link such that the included total traffic satisfies a required pulse loss probability. We should also determine the number of connected RBSs, that is, the number of radio zones, such that a required call blocking probability is satisfied, although this number depends on the standard deviation of traffic distribution.

Figure 6.65 shows the total accommodated traffic versus the required pulse loss probability for different values of the frequency sweep range of chirp multiplexing transform, ΔF. When the required pulse loss probability is reduced, the total traffic that can be accommodated by a bus link decreases. However, enlarging ΔF can improve the pulse loss probability or the accom-

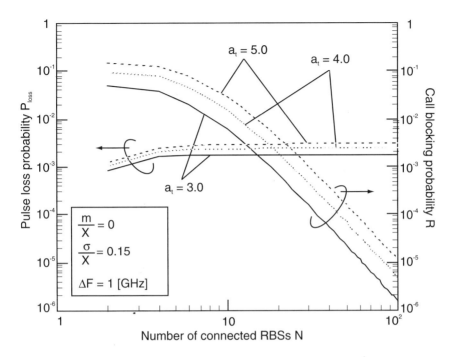

Figure 6.64 Pulse loss probability and call blocking probability versus the number of connected RBSs.

modated total traffic, a_t. For example, if we assume 10^{-3} of pulse loss probability is required, the networks have to be constructed in such a manner that a bus link includes 1.6 traffic when $\Delta F = 1$ GHz.

However, the reduction of the pulse duty of the pulse amplitude modulated/optical intensity modulated signal, that is, the enlargement of ΔF, causes the degradation of C/N signal performance detected at the RCS. The relationship between the pulse duty and the received C/N was theoretically analyzed in Section 6.5.2, and in the case of $\Delta F = 1$ GHz and $B_{RF} = 300$ kHz, the pulse duty is reduced to 3.0×10^{-4} from (6.202), but a C/N of more than 50 dB is obtained at the pulse duty as shown in Section 6.5.2.

Figure 6.66 shows the required number of RBSs to satisfy the call blocking probability of 10^{-2} versus the normalized standard deviation σ/X. From Figure 6.66 we can see that the required number of RBSs is very dependent not only on the total traffic but also on the standard deviation. For example, under the conditions of $\sigma/X = 0.15$ and $a_t = 3.0$, connection of more than 10 RBSs is required.

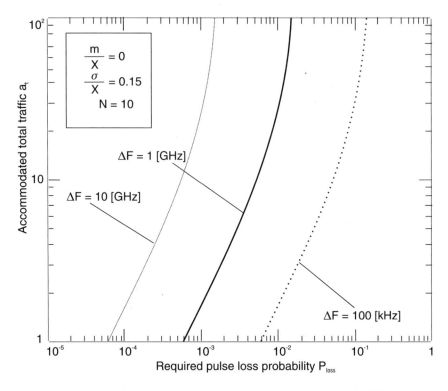

Figure 6.65 Accommodated total traffic versus required pulse loss probability.

6.5.4.4 Summary

This section proposed an asynchronous TDMA bus link system that uses the chirp multiplexing transform. The performance of the proposed system was investigated by theoretically analyzing the pulse loss probability and the call blocking probability considering the traffic partiality in the area covered by a fiber optic bus link.

The relationship between pulse loss probability and call blocking probability, the effect of traffic partiality on them, the total allowable traffic covered by a bus link, and the required number of connected RBSs were investigated and the following results were obtained:

- An asynchronous TDMA bus link system can be constructed that uses the chirp multiplexing transform technique. Because the type of radio service can be distinguished by its transmission time in the proposed system, the switching of radio signals according to type

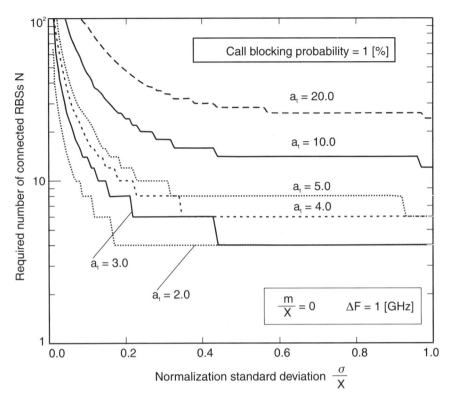

Figure 6.66 Required number of RBSs versus normalized standard deviation of traffic distribution.

of service can be performed at the optical stage by using a photonic time switch.

- The pulse loss probability performance depends only on the total traffic in the area covered by a bus link, but is independent of the traffic partiality, the position of peak traffic intensity, and the number of connected RBSs.

- The call blocking probability performance is improved by increasing the number of connected RBSs; thus, a number at which the call blocking probability becomes smaller than the pulse loss probability exists. This system should be operated using that number of RBSs.

- The pulse loss probability can be improved by increasing the bandwidth of the chirp multiplexing transform or decreasing the total traffic in the area covered by a bus link.

- One of the ways to determine the area size covered by a bus link and the number of connected RBSs was introduced. The area size covered by one bus link should be designed in a manner such that the covered traffic satisfies the required pulse loss probability performance. However, the number of connected RBSs should be determined in a manner that satisfies the required call blocking probability considering the traffic partiality.

6.6 Photonic WDM and Routing

6.6.1 Outline of the Photonic Network Technologies

Extensive research on photonic networks is rapidly progressing to the point where we may be able to realize flexible and seamless high-speed networks. The core technology is dense WDM and wavelength routing technologies.

Now, these technologies apply mainly to high-capacity baseband transmission. These systems transmit plural low-speed bit streams to prevent baseband pulse waveform deterioration due to dispersion of fibers and connectors. This technology was developed and introduced in North America as long-line transmission systems in 1996, and 16 WDM of a 2.5-Gbps stream system is mainly used now. And 30 to 40 WDM of a 10-Gbps system is being introduced and a more than 100-WDM system is under study.

To support the rapidly growing Internet traffic, IP over WDM and IP routing technologies are also under study. High-speed IP routing using dense WDM technologies is now considered to be one of the solutions as shown in Figure 6.67. These technologies are called *photonic network technologies* [44]. These core technologies are also applicable to the radio over fiber routing technologies that are shown in this section.

6.6.2 Photonic Networks and Wavelength Switching

WDM photonic network configurations are classified as shown in Figure 6.68. The various types of wavelengths assigned to radio signals are shown in Figure 6.69.

Most simple networks are point-to-point bulk transmission systems [Figure 6.68(a)]. In the system, one fiber is used to multiplex multiple wavelengths. All of the wavelengths are amplified periodically and transmitted to one destination. In this system, every wavelength is assigned to transmit every radio signal from RBSs in a small area, like a passive double star. For another application, every wavelength is assigned to every radio signal from

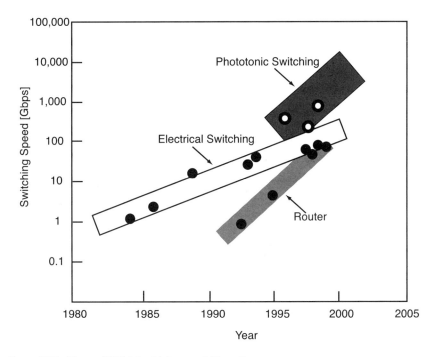

Figure 6.67 IP over WDM for high-speed IP routing.

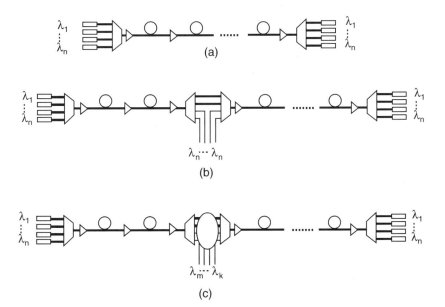

Figure 6.68 Various WDM configurations: (a) WDM bulk connection, (b) fixed WDM add drop, and (c) dynamic WDM add drop.

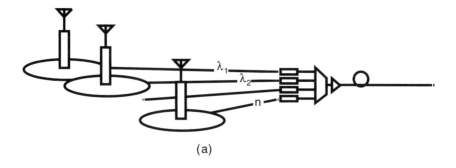

(a)

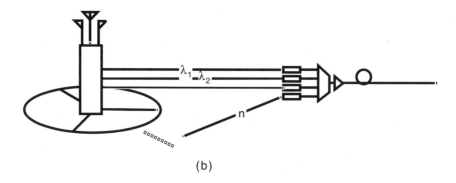

(b)

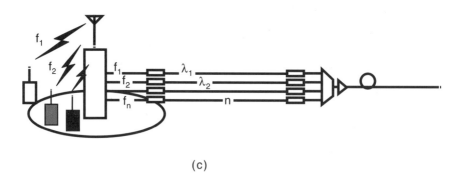

(c)

Figure 6.69 Various wavelength assignment methods: (a) wavelength assigned to RBSs, (b) wavelength assigned to sectors/diversity, and (c) wavelength assigned to radio service/frequency bands.

one of the sectors of RBSs. This configuration works for single-operator single RCS system applications.

The next configuration, shown in Figure 6.68(b), is the fixed WDM add drop point system. For example, the wavelength of this system is assigned to every RBS that is operated at a different place, and they are bus connected in one fiber. Using this topology, beat noise among laser diodes can be reduced. For another example, the wavelength of this system is assigned to every radio service in a single RBS and these services are add dropped at the intermediate node and rerouted to another place where the different service is operated. Usually the different service has a different frequency band, so the radio FDM is translated directly to the optical wavelength.

Another configuration is the dynamic WDM add drop configuration shown in Figure 6.68(c). In this system, the wavelength is dynamically rerouted to different destinations. A peer-to-peer dynamic connection can be supported by this configuration, and one radio application for this type might be peer-to-peer connection of radio LAN packets.

6.6.3 Photonic Switching Technology

If we use photonic switching, the bulk of the radio signals will be reroutable, so all types of radio signals will be reroutable. The amount of radio signal equipment required is reduced severely as is the delay time due to electronics circuits; for instance, modulation and demodulation are also reduced. One example of the wavelength switching network is shown in Figure 6.70, and the add-drop switch is shown in Figure 6.71. As shown in these figures, it is important for optical network devices to be fabricated with compact size and a high on-off ratio for the optical power in mind.

6.6.4 Summary

In this section, a photonic network based on dense WDM technologies was described, and these technologies are applied to the simple point-to-point bulk transmission link that has no node between source and destination. The applied signal format is baseband optical pulses. Extensive research is being conducted on these technologies, and they may be applicable to radio over fiber signal routings.

Acknowledgments

Shozo Komaki would like to thank Associate Professor K. Tsukamoto, Dr. Y. Shoji, and Dr. S. J. Park for the preparation of this chapter.

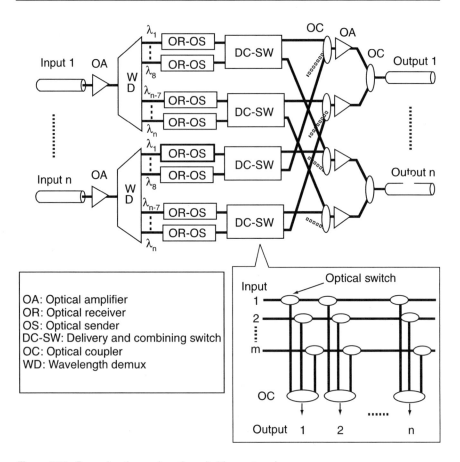

Figure 6.70 Example of wavelength switching networks.

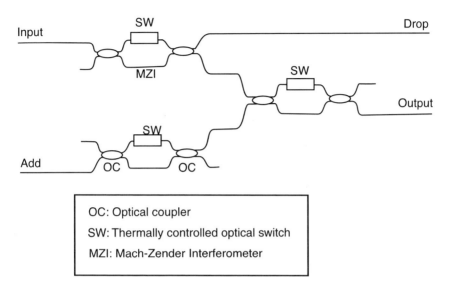

Figure 6.71 Configuration of the add-drop switch.

References

[1] Way, W. I., "Subcarrier Multiplexed Lightwave System Design Considerations for Subcarrier Loop Applications," *J. Lightwave Tech.*, Vol. 7, No. 11, Nov. 1989, pp. 1806–1818.

[2] Komaki, S., et al., "Proposal Fiber and Radio Extension Link for Future Personal Communications," *Microwave and Optical Technology Lett.*, Vol. 6, No. 1, Jan. 1993, pp. 55–60.

[3] Namiki, J., et al., "Optical Feeder Basic System Design for Microcellular Mobile Radio," *IEICE Trans. Commun.*, Vol. E76-B, No. 9, Sept. 1993, pp. 1069–1077.

[4] Ogawa, H., "Microwave and Millimeter-Wave Fiber Optic Technologies for Subcarrier Transmission Systems," *IEICE Trans. Commun.*, Vol. E76-B, No. 9, Sept. 1993, pp. 1078–1090.

[5] Way, W. I., "Optical Fiber-Based Microcellular Systems: An Overview," *IEICE Trans. Commun.*, Vol. E76-B, No. 9, Sept. 1993, pp. 1091–1102.

[6] Komaki, S., et al., "Proposal of Radio Highway Networks for Future Multimedia-Personal Wireless Communications," *IEEE 1994 Intl. Conf. Personal Wireless Communications (ICPWC '94)*, Bangalore, India, Aug. 1994, pp. 204–208.

[7] Komaki, S., and E. Ogawa, "Trends of Fiber-Optic Microcellular Radio Communication Networks," *IEICE Trans. Electron.*, Vol. E79-C, No. 1, Jan. 1996, pp. 98–104.

[8] Taub, H., and D. L. Schilling, *Principles of Communication Systems,* New York: McGraw-Hill, 1986.

[9]   Harada, H., et al., "TDM Intercell Connection Fiber-Optic Bus Link for Personal Radio Communication Systems," *IEICE Trans. Commun.*, Vol. E78-B, No. 9, Sept. 1995, pp. 1287–1294.

[10] Shoji, Y., et al., "Fiber-Optic Radio Access Networks Using Photonic Self-Synchronized TDM Bus Link," *Proc. 1st Intl. Symp. Wireless Personal Multimedia Communications (WPMC'98)*, Yokosuka, Japan, Nov. 1998, pp. 145–150.

[11] Shoji, Y., K. Tsukamoto, and S. Komaki, "Proposal of the Radio Highway Networks Using Asynchronous Time Division Multiple Access," *IEICE Trans. Commun.*, Vol. E79-B, No. 3, March 1996, pp. 308–315.

[12] Shoji, Y., K. Tsukamoto, and S. Komaki, "Proposal of Chirp Multiplexing Transform/Intensity Modulation/Direct Detection System for Radio Highway Networks," *IEICE Trans. Fundamentals,* Vol. E81-A, No. 7, July 1998, pp. 1396–1405.

[13] Prucnal, P. R., M. A. Santoro, and T. R. Fan, "Spread Spectrum Fiber-Optic Local Area Network Using Optical Processing," *J. of Lightwave Tech.*, Vol. LT-4, No. 5, May 1986, pp. 547–554.

[14] Prucnal, P. R., M. A. Santoro, and S. K. Sehgal, "Ultrafast All Optical Synchronous Multiple Access Fiber Networks," *IEEE J. on Selected Areas in Commun.*, Vol. SAC-4, No. 9, Dec. 1986, pp. 1484–1493.

[15] Santoro, M. A., and P. R. Prucnal, "Asynchronous Fiber Optic Local Area Network Using CDMA and Optical Correlation," *Proc. IEEE,* Vol. 79, No. 9, Sept. 1987, pp. 1336–1338.

[16] Kwong, W. C., P. A. Perrier, and P. R. Prucnal, "Performance Comparison of Asynchronous and Synchronous CDMA Techniques for Fiber-Optic Local Area Networks," *IEEE Trans. Commun.*, Vol. 39, No. 11, Nov. 1991, pp. 1625–1634.

[17] Holmes, A. S., and R. R. A. Syms, "Al Optical CDMA Using Quasi-Prime Codes," *J. of Lightwave Tech.*, Vol. LT-10, No. 2, Feb. 1992, pp. 279–286.

[18] Salehi, J. A., A. M. Weiner, and J. P. Heritage, "Coherent Ultrashot Light Pulse Code-Division Multiple Access Communication Systems," *J. of Lightwave Tech.*, Vol. 7, No. 3, March 1990, pp. 478–491.

[19] Zaccarin, D., and M. Kavehrad, "All Optical CDMA System Based on Spectral Encoding of LED," *IEEE Photon. Tech. Lett.*, Vol. 4, No. 4, April 1993, pp. 479–482.

[20] Benedetto, S., and G. Olmo, "Performance Evaluation of Coherent Optical CDMA," *Electron. Lett.,* Vol. 27, No. 22, Oct. 1991, pp. 2000–2002.

[21] Foschini, G. J., and G. Vannucci, "Using Spread-Spectrum in a High-Capacity Fiber-Optical Network," *J. of Lightwave Tech.*, Vol. LT-6, No. 3, March 1988, pp. 370–379.

[22] Vannucci, G., and S. Yang, "Experimental Spreading and Despreading of the Optical Spectrum," *IEEE Trans. Commun.*, Vol. 37, No. 1, Jan. 1994, pp. 63–65.

[23] Park, S, J., K. Tsukamoto, and S. Komaki, "Proposal of CDMA Cable-to-the-Air System," *Proc. IEICE Spring Conf.,* Vol. SB-5-8, March 1996, pp. 665–666.

[24] O'Farrell, T., and S. I. Lochmann, "Performance Analysis of an Optical Correlator Receiver for SIK DS-CDMA Communication Systems," *Electron. Lett.,* Vol. 30, No. 1, Jan. 1994, pp. 63–65.

[25] O'Farrell, T., and S. I. Lochmann, "Switched Correlator Receiver Architecture for Optical CDMA Networks with Bipolar Capacity," *Electron. Lett.,* Vol. 31, No. 11, May 1995, pp. 905–906.

[26] Khaleghi, F., and M. Kavehrad, "A New Correlator Receiver Architecture for Noncoherent Optical CDMA Networks with Bipolar Capacity," *IEEE Trans. on Commun.,* Vol. 44, No. 10, Oct. 1996, pp. 1335–1339.

[27] Nguyen, L., B. Aazhang and J. F. Young, "All-Optical CDMA with Bipolar Codes," *Electron. Lett.,* Vol. 31, No. 6, March 1995, pp. 469–470.

[28] Park, S. J., K. Tsukamoto, and S. Komaki, "Proposal of Direct Optical Switching CDMA for Cable-to-the-Air System and Its Performance Analysis," *IEICE Trans. on Commun.,* Vol. E81-B, No. 6, June 1998, pp. 1188–1196.

[29] Park, S. J., K. Tsukamoto, and S. Komaki, "Performance Analysis of Direct Optical Switching CDMA Cable-to-the-Air Network," *Proc. SPIE Optical Fiber Communication,* Vol. 3420-04, Taiwan, July 1998, pp. 20–29.

[30] Park, S. J., K. Tsukamoto, and S. Komaki, "Proposal of Radio Highway Network Using a Novel Direct Optical Switched CDMA Method," *Proc. MWP (Microwave Photonics),* Japan, Vol. TU4-6, Dec. 1996, pp. 77–80.

[31] Park, S. J., K. Tsukamoto, and S. Komaki, "Polarity-Reversing Type Photonic Receiving Scheme for Optical CDMA Signal in Radio Highway," *IEICE Trans. on Electron.,* Vol. E81-C, No. 3, March 1998, pp. 462–467.

[32] Park, S. J., K. Tsukamoto, and S. Komaki, *Theoretical Analysis on Connectable RBS Number and CNR in Direct Optical Switching CDMA Radio Highway,* IEICE Tech. Report, MWP97-23, Jan. 1998, pp. 80–85.

[33] Tamura, S., S. Nakano, and K. Okazaki, "Optical Code-Multiplex Transmission by Gold Sequence," *J. of Lightwave Tech.,* Vol. LT-3, No. 1, Feb. 1985, pp. 121–127.

[34] Yokoyama, M., *Spectrum Spreading Communication System,* Tokyo, Japan: Science Technology Inc., 1988, pp. 401–409.

[35] Banat, M. M., and M. Kavehrad, "Reduction of Optical Beat Interference in SCM/WDM Networks Using Pseudorandom Phase Modulation," *J. of Lightwave Tech.,* Vol. LT-12, No. 10, Oct. 1994, pp. 1863–1868.

[36] Kajiya, S., K. Tsukamoto, and S. Komaki, "Proposal of Fiber-Optic Radio Highway Networks Using CDMA Method," *IEICE Trans. Electronics,* Vol. E79-C, No. 1, Jan. 1996, pp. 111–117.

[37] Park, S. J., K. Tsukamoto, and S. Komaki, "Proposal of Reversing Optical Intensity CDMA Routing Method in Radio Highway for CDMA Radio," *Proc. CIC CDMA Intl. Conf.,* Korea, Oct. 1998, pp. 269–272.

[38] Mervyn, A. J., P. M. Grand, and J. H. Collins, "The Theory, Design, and Applications of Surface Acoustic Wave Fourier-Transform Processors," *Proc. IEEE,* Vol. 68, No. 4, April 1980, pp. 450–468.

[39] Shoji, Y., et al., "Proposal of Radio Highway Networks Using Chirp Multiplexing Transform," *Intl. Topical Meeting on Microwave Photonics (MWP'96),* Kyoto, Japan, Sept. 1996, pp. 37–40.

[40]   Shoji, Y., et al., "Fiber-Optic Virtual Radio Free Space Network Using Chirp Multi-
       plexing Transform for Multiband Operation of Multimedia Mobile Radio," *8th Intl.
       Workshop on Optical/Hybrid Access Networks,* Atlanta, GA, March 1997, p. 16.

[41]   Shoji, Y., K. Tsukamoto, and S. Komaki, "A Consideration on Radio Highway
       Networks Using FDM-TDM Conversion Scheme," *Intl. Topical Meeting on Microwave
       Photonics (MWP'97),* Duisburg, Germany, Sept. 1997, pp. 219–222.

[42]   Kobayashi, K., T. Kumagai, and S. Kato, "A Group Demodulator Employing Multi-
       Symbol Chirp Fourier Transform," *IEICE Trans. Commun.,* Vol. E77-B, No. 7,
       July 1994, pp. 905–910.

[43]   Kumagai, T., and K. Kobayashi, "A New Group Demodulator with Multi-Symbol
       Chirp Fourier Transform for Mobile Communication Systems," *Proc. IEEE
       ICUPC'95,* Nov. 1995, pp. 397–401.

[44]   Sato, K., "Photonic Networks," *NTT R&D,* Vol. 49, No. 1, 2000.

7

Radio over Fiber Technology: Current Applications and Future Potential in Mobile Networks—Advantages and Challenges for a Powerful Technology

Alan Powell

7.1 Historical Background and Evolution of Key Factors in 2G Radio over Fiber Networks

Radio over fiber (or high-frequency analog fiber optic links) was first developed in the early 1980s in the United States for military applications. RoF technology was used to distance the radar emitters (dish) far from the control electronics and personnel, because of the development of radar-seeking missiles (antiradiation missiles). In the case of lower frequency radars, this was done at the actual carrier frequency; but for higher frequency radars, the intermediate frequency was carried instead. At that time, it was necessary to use linear or analog systems because digital systems did not have the necessary speed, nor could they resolve the detail required to preserve target information.

To this end a number of research programs were funded by the U.S. government to develop high-frequency analog fiber optic systems. Competing technologies for this were externally modulated lasers, directly modulated lasers, and semiconductor lasers. Eventually directly modulated semiconductor lasers proved to be the transmitter of choice after overcoming problems of reliability, temperature stability, and coupling to the fiber. Very soon it was possible to modulate diode lasers up to 2 GHz and these started to be used in increasing volumes. In time the military requirement switched to much higher frequencies, and manufacturing techniques evolved to handle the geometries required, making the lower frequency lasers easier to manufacture and lower in price.

Finally in the late 1980s, lasers and photodetectors were transferred to industry-scale production. At the same time the radio industry was evolving and the need for wide-area coverage was increasing. Coverage could not be provided by a single base station so techniques were evolved that took advantage of multiple transmitters, all broadcasting on or near the same frequency. This technology was called *Quasi-Sync* in Europe and *Simulcast* in America. But these systems were notorious for frequency drift and had to use very expensive frequency standards with constant maintenance. In the receiving path, it was necessary to use voting systems and complex audio equalizers to select the most suitable receiver to detect the mobile transmitter. These were complex to set up and slow to respond; often the first part of a conversation would be lost.

Radio over fiber became an obvious technology to use to provide synchronous transmission and near-instantaneous reception of radio signals. When attached to a transceiver, they function as multiple antennas connected to the antenna socket, all radiating the same signal, and because there is only one receiver, voting is not required. The earliest practical systems were UHF and 800-MHz trunking systems in the United Kingdom at Gatwick Airport and Baltimore Airport Tunnel in the United States.

7.1.1 Basic RoF Systems

All RoF systems have the same basic configuration: A two-way interface contains the analog laser transmitter and photodiode receiver, which connects the base station transmitters and receivers to a pair of single-mode optical fibers. At the other end of the fibers is a remote unit that uses a similar photodiode receiver and analog laser transmitter to convert the optical signals to and from an antenna suitable for radiating the signals in question, as shown in Figure 7.1. More complex systems include multiple remote units, dual-band units, and comprehensive alarms. Some systems use optical wavelength-division multiplexers to allow bidirectional use of a single optical fiber.

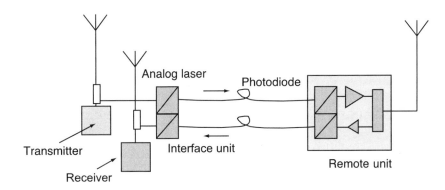

Figure 7.1 Basic RoF system.

7.1.1.1 Operation

Engineers used to conventional transmitters and receivers sometimes have trouble with RoF systems, because output powers are very low, input dynamic ranges are half that of a receiver, and the noise figures can horrify those not familiar with them. However, if the systems are viewed as a larger part of the transmitter-to-receiver link, their advantages can be seen. RoF systems can be broken down into separate transmitting and receiving paths to allow easier understanding of their operation and limitations.

7.1.1.2 The Transmitting Path

The transmitting path (known in the industry as the downlink) shown in Figure 7.2 shows multiple carriers brought together in an ideal combiner. From there they pass as a single complex carrier to the laser where they amplitude modulate a semiconductor laser that is biased into the middle of its operating range. The optical output of this laser is coupled via a fiber to

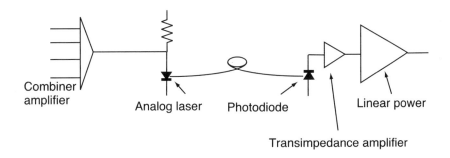

Figure 7.2 The transmitting path.

a photodetector where the optical signals are converted back to a multicarrier signal and amplified in a linear amplifier.

All components in this link are operating in a linear mode and, when carrying multiple carriers, are capable of producing intermodulation products. The amplitude of these intermodulation products depends on the third-order intercept point of the component and the amplitude of the signals passing through it. Any intermodulation products produced early on in the link will be amplified by any subsequent amplifier. Normally this is not a problem in the transmitting path (downlink) because signals in the optical components are low in relation to the third-order intercept of the laser and photodiode. Intermodulation products are normally generated in the linear power amplifier where the signals are high in relation to its third-order intercept point. (More details are given in Chapters 1, 2, and 3.)

Throughout the world regulatory authorities specify the level of inter-modulation products that are allowed to be generated by a radio system; sometimes this is an absolute level (−36 dBm in the case of CEPT), or sometimes it is a level relative to the carrier (−45 dBc in the case of the United Kingdom's MPT1380). This defined intermodulation level combined with the third-order intercept of the linear power amplifier, coupled with the number of carriers, limits the power per channel that the remote repeater can generate. Third-order intermodulation products are the most trouble-some: Because they fall on or very close to the signal frequencies, they also rise three times as fast as the signal amplitude increases. A quick way of working out for two carriers what signal level produces a given level of intermodulation in an amplifier can be described as follows:

1. Given: The intermodulation level is twice as far below the carrier as the carrier is below the third-order intercept point of the amplifier.

2. So an amplifier with a third-order intercept point of +60 dBm (1,000W)
   And a per-carrier level of +28 dBm
   Gives a difference of 32 dB.

3. Double this to get 64 dB.

4. So the intermodulations are 64 dB below the carrier (+28 dB), to give a level of −36 dBm for two carriers.

We can see from this explanation that a very big (expensive) amplifier is required to produce carriers of only 800 mW. Note that, as a general rule of thumb, an amplifier with a third-order intercept point of +60 dBm has a 1-dB compression point of about 50 dBm or 100W. Techniques exist, such as feed-forward correction, which can lift the third-order intercept of an amplifier relative to the 1-dB compression point by 20–30 dB.

We can easily infer from the preceding discussion that linear amplifiers and therefore RoF systems are not the best choice for covering large areas. But they do not need to cover large areas. They use the fibers to cover the majority of the distance between the base station and the mobile.

7.1.1.3 The Receiving Path

The receiving path, or uplink, shown in Figure 7.3 appears in front of the receivers as a wideband linear amplifier capable of handling signals across the whole of the receiving band (which is defined by the input filter). In a conventional receiver, automatic gain control is applied to, or very near, the first stages of gain in the receiver. This allows the receiver to compensate for a very large range of input signal levels (90–110 dB) for the single signal they are receiving. Automatic gain control techniques cannot be used in an amplifier (and by inference RoF systems) that is handling the whole of the receiving band, because it may be passing signals at a high level and at a low level at the same time. If automatic gain control were to be adjusted to reduce the big signals, it would also reduce the small signals and probably kill them. Good low-noise amplifiers without automatic gain control will have a dynamic range of about 60–70 dB and it is the same for RoF uplink paths.

To an engineer used to conventional radio receivers, limiting the input dynamic range to 60 dB is a horrifying prospect. However, we must point out that a conventional receiver will have to take signals from the base of the mast right out to signals generated on the horizon. RoF systems are not intended for this range of signals; they instead use the unlimited length of

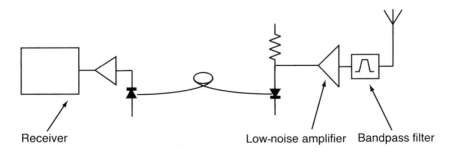

Figure 7.3 Off-air repeater with an RoF distribution system.

fiber to take the receiving antenna out among the portable radios where such ranges for input signals are not experienced.

The noise figure of the laser/photodiode combination is extremely high at around 30–40 dB, and for short lengths is at an extreme disadvantage against coaxial cable. However, as the distance increases, the coaxial losses go up remorselessly, whereas fiber losses are <0.5 dB/km [1]. In a link consisting of a series of mixed amplifiers and attenuators, it can be shown that if the amplifier is placed at the front and it has higher gain and lower noise figure than the parameters of this will dominate the whole chain.

The formula known as Friis's noise equation for such a chain is

$$F_c = F_1 + \frac{F_2 - 1}{G_1} + \frac{F_3 - 1}{G_2} + \ldots \frac{F_n - 1}{G_1 G_2 \ldots G_n} \qquad (7.1)$$

where F_1 and F_2 are the noise figures in numeric form and G_1 and G_2 are the gain of each stage in numeric form. It can be easily seen that the noise for the whole cascade chain is dominated by the noise contribution of the first stage so by using a good low-noise amplifier with a moderate amount of gain we can overcome the noise generated by the laser/photodiode.

7.1.2 Products and Manufacturers

Typical of such modern equipment is the Tekmar Sistemi Britecell single- and dual-band cellular remote repeater system as shown in Figures 7.4 and 7.5.

7.1.3 Some Early Players in the RoF Market

Decibel Products, of Dallas, Texas, a manufacturer of antennas and antennas systems, produced the world's first range of lamppost-mounted microcells, which were very popular in Australia. They also manufactured a number of custom-made systems, typical of which was a citywide hybrid system comprised of RoF and leaky feeders for the underground metro in Newcastle, United Kingdom. AT&T experimented with RoF in microcells using their own in-house laser fabrication facility. Ortel, Inc., of Alhambra, California, was the original manufacturer of high-frequency analog links for the U.S. military. As they moved to vertically integrate the company they started to manufacture systems mostly of a standard nature for cellular applications. Bosch GmbH in Berlin was the first company to manufacture an RoF system for the Inter Continental Express (ICE) in Germany.

Figure 7.4 Britecell remote repeater. (Photograph courtesy of Tekmar Sistemi S.R.L.)

Figure 7.5 Britecell local interface unit. (Photograph courtesy of Tekmar Sistemi S.R.L.)

Manufacturers with more advanced knowledge of cellular engineering eventually began to make RoF systems. Companies such as Mikom in Germany and Tekmar Sistemi in Italy were able to make systems that were easy to integrate into conventional GSM systems and were able to offer the advantages of additional capacity, signal control, and improved handover capacity. Other companies have appeared that use fiber for conveying the signal but use downconversion and upconversion to transport the signals. This allows for the use of cheaper multimode fibers and connectors but at the expense of increased equipment costs and more complex filtering to cope with delay issues.

7.2 Market Acceptance and Competitive Analysis of 2G Networks

7.2.1 RoF Market Drivers

Initially, site difficulties, timescales, and the costs associated with installing new radio base stations drove the RoF market, although sometimes it was the only way of providing coverage in very sensitive areas where revenue was not as important as seamless coverage. Now in the cellular market it is the cost-effective coverage of gaps or dead spots (e.g., underground parking garages) that can add revenue and increasingly the control and distribution of high capacity traffic that also drives the installation of RoF systems.

These days it is the cost of installation that drives the use of RoF systems. For instance, it is much easier to pull a 0.25-inch fiber optic cable through a roof space than 1-inch-diameter coaxial cable with a minimum bend radius of 3 feet. The remote repeaters themselves are becoming smaller and lower power; they are simple to mount with a few screws and their indifference to fiber length means it is possible to keep a coil of fiber stored in a suitable spot in case the repeater needs to be moved.

In harsh environments the simplicity of the RoF remote units allows increased reliability compared to a base station. An RoF unit can be mounted alongside a railway, for instance, where the environment is full of iron dust mixed with grease and graphite, which would very quickly clog up the fans in a base station. Access to remote equipment is often limited and a complex base station would be difficult to maintain, whereas an RoF remote unit is easy to maintain because it has so few parts.

Finally, the higher carrier frequencies associated with modern cellular systems are tipping the balance in favor of RoF systems and away from conventional coaxial distribution systems because of the coax cable losses. The negative side of using RoF systems is the need for dark fiber or an effective way to run a fiber cable. This can change dramatically from country to country depending on local telecommunications regulations and monopolies versus the competition scenario.

7.2.2 Competitive Scenario

RoF has always been perceived by network infrastructure manufacturers as a competing technology to conventional base stations, but it should really be seen as a complementary solution to mobile network infrastructure, because it does not in itself add to the capacity of a network. In dense environments, however, it can increase capacity by offering greater frequency reuse by tighter control of cell boundaries.

The larger original equipment manufacturers including Motorola, Ericsson, and Nokia analyzed RoF in their research and development laboratories, and although it may have reached the internal validation and technical acceptance stage, the next step toward becoming part of a main network's product development stream was never reached.

This situation transferred itself to mobile network operators, who, although they were often well aware of existing RoF products and potential advantages connected with their deployment, never included them in their early network planning, and only at a more mature rollout stage start did they begin to consider this technology as capable of solving some very specific cases where adding an RBS was clearly too costly, impossible, or unpractical.

7.2.3 Applications Overview

Applications for RoF in mobile telephone networks are split into two types: radio coverage extension and capacity distribution and allocation.

7.2.3.1 Radio Coverage Extension

This first case includes applications for railway and motorway tunnels, canyons, and other similar dead spot areas that because of their very specific needs often do not justify the costs of installing and operating new RBSs. In this case the RoF solution extends the reach of an existing base station to one or more locations that are shadowed or otherwise obstructed in terms of propagation of radio waves. The RoF system simply becomes part of the base transceiver station antenna system; one or more remote units are connected to an existing base station via a fiber cable that can easily be laid due to the linear system geometry and availability of cable ducts.

In some cases where frequency planning is not a limiting factor, the RoF coverage extension system may be interfaced to the donor base station via an off-air repeater.[1] The donor antenna is located in a suitable place where radio coverage from the donor base station is still reliable; the RF repeater amplifies both forward and reverse path signals to the appropriate level in order to drive the RoF system as shown in Figure 7.6.

Examples of applications belonging to this case (radio coverage extension) have been reported worldwide during system rollout since the early days of RoF technology and do not compete with but rather complement the traditional approach of new base stations. Several cellular service providers have deployed RoF as the standard solution for highway and railway tunnels in European countries (i.e., Germany, Switzerland, Austria, and Italy) to complement and enhance the solutions based on RF repeaters only.

7.2.3.2 Capacity Distribution and Allocation

This second case includes applications for subway stations, exhibition grounds, airports, downtown street levels, and other densely populated environments where the traffic requirement leads to dedicated radio channels

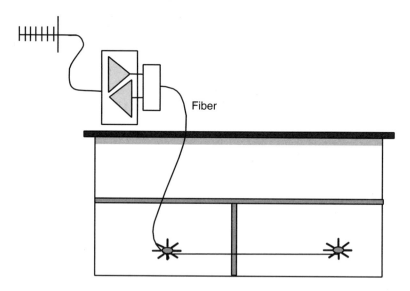

Figure 7.6 Off-air repeater with RoF distribution.

---

1. Off-air repeaters are bidirectional linear amplifiers that boost radio signals to compensate for the radio path loss. Their RF power gain normally ranges from 50 to 80 dB, and output power from 0.5 to 5W (1-dB compression); they often include band-specific or sub-band-specific selectivity so as not to create undesired interference in adjacent bands belonging to other services.

that need to be distributed appropriately. Such a system requires a thorough evaluation of network integration constraints, and technical designers need to consider many parameters such as cell boundaries, handover capacity, and people movement in order to accomplish the required targets, which normally include network decongestion, interference reduction, and quality of service improvement.

In this scenario, many remote units are connected to a dedicated base station via fiber cables, and the system normally has a star configuration. Sometimes operators share the RoF system. This application is typical of mature networks with high subscriber penetration where the cellular operator aspires to offer enhanced products such as private virtual networks.

Examples of applications belonging to this second case (capacity distribution and allocation) are reported in situations such as airports and shopping malls and are frequently shared by several network operators. RoF here is less generally accepted as a standard solution because of its direct competition with micro base stations and because of the plurality of surrounding network conditions that often lead planning engineers to take distinct approaches case by case.

In-building capacity distribution often uses passive coaxial technology, with RoF considered as a viable option in case of long cable runs or difficult environments where the bulky coaxial cable may not be practical to lay, or where Earth loops and lightning protection have a significant part to play in other buildings' electrical systems. Almost all European and North American mobile network operators have RoF systems, some of which are mentioned next.

On a worldwide basis, only Australia and South Korea have widely used RoF for outdoor capacity applications; Telstra (incumbent Australian landline and mobile operator) has used RoF systems since 1993 for their Advanced Mobile Phone System (AMPS) network and further migration to GSM. They deployed a real outdoor street-level microcellular layer in Melbourne and Sydney, with more than 300 remote units mounted on light poles in downtown areas, as shown in Figure 7.7. Similarly, the mobile network operators of South Korea adopted RoF around the mid-1990s to distribute the capacity of their CDMA systems at 800 and 1,800 MHz with medium- and high-power fiber remote units. In both cases the key driving factors were the high-cost and cumbersome base stations relative to RoF, and fiber cable availability at favorable or no cost.

7.2.4 Market Scenario

As outlined previously, three key factors contributed to the partially successful market positioning for RoF in 2G networks:

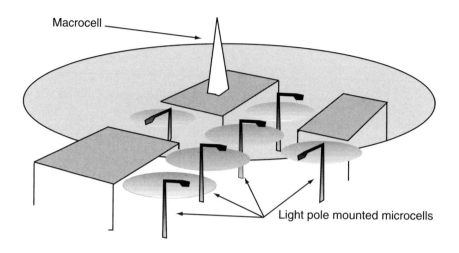

Figure 7.7 Australian microcell application.

1. Lack of support from network infrastructure manufacturers;
2. Late-coming RoF maturity compared to 2G network technology;
3. The availability of dark fiber.

An objective assessment of the success of RoF is rather difficult, considering the current competitive environment and the consequent difficulty in retrieving reliable information. An attempt that does not call for precise numbers, but rather tries to set a frame for the magnitude of success for RoF, would first need to quantify its market potential. We may take Australia as a reference of a successful case for which we have relatively consistent data availability. Here the investment made in RoF technology by one single operator lies at around 5% of the overall mobile network infrastructure, with more than 300 fiber microcells and 50 in-building systems; in all, the author estimates that they probably carry more than 10% of total mobile calls on a daily basis (see Table 7.1).

7.2.5 Market Size Estimate Versus Overall Mobile Radio Infrastructure

According to the Australian case, the potential for RoF in any 2G network may account for around 5% of the radio infrastructure, for an average figure of $20 to $100 million worth per network. Assuming 50 operators on a worldwide basis would represent most of the 2G market, our estimation for

Table 7.1
Australian Cellular Market

Network coverage is 96% of population
Network coverage is 4% land mass
66% of population live in the five capital cities
50% of population live in Sydney and Melbourne
1,800-MHz/800-MHz auction in 1998
Three new operators, four new networks
3 × 800-MHz CDMA, 1 × 1,800-MHz GSM

the overall potential for RoF systems would be projected to be anything between $1 and $5 billion.

With the exception of a few cases (Australia and the United States), RoF is mostly used for in-building and confined environment solutions (also called special projects), where it has to compete with a number of different technologies (radiating cable, passive distributed antenna systems, RF repeaters, micro/pico base stations) and where, because of the cost and the difficulties associated with accessing the site, most operators refrain from making the required decision and investment. Special cases with third-party investment (building owner, service management companies) are fraught with problems such as who owns the frequencies and what happens if the system fails.

7.2.6 Issues with Regulatory Bodies

Most regulatory bodies agree that the benefits of RoF (low RF power, close control of coverage, and handover capacity benefits) are very desirable. Yet with the exception of a few countries, the regulatory bodies do not have a framework around which to approve such equipment so they normally put the burden on the cellular operators. For large-scale citywide systems, dark fiber is another area of difficulty. In some countries the regulatory authorities force the incumbent telecommunications company to rent dark fiber to the cellular operators. Unfortunately, dark fiber is considered to have almost infinite bandwidth and it is necessary to restrict the bandwidth using complex filtering to satisfy the telecommunications company. The situations that have been the most successful usually occur when the telecommunications company also owns the cellular company.

Licensing has been an issue in some cases where the owner has a local radio system (rather than national) and wants to extend the coverage. The low-loss nature of the fiber means remote repeaters can be situated tens of kilometers away from the host site and normally this application requires a second license.

7.3 The 3G Technological Challenge

7.3.1 Radio Link

The radio link between the mobile unit and the base station is subject to fading, distortion, and interference, whereas signals in optical fiber are immune to these effects. So where an RoF system is used it is logical to take the fiber as close to the mobile as possible and make the radio part of the link as short as practical, the only purpose of the radio portion being to provide a flexible link.

Given the desirability of a short radio link, it is necessary to have many radiating points so that a mobile is never very far from an antenna. Short-range links require lower RF power, which in turn leads to lower cost amplifiers. Systems will be configured to have single antennas or batches of antennas linked back to the base transceiver station. Allocation of the remote antennas to sectors will be dynamic, based on the capacity requirements at that moment. Switching of this capacity could possibly be done at the optical level. The phase/time delays in fiber are constant, and switching between fibers could easily be compensated for either in the software or by the simple addition of fiber optic delay lines in the shortest fibers.

RoF systems are very linear and are completely transparent to any modulation scheme; power control is unaffected and is simply passed back to the base transceiver station. The simple linear nature of RoF systems means that mixing of 2G and 3G signals on the same fibers is possible. This means that only one system need be fitted to a building with the only modification required being an upgrade of the filtering system.

RoF systems offer the operators the ability to install many radiating points that can pass broadband signals at economical prices throughout a coverage area with a great deal of macrodiversity (see Section 7.3.2). This coincides with the trend of cell shrinkage associated with GPRS/Enhanced Data Rates for Global Evolution (EDGE) and the tougher requirements on the carrier-to-interference ratio (C/I) (i.e., smaller cells and many more of them).

7.3.2 Macrodiversity

Diversity in cellular systems is well understood, whether it is transmitting or receiving diversity. Typically, when a number of paths are available on which a signal can propagate, the best is selected. Probabilities of achieving low bit error rates improve when the number of signal paths increases. In an outdoor environment there is little distribution between the best and worst paths between a mobile and the base station, and the effective gain improvement that results from using diversity is only a few decibels. With macrodiversity, great differences are seen between the best and worst paths; this is akin to using two separate base stations with an associated handover procedure. (More details are given in Chapter 4.)

If an RoF system is installed in a building with many distributed antennas, the path between a mobile and one system antenna may be attenuated by 100 dB by a metal partition; yet if it is in sight of another antenna with a path loss of only several tens of decibels, then macrodiversity gain may be viewed as the difference between the two paths and could be as much as, say, 50–60 dB. Viewed from the other side of the partition, the situation will be reversed but the system still sees a macrodiversity gain of 50–60 dB.

The "organic" nature of buildings means that a system that works today might not work tomorrow. For example, a great deal of money might be spent on providing coverage within a building, only to be wasted when the occupant decides to erect metal partitions or wired glass windows. By employing a multiplicity of antennas in a building, all linked back to the BTS by RoF systems, it will become almost impossible to break or degrade the base-mobile-base link, RF power will be low, and coverage will be limited to its intended area—a "perfect radio channel."

Reference

[1]   *RF/Microwave Fiber Optic Link Design Guide,* Alhambra, CA: Ortel Corporation, 1989.

8

Radio over Fiber Multiple-Service Wireless Communication Systems
Masayuki Fujise

8.1 Introduction

The Communications Research Laboratory is developing a road-to-vehicle multiple-service communication system based on RoF technology in the millimeter-wave frequency region of 36–37 GHz. In the experimental system, the vehicle can receive three wireless services such as the personal handy-phone system, electronic toll collection system, and a broadcasting satellite. In this chapter, the system concept and experimental system configuration are explained. In addition, several concepts for other multiple-service wireless systems based on RoF are discussed.

As one form of the ITS information communication system, a road-to-vehicle communication system using the 36- to 37-GHz millimeter-wave band based on RoF technology has been studied and the experimental test bed has been under construction [1–3]. The experimental system has adopted a scheme that can provide multiple wireless services through the integration of mobile communications and broadcasting services other than the electronic toll collection system, which is one of the dedicated ITS services. Here, various concepts for future multiple-service radio communications based on RoF are proposed, and the experimental system configuration being developed is introduced.

RoF technology has been practically used in cellular phones as the blind zone cover technology, because it is difficult for radio waves to reach the back of a tunnel or into an underground arcade [3]. RoF technology has been considered for millimeter-wave transmission for relatively long distances, because transmission losses are great when high-frequency radio waves such as millimeter waves are transmitted through the cable and the waveguide [4].

Because the lightwave modulated by an RF signal is transmitted through an optical fiber, the interference that is normally generated when a radio wave propagates through air is not generated. Furthermore, the analog intensity modulation can make the transmission system very simple. However, the external modulator should have the characteristics of good linearity and fast response and the amplifier in the RF unit should also exhibit good linearity. The costs of these components need to be low if they are going to be of practical use. At present, research and development to overcome these problems have been extensively carried out and the optical fiber as a new transmission medium for radio waves is expected to be practically utilized.

8.2 Application for ITS

People, their vehicles, and the roads on which they travel are considered to be an integrated system in the ITS concept and we are pushing forward with a national project to realize a safer, more comfortable traffic environment in the twenty-first century.

The Communications Research Laboratory, Ministry of Posts and Telecommunications Japan, has been working on the research and development of ITS telecommunications. Research and development of RoF road-to-vehicle communications are one such ITS research area. In this chapter, the current status of RoF technology is introduced and the possibility for its practical use is discussed.

Figure 8.1 shows the radio environment that surrounds the vehicle during its travels. As shown in this figure, the vehicle runs while transmitting and receiving various kinds of radio waves. Examples of ITS services in the vehicle include the vehicle information and communication system (VICS) and electronic toll collection (ETC) system. The vehicle information and communication system service is offered by a 2.5-GHz-band radio beacon, by a light beacon, or by an FM multiplexed broadcast, whereas the electronic toll collection system service is offered as a form of dedicated short-range communication using the 5.8-GHz band. However, mobile telephones

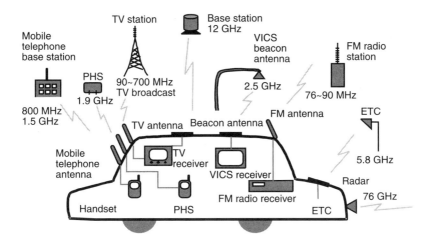

Figure 8.1 Current surroundings for vehicle communications, broadcasting, and radar. (PHS: personal handy-phone system.)

communicate in the 800-MHz and 1.5-GHz bands, and IMT 2000 will use the 2.1-GHz band, which started up on October 1, 2001.

Although the car is enjoying the various radio services mentioned above, it might become a hedgehog from the point of view of the limitations on space for mounting antennas, especially as new services are introduced one after another in the future. Moreover, the interior of the vehicle would be gradually compressed if radio communication equipment and terminals are installed individually for each service. To cope with such a problem, the concept of ITS road-to-vehicle communication using RoF technology has been proposed, and the research and development of such technology has been attempted by the Communications Research Laboratory.

The basic concept under development is shown in Figure 8.2. Various RF signals such as an existing mobile telephone or next-generation mobile telephone and ITS communication and satellite broadcasting are fed into the integrated base station from each service base station (BS). Then each RF is converted and integrated into some common-use frequency band. Furthermore, the lightwave emitted from a laser diode is modulated by the integrated radio signal and is transmitted to the local base station through optical fiber cable. At the local base station, the received optical signal is converted into a radio signal, which is amplified and emitted into the air from an antenna. In this process, FDM can be employed as a multiplexing technique.

Here, to transmit such an integrated broadband radio signal, the microwave or millimeter-wave frequency band is preferred. At the vehicle, however,

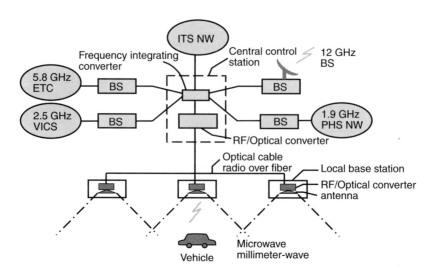

Figure 8.2 Concept of the ITS multiple-service network (NW) based on a common frequency band that uses RoF technology.

the integrated radio signal is received by a broadband antenna and divided into several original radio service signals. By setting up the downlink of the multiple service in this manner, radio interface between vehicle and local base station can be unified. The uplink from the vehicle to the integrated base station can be realized by a scheme that is the reverse of the downlink. Furthermore, integration of the mobile terminals mounted on the vehicle is desired and the application of software-defined radio technology is expected, as shown in Figure 8.3.

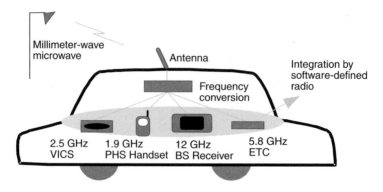

Figure 8.3 Integration of mobile communication links in a vehicle.

8.3 Multiple-Service Transmission Experiment

To confirm the feasibility of multiple-service RoF transmission, three kinds of radio signal multiplexing transmission experiment were tested using the experimental system shown in Figure 8.4. Radio signals for current services such as personal digital cellular (PDC), personal handy-phone system, and IS-95 were employed as a transmitted signal and integrated into a common-use frequency band at 5.8 GHz by means of FDM. Such an integrated optical signal is converted into the RF signal in the photodiode, after being transmitted through 5 km of optical fiber, and it is observed by the spectrum analyzer.

The spectrum shown in Figure 8.5 is obtained after the transmission. It reveals good features and a 40-dB dynamic range. The result of the modulation analysis is shown in Table 8.1 where EVM is the error vector magnitude and ρ is the rho factor, which is a measure of the quality of digital modulation similar to the EVM. The rho factor is determined by measurement of the normalized correlated power between the measured signal and reference signal. This experiment has successfully demonstrated the feasibility of common frequency band RoF transmission of the multiple-service signals.

8.4 The 5.8-GHz-Band Dual-Service Prototype System

Next, the Communications Research Laboratory attempted to design a dual-service communication system that integrates the electronic toll collection system and the personal handy-phone system in the 5.8-GHz band, based on the basic experimental results of the multiple-service transmission. As a

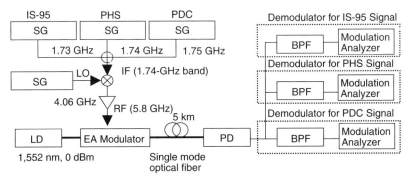

Figure 8.4 Configuration of the experimental setup for triple-service transmission. (BPF: bandpass filter; SG: signal generator; LO: local oscillator.)

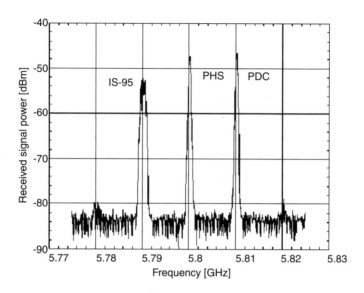

Figure 8.5 Measured spectrum of triple-service radio transmission by common frequency band RoF.

Table 8.1
Measurement Results of Transmission Qualities

Fiber Length	IS-95 (ρ)	Personal Handy-Phone System (EVM:%)	Personal Digital Cellular (EVM:%)
5 km	0.99764	2.30	2.28
20 km	0.99750	3.07	2.29

result, we confirmed that such a dual-service transmission can be realized by using RoF technology. Figure 8.6(a) shows the configuration for the downlink of the electronic toll collection system and personal handy-phone system dual-service RoF transmission system in the 5.8-GHz band, and Figure 8.6(b) shows the frequency allocation of the system.

The entire view of the prototype system is shown in Figure 8.7. The RF unit and antenna used for the practical electronic toll collection system in the 5.8-GHz band are also employed for this prototype system. An existing base station, switching system, handset for the personal handy-phone system, and a mounted mobile terminal and controller for the electronic toll collection system service are used. In this system, voice calls can be made through the personal handy-phone system link, while a charge notice announcement,

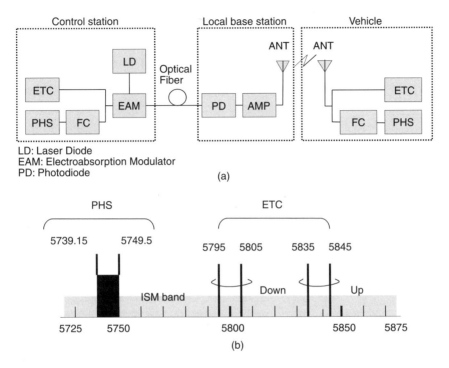

LD: Laser Diode
EAM: Electroabsorption Modulator
PD: Photodiode

(a)

(b)

Figure 8.6 (a) Configuration of developed prototype system and (b) frequency allocation of developed prototype system. (FC: frequency converter; ISM: industrial, science, and medical.)

which simulates electronic toll collection system service, is carried out. As shown in Table 8.2, electronic toll collection system and personal handy-phone system have different modulation schemes of ASK and $\pi/4$-DQPSK and transmission rates of 1.02 Mbps and 384 Kbps, respectively. It has been verified that this prototype system is able to transmit integrated different services, while maintaining good quality for each service.

8.5 The 36- to 37-GHz-Band Multiple-Service Experimental Facility

The realization of a system that can provide large-capacity multimedia information in a short time is desired for future ITS telecommunications systems. This led to consideration of a millimeter-wave RoF road-to-vehicle communications system that ensures such a broadband transmission capability. As shown in Figure 8.8, we are establishing an experimental facility based on

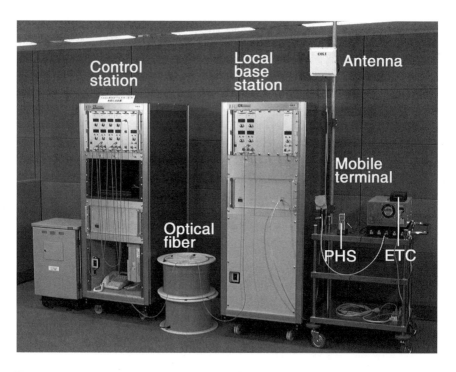

Figure 8.7 The 5.8-Hz-band electronic toll collection (ETC) system and personal handy-phone system (PHS) dual-service prototype system.

Table 8.2
Specification for the Developed System

Modulation/access	Personal handy-phone system	$\pi/4$ DQPSK	384 Kbps	TDMA-TDD
	Electronic toll collection system	ASK	1.024 Mbps	Slotted ALOHA
Frequency	Personal handy-phone system	5,739.15–5,749.95 MHz		
	Electronic toll collection system	5,795, 5,805 down; 5,835, 5,845 up MHz		
Antenna gain	Road	18 dBi		
	Vehicle	5 dBi		
RF output power	Road	3 dBm		
	Vehicle	10 dBm		
Optical fiber length		1 km		

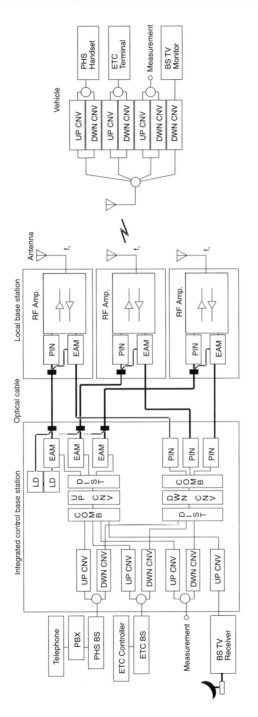

Figure 8.8 Configuration of the experimental facilities for a multiple-service common frequency band-RoF transmission system in the millimeter-wave band. (PBX: private branch exchange; PIN: p-i-n photodiode.)

street cell RoF multiple-service transmission in a millimeter-wave frequency region. This experimental facility consists of an integrated control base station, optical cables, a local base station, and a mobile terminal on the vehicle. The integrated control base station is installed in a laboratory room in the center building at Yokosuka Research Park. The local base stations are also located here along with the ITS test course, which is about a 240-m-long public road at Yokosuka Research Park. The section between the integrated control base station and the local base station is connected with the optical fiber cables. The distance between the local base stations is 20m, and six local base stations have been established at present.

A complete view of the experimental test course is shown in Figure 8.9. In this experimental system, the RF signal radio service is converted into a millimeter-wave common frequency band and the integrated RF modulates the lightwave, which is transmitted through the optical fiber cable. Then, the local base station receives the optical signal and converts it into the radio signal, which is emitted toward the street after being amplified. By adopting such a configuration, unification of the radio interface can be obtained.

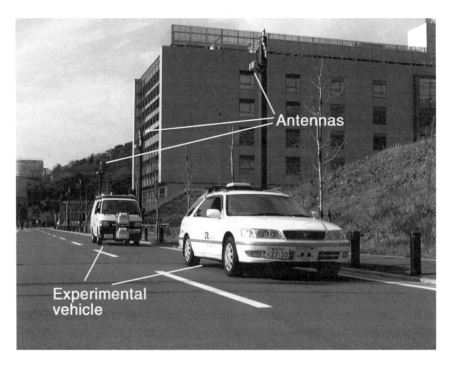

Figure 8.9 Experimental test course at Yokosuka Research Park.

The fact that multiple service can be provided through only one antenna and an RF unit mounted on the vehicle is extremely advantageous. The frequency allocation of the multiple service in the common frequency band is shown in Figure 8.10. The common frequency bands of 36.00–36.50 and 36.75–37.25 GHz are used for the downlink and the uplink, respectively. In this system, we have introduced the electronic toll collection (ETC) system as one of the typical ITS services, personal handy-phone system (PHS) as one of the mobile communication services, and BS as the broadcasting service. Two major schemes are available for generating 36-GHz radio signals. One scheme transmits an optical signal that has been modulated directly by the 36-GHz radio signal. The other scheme transmits an optical signal that has been modulated by the integrated IF signal of each service. In this case, the transmitted IF signal is converted into an RF signal at the local base station. Our experimental system adopts both schemes.

As part of the integrated technology for the mobile terminal, research on a multimode terminal that uses software-defined radio technology was carried out [5]. This experiment also promoted development of an ITS broadband radio access link called the multimedia lane and station [6].

8.6 ACTS FRANS Project

Another field trial of the RoF system was conducted. The ACTS FRANS is an RoF-based ATM network and services project [7]. In the trial system, video-on-demand, high-speed Internet, visiophony, and so forth are delivered to customers. The scheme for the demonstration is shown in Figure 8.11. The ATM interfaces to the RoF network are 622 Mbps downstream and 40 Mbps upstream, respectively. The frequency bands for the free-space

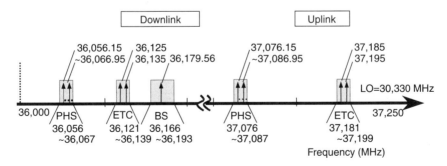

Figure 8.10 Frequency allocation of multiple-service RoF transmission system in the 36- to 37-GHz band.

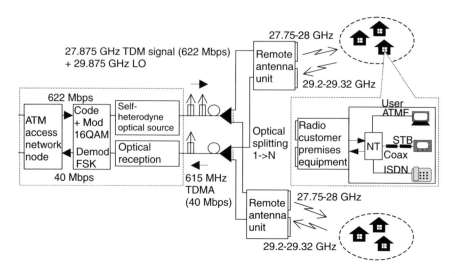

Figure 8.11 The RoF ATM passive optical network concept [7]. (NT: network termination; ATMF: ATM Forum; STB: set-top box.)

downlink and uplink are 27.75–28 and 29.2–29.32 GHz, respectively. For the transmission of RoF, an optical millimeter-wave source is generated based on the self-heterodyne technique. The coherent optical signals are received and demodulated at a high-speed photodiode. The FRANS project successfully demonstrated multiple-service transmission through an RoF network.

8.7 Application to Mobile CATV

CATV is very useful not only for broadcasting services but also for providing convenient telecommunications services such as telephony and Internet access. RoF technology has a good affinity for the CATV system and it has also been used practically in the CATV system. Wireless CATV is considered for the fixed CATV network. If we convert this fixed wireless CATV concept to that of mobile wireless CATV, mobile users can then be provided with CATV services.

Within the same CATV network service area, the CATV network can provide seamless service between indoor environments and outdoor environments. The user can obtain both the fixed service and the mobile service only if a mobile terminal with a set-top box function and user identification function is prepared. Furthermore, the deployment of mobile CATV service on a nationwide scale is also expected if each local CATV network is connected with each other and the roaming control is done when the vehicle goes into an area under the control of a different local CATV network. The concept of mobile CATV based on RoF technology is shown in Figure 8.12. More applications were explained in Chapter 7.

8.8 Fixed Mobile Communication Service

Cellular phones and personal handy-phone systems are used not only in mobile environments but also in office environments. In the back of a room in a building, handset terminals may not be able to receive signals. Even if handset terminals can receive signals, it may be difficult to connect handsets to the mobile network, because congestion may be caused by a large number of users. In such a situation, dedicated transmission using RoF technology is attractive because it can alleviate such a problem in mobile communication service. The basic concept is shown in Figure 8.13. This may be called a kind of seamless communication service between the outdoor environment and indoor environment.

No problem exists fundamentally if the delay time due to the propagation in the total section between base station and the terminal including the optical fiber section is less than that originally allowed for the system. This system concept results in the relaxation of congestion in the network, because we can utilize a different frequency band, such as the millimeter-wave frequency region. The existing optical fiber cables installed for the local-area

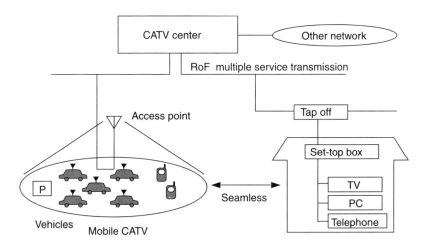

Figure 8.12 Mobile CATV concept.

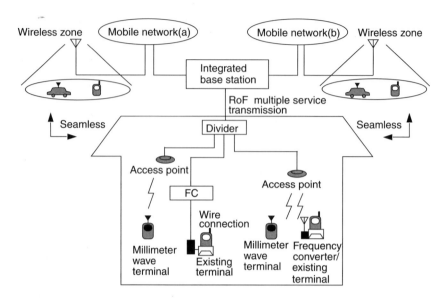

Figure 8.13 Concept of fixed mobile communications service.

network and so forth can be shared with such types of fixed services of the mobile communications. The adoption of WDM technology or of transmission technology that combines an RoF signal with a digital optical signal may enable the utilization of the existing installed fibers [8].

8.9 Integrated Multiple-Service Mobile Communication

From the viewpoint of flexibility and expandability, it is well known that RoF transmission technology makes a great contribution to the realization of a new mobile communication infrastructure that includes existing mobile communication services and broadcasting services, harmonized with the progress of the software-defined radio technology. An example of the mobile communication network and its backward compatibility is shown in Figure 8.14.

By introducing the software-defined radio technology and rewriting the software in the digital signal processor housed in the terminal, a multimode terminal that can access various kinds of mobile communication systems can be obtained. By using the multimode terminal and the RoF network shown in Figure 8.14, a flexible and convenient infrastructure may be realized that seamlessly integrates not only new services but also existing services.

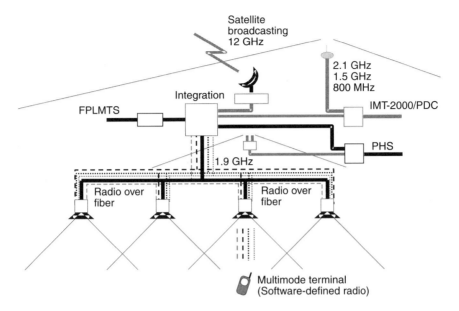

Figure 8.14 Concept of a multiple-service wireless communication system realized by means of RoF technology. (FPLMTS: Future Public Land Mobile Telecommunications System.)

The broadband transmission capability of the optical fiber will remain suitable in case a new service is introduced.

In this proposed network structure, existing facilities such as the integrated control station and the local base station can be used almost as is, if the transmission bandwidth and the gain of the RF units and the antenna are good enough. Per the features of the RoF system introduced here, the local base station relays the RF signal and handover between cells is not required because the base station of each individual service has the handover function.

8.10 Conclusions

In RoF transmission technology, it is necessary to pay attention to the optical fiber length, the emission wavelength of the laser, and the transmission frequency for the RF signal, since the optical fiber has the responsibility for chromatic dispersion. In addition, robustness adequate for the application in real outdoor field environments and cost reduction of the devices for the millimeter-wave signals and optical signals are desired so that such a system may spread widely. At the same time, it is also necessary to create killer applications that are very attractive to users.

The RoF network structure is more flexible and easy to expand when compared with conventional network structures. Although various problems still exist, RoF may be called a core technology for telecommunications in the twenty-first century and it has the potential to open the door for a new style in mobile communications and ITS communications.

References

[1] Fujise, M., and H. Harada, "Multimode DSRC by Radio on Fiber," *Proc. 1998 Communication Society Conference of IEICE,* Yamanashi, Japan, No. SAD-2-8, Sept. 1998, pp. 32–33.

[2] Fujise, M., et al., "ITS Multi Service RVC Test Bed Based on the ROF Technology (1)," *IEICE General Conference,* Hiroshima, Japan, No. A-17-33, March 31, 2000, p. 389.

[3] Ebine, Y., "Development of Fiber-Radio Systems for Cellular Mobile Communications," *MWP'99 Digest,* No. F-10.1, IEEE, Melbourne, Australia, Nov. 17–19, 1999, pp. 249–252.

[4] Kuri, T., et al., "Fiber-Optic Millimeter-Wave Downlink System Using 60 GHz-Band External Modulation," *J. Lightwave Technology,* Vol. LT-17, No. 5, May 1999, pp. 799–806.

[5] Kojima, F., et al., *Proposal for Multimedia Lane and Station Application for ITS Road-to-Vehicle Communication System,* IEICE Technical Report, ITS, Feb. 2000.

[6] Harada, H., Y. Kamio, and M. Fujise, *A New Multi-Mode and Multi-Service Software Radio Communication System for Future Intelligent Transport Systems,* IEICE Technical Report SR99–21, Nov. 1999.

[7] Deborgies, F., et al., "Progress in the ACTS FRANS Project," *MWP'99 Digest,* No. T-7.1, IEEE, Melbourne, Australia, Nov. 17–19, 1999, pp. 115–118.

[8] Harada, H., N. Mineo, and M. Fujise, "A Feasibility Study on Fiber-Optic Radio Transmission System over IM/DD Digital Optical Transmission Network," *IEICE General Conference,* Tokyo, Japan, No. B-5-258, March 31, 1999, p. 609.

List of Acronyms

ACLR	adjacent channel power leakage ratio
AM	amplitude modulation
BER	bit error ratio
CATV	cable television
CDMA	code-division multiple access
CIM	carrier-to-intermodulation ratio
CIR	carrier-to-receiver ratio
CINR	carrier-to-interference plus noise ratio
CSIR	carrier signal-to-self-interference plus noise ratio
C/N	carrier-to-noise ratio
DECT	digital enhanced telephone
DPCCH	dedicated physical control channel
DPDCG	dedicated physical data channels
EAM	electroabsorption modulator
EDGE	Enhanced Data Rates for Global Evolution
ETSI	European Telecommunications Standards Institute
EVM	error vector magnitude
FEC	forward error code
FM	frequency modulation
GPRS	General Packet Radio Service
GSM	Groupe Spéciale Mobile or Global System for Mobile-communications
HDTV	high-definition television
IIP	input intercept point

IM3	third-order intermodulation distortion
IM-DD	intensity modulation-direct detection
IMT 2000	International Mobile Telecommunications 2000
ITS	intelligent transport systems
ITU	International Telecommunication Union
M-QAM	multilevel quadrature amplitude modulation
OIP	output intercept point
OPRC	optical polarity-reversing correlator
pdf	probability density function
PM	phase modulation
PN	pseudonoise
QAM	quadrature amplitude modulation
RAU	remote antenna unit
RBS	radio base station
RF	radio frequency
RIN	relative intensity noise
RoF	radio over fiber
ROI-CDMA	reversing optical intensity carrier division multiplexing access
SCM	subcarrier multiplexing
SDH	synchronous digital hierarchy
SFDR	spurious free dynamic range
S/N	signal-to-noise ratio
STDMA	synchronous time-division multiple access
TACS	total access communication system
TDM	time division multiplexing
TDMA	time-division multiple access
TOI	third-order intercept point
UMTS	Universal Mobile Telecommunication System
UTRA	UMTS Terrestrial Radio Access
WCDMA	wideband CDMA
WDM	wavelength division multiplexing

About the Authors

Hamed S. Al-Raweshidy received his B.Sc. and M.Sc. in electrical engineering from the University of Technology in Baghdad, in 1977 and 1980, respectively. He has worked for the Military Engineering College, the Space and Astronomy Research Centre in Baghdad, CARL ZEISS in Germany, and Perkin Elumer in the United States. In 1987 he completed a postgraduate course in optical information processing at Glasgow University and subsequently obtained his Ph.D. in the spread spectrum multiplexing technique for communication networks from Strathclyde University in Glasgow. Dr. Al-Raweshidy has also worked at the University of Oxford; at British Telecom Laboratories; at the Centre for Communication Networks Research and the Department of Electrical and Electronic Engineering at Manchester Metropolitan University; and at the Communication Systems Division/Electronic Engineering Department at the University of Kent. Dr. Al-Raweshidy is a member of the IEE and the New York Academy of Sciences and a senior member of the IEEE. He is a guest editor for the *International Journal of Wireless Personal Communications,* and is on editorial boards for *Communications and Mobile Computing* and *Wireless Personal Communications.* In addition, Dr. Al-Raweshidy is a member of several conference committees and has organized several workshops and tutorial sessions on RoF for 3G mobile communication networks and Mobile IP in Europe and Japan. At present, in the Communications Systems Division in the Department of Electronics at the University of Kent, he is leading a group researching Mobile Internet Protocols (MIP) and RoF for 3G (UMTS) networks and beyond. Dr. Al-Raweshidy is currently the director of IIT Telecom Ltd.

Shozo Komaki received his B.E., M.E., and Ph.D. in electrical communication engineering from Osaka University in 1970, 1972, and 1983, respectively. In 1972, he joined the NTT Radio Communication Labs, where he was engaged in repeater development for a 20-GHz digital radio system and 16- and 256-QAM digital microwave systems. In 1990, he began working at Osaka University in the Faculty of Engineering, and began research on RoF communication systems, mobile communications, radio agents, and wireless IP communications. He is currently a professor at Osaka University. He is the coauthor of the *Encyclopedia of Telecommunications,* published by Marcel Dekker.

István Frigyes graduated from the Budapest Technical University with a degree in electrical engineering in 1954. He later received his Ph.D. jointly from the Budapest Technical University and the Hungarian Academy of Sciences in 1979. In 1995 Dr. Frigyes received his Ph.D. Habil. He also received a D.Sc. from the Hungarian Academy of Sciences in 1996. After various positions in industry and in industrial research and development, Dr. Frigyes began working for the Budapest University of Technology, where he is currently a professor. His main interest is in digital wireless communications, including RoF technology. He has taught courses all over the world. He is the coauthor of seven books, five of which are in English and one of which is *Digital Microwave Transmission,* published by Elsevier.

Masayuki Fujise received his B.S., M.S., and Dr. Eng. in communication engineering from Kyushu University in Fukuoka, Japan, in 1973, 1975, and 1987, respectively, and his M. Eng. in electrical engineering from Cornell University in 1980. He joined Kokusai Denshin Denwa Co. Ltd. (KDD) in 1975 and worked in the research and development laboratories researching optical fiber measurement technologies for optical fiber transmission systems. In 1990, he joined Advanced Telecommunications Research Institute International (ATR) as the department head of optical and radio communications research laboratories in Kyoto, Japan. There he managed research on optical intersatellite communications and active array antenna for mobile satellite communications. In 1997 Dr. Fujise joined the Communications Research Laboratory (CRL) of the Ministry of Posts and Telecommunications in Japan, where he is a group leader of the mobile communications group. Currently, CRL is an independent administrative institution in Japan. Dr. Fujise is interested in RoF, communications, software-defined radio technology, and ITS.

Dr. Fujise was the recipient of the Jack Spergel Memorial Award of the 33rd International Wire & Cable Symposium in 1984 and he is currently a member of the IEICE Japan and the IEEE.

Stephan Hunziker received his Ing. HTL from FH Aargau, Brugg-Windisch, Switzerland, and his Dipl. Ing. and Ph.D. from ETH Zurich, Switzerland, all in electrical engineering. His Ph.D. thesis was on the analysis and optimization of transparent fiber optic links.

Dr. Hunziker has worked with Synchronous Communications, Inc. in San Jose, California, on fiber optic transceiver design, and with Contraves Space/Unaxis in Zurich, on coherent optical intersatellite links. He currently works for Avalon Photonics Ltd. in Zurich, where he is product manager for datacom VCSELs. Since 1991, he has also been a lecturer at FH Aargau.

Dr. Hunziker's current research interests include high-speed laser-diode modeling, microwave circuit design, fiber optic link analysis and design, optoelectronic packaging, VCSEL, and high-speed digital and analog modulation and characterization. He is also a member of the IEEE and SPIE.

Mohsen Kavehrad received his B.Sc. in electronics from Tehran Polytechnic in 1973, his M.Sc. from Worcester Polytechnic Institute (WPI) in Massachusetts in 1975 and his Ph.D. from Polytechnic University in Brooklyn, New York, in 1977, both in electrical engineering. He has worked for Fairchild Industries, GTE, and Bell Laboratories. In 1989, Dr. Kavehrad became a professor at the Department of Electrical Engineering at the University of Ottawa. Simultaneously he became the director of the broadband communications research laboratory at the University of Ottawa. He was also the director of the Ottawa-Carleton Communications Center for Research (OCCCR). Dr. Kavehrad has consulted for NTT Laboratories, NORTEL, and Lucent (Bell Labs). In 1997, he joined the Department of Electrical Engineering at Pennsylvania State University as the AMERITECH (W. L. Weiss) Chair Professor of Electrical Engineering. In addition, Dr. Kavehrad was also the founding director of the Center for Information and Communications Technology Research (CICTR) and the chief technology officer and a vice president at TeleBeam, Inc. in Pennsylvania.

Dr. Kavehrad's current research interests are in broadband wireless and optical fiber communications systems and networks. He has published over 250 refereed papers and contributed to several books. He also holds several issued patents in these areas. Dr. Kavehrad is a Fellow of the IEEE and has received several awards. He was a technical editor for *IEEE Transactions on Communications, IEEE Communications Magazine,* and *IEEE Magazine on*

Lightwave Telecommunications Systems, and is currently on the editorial board of the *International Journal of Wireless Information Networks.* He has chaired, organized, and been on advisory committees for several international conferences and workshops.

Alan Powell is an engineer working in the radio industry. In 1988 he built and patented the first RoF systems in the world and has subsequently designed and installed many of these systems around the globe. He has worked for Philips, RCA Semiconductors, and, most recently, Allen Telecomms. Mr. Powell's background is in systems engineering in indoor and tunnel radio systems. He was awarded a patent on RoF in the United Kingdom and the United States in 1988.

David Wake is currently a senior research fellow at University College London, where he is working on low-cost, fiber-radio distributed antenna systems for in-building coverage of current and emerging mobile communications networks. Dr. Wake has spent most of his career working at BT Laboratories in the United Kingdom, where he pioneered many new device and system concepts for fiber-radio access networks. Dr. Wake has published extensively in this subject area and has contributed to several books. He also holds several patents. Dr. Wake has served on a number of committees associated with the International Topical Meetings on Microwave Photonics and is the 2001 meeting workshop chairman.

Index

Recent Titles in the Artech House Mobile Communications Series

John Walker, Series Editor

For further information on these and other Artech House titles, including previously considered out-of-print books now available through our In-Print-Forever® (IPF®) program, contact:

Artech House
685 Canton Street
Norwood, MA 02062
Phone: 781-769-9750
Fax: 781-769-6334
e-mail: artech@artechhouse.com

Artech House
46 Gillingham Street
London SW1V 1AH UK
Phone: +44 (0)20 7596-8750
Fax: +44 (0)20 7630-0166
e-mail: artech-uk@artechhouse.com

Find us on the World Wide Web at:
www.artechhouse.com